Introduction to
High Energy Physics

DONALD H. PERKINS
University of Oxford

ADDISON-WESLEY PUBLISHING COMPANY
Reading, Massachusetts · Menlo Park, California
London · Amsterdam · Don Mills, Ontario · Sydney

This book is in the
ADDISON-WESLEY SERIES IN ADVANCED PHYSICS

Morton Hamermesh
Consulting Editor

ISBN 0-201-05785-9
HIJKLMNOPQ-MA-89876543210

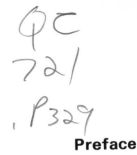

Preface

The main intention behind this book has been to present the more important aspects of the field of high energy physics at an elementary level. The content is based on courses of lectures given to Oxford physics undergraduates in their second and final years. The book would also serve as an introductory text for first-year graduate students specializing in nuclear physics. I have tried to aim at as broad a coverage of the subject as possible while still keeping the text to a reasonable length.

The material presented is more than would normally be covered in an undergraduate course. For this reason, the discussion of the basic introductory ideas, experimental methods, and invariance principles and symmetries, which form the backbone of the subject, are covered almost entirely in the first three chapters, which could well form the basis for a shorter course. The remaining chapters deal principally with the detailed features and models of weak, electromagnetic, and strong interactions. An attempt has been made to keep individual chapters reasonably self-contained, although some cross reference has been unavoidable. Some important aspects of the subject, such as hypernuclei, photomeson production, and cosmic rays, have been omitted completely in the interests of space.

It has been assumed that the reader has a working knowledge of elementary quantum mechanics. Wherever possible, however, the material has been presented from the empirical viewpoint, with the minimum of formalism. At no point have I hesitated to sacrifice mathematical rigor in the interests of clarity and of keeping the discussion at a simple level. Mathematical details, which can often as easily obscure as illuminate the basic physical principles, have been omitted completely or relegated to appendixes.

The material has been compounded from the many excellent advanced texts on the subject, from review articles, and from original papers. I have included a short bibliography at the end of each chapter, for further and deeper reading. References to original papers have generally been given where results from particular experiments are quoted, but it has not been possible to make this list either comprehensive or fully representative.

The book is intended to form part of a four-volume course in modern physics published by Addison-Wesley. The companion volume, *Introduction to Nuclear Physics*, by H. A. Enge, appeared in 1964. The two volumes together can be used to cover the whole field of nuclear and particle physics. The set will be completed shortly by a text on atomic physics by H. A. Enge, and one on the subject of solid-state physics.

ACKNOWLEDGMENTS

For permission to reproduce various photographs, figures, and diagrams, I am indebted to the authors cited in the text, and to the following laboratories and publishers:

Brookhaven National Laboratory, Long Island: Fig. 6.6.

CERN Information Services, Geneva: Figs. 1.4, 2.13, 2.14, 2.16, 3.5, 3.6, 3.15, 4.24, 4.25, 4.26, 4.40, 5.26, 5.28, and 7.27.

Rutherford High Energy Laboratory, England: Fig. 2.18.

Stanford Linear Accelerator Laboratory, Calif.: Fig. 2.23.

The American Institute of Physics, New York, publishers of *Physical Review*, *Physical Review Letters*, and *Reviews of Modern Physics*: Figs. 1.2, 2.24, 2.25, 2.27, 3.2, 3.13, 3.14, 3.17, 3.19, 3.20, 4.3, 4.16, 4.34, 4.35, 4.36, 4.37, 5.3, 5.16, 5.20, 5.22, 5.25, 6.8, 7.4, 7.5, 7.6, 7.7, 7.8, 7.9, 7.14, 7.17, 7.24, 7.26, 7.35.

Annual Reviews Incorporated, Palo Alto, Calif., publishers of *Annual Reviews of Nuclear Science*, for Fig. 4.10.

The Italian Physical Society, Bologna, publishers of *Il Nuovo Cimento*, for Figs. 7.19 and 7.21.

The North-Holland Publishing Co., Amsterdam, publishers of *Physics Letters*, for Figs. 2.27, 3.13, 3.14, 5.3, 5.20, 5.22, 5.25, 6.8, 7.14, 7.24, and 7.35.

Pergamon Press, Ltd., London, for Fig. 1.3.

The Physical Society, London, publishers of *Proceedings of the Physical Society*, for Fig. 2.2.

I have benefited from the help and advice of many colleagues during the preparation of the text. I am especially indebted to Dr. M. G. Bowler and Dr. D. Radojicic, for reading the manuscript and their numerous suggestions for amendments and improvements. Finally, I should like to thank Mrs. Jill Hudson and Mrs. Janet Caldwell for their careful typing of the manuscript.

Oxford, England   D. H. P.
November 1971

Contents

Chapter 7   Strong Interactions II—Dynamical Models

Tables

Appendixes

TABLE OF LONG-LIVED PARTICLES, STABLE OR DECAYING BY WEAK OR ELECTROMAGNETIC TRANSITIONS†

(for explanation of symbols, see footnote)

	Particle	J^P	I^G	Mass MeV	Mean life sec	Mode	Decay Fraction	p_{max}(MeV/c)
	γ	1^-	—	0	stable	—	—	—
LEPTONS	ν_e	$\frac{1}{2}$	—	$0(<50\text{ eV})$	stable	—	—	—
	ν_μ	$\frac{1}{2}$	—	$0(<1\text{ MeV})$				
	$e^\pm$	$\frac{1}{2}$	—	$0.511004(\pm 2)$	stable	—	—	—
	$\mu^\pm$	$\frac{1}{2}$	—	$105.660(\pm 1)$	$2.198(\pm 1) \times 10^{-6}$	$e\nu\bar{\nu}$	100%	53
MESONS	$\pi^\pm$	0^-	1^-	$139.58(\pm 1)$	$2.603(\pm 6) \times 10^{-8}$	$\mu\nu$	$\sim 100\%$	30
						$e\nu$	1.24×10^{-4}	70
						$\mu\nu\gamma$	1.24×10^{-4}	30
						$\pi^0 e\nu$	1.02×10^{-8}	5
						$e\nu\gamma$	3×10^{-8}	70
	π^0	0^-	1^-	$134.97(\pm 1)$	$0.9(\pm 2) \times 10^{-16}$	$\gamma\gamma$	98.8%	67
						$\gamma e^+ e^-$	1.17%	67
	$K^\pm$	0^-	$\frac{1}{2}$	$493.8(\pm 1)$	$1.235(\pm 4) \times 10^{-8}$	$\mu^\pm \nu$	$63.8\% \,(K_{\mu 2})$	236
						$\pi^\pm \pi^0$	$20.9\% \,(K_{\pi 2})$	205
						$\pi^\pm \pi^+ \pi^-$	$5.6\% \,(K_{\pi 3})$	126
						$\pi^\pm \pi^0 \pi^0$	1.7%	133
						$\mu^\pm \pi^0 \nu$	$3.2\% \,(K_{\mu 3})$	215
						$e^\pm \pi^0 \nu$	$4.9\% \,(K_{e3})$	228
						$e^\pm \nu,$	$1.2 \times 10^{-5}\,(K_{e2},$	247
							and others	
	$K^0, \overline{K^0}$	0^-	$\frac{1}{2}$	$497.8(\pm 2)$	$50\% \, K_S, 50\% \, K_L$			
	K_S				$0.862(\pm 6) \times 10^{-10}$	$\pi^+ \pi^-$	68.7%	206
						$\pi^0 \pi^0$	31.3%	209
				$(m_{K_L} - m_{K_S}) = \dfrac{0.47}{\tau_S}$		$\pi^+ \pi^- \gamma$	3×10^{-3}	206
	K_L				$5.17(\pm 4) \times 10^{-8}$	$\pi^0 \pi^0 \pi^0$	21.5%	139
						$\pi^+ \pi^- \pi^0$	12.6%	133
						$\pi\mu\nu$	26.8%	216
						$\pi e\nu$	38.8%	229
						$\pi^+ \pi^-$	1.57×10^{-3}	206
							and others	
	η	0^-	0^+	$548.8(\pm 6)$	$(\text{Width} = 2.6 \text{ keV})$	$\gamma\gamma$	39%	274
						$\pi^0 \gamma\gamma$	3%	258
						$3\pi^0$	30%	179
						$\pi^+ \pi^- \pi^0$	23%	174
						$\pi^+ \pi^- \gamma$	5%	236
	η' (or X^0)	0^-	0^+	$958(\pm 1)$	$(\text{Width} <4 \text{ MeV})$	$\eta\pi\pi$	66%	—
						$\rho^0 \gamma$	30%	—
						$\gamma\gamma$	$\sim 5\%$	—

† Adapted from the review of particle data by A. Barbaro-Galtieri, S. E. Derenzo, L. R. Price, A. Rittenberg, A. H. Rosenfeld, N. Barash–Schmidt, C. Bricman, M. Roos, P. Söding, and C. G. Wohl, *Reviews of Modern Physics* **42**, No. 1 (1970).

Footnote

1. J^P, spin and parity of particle.
2. I^G, isospin and G-parity.
3. Errors given in a bracket [thus (± 2)] refer to the last decimal place.
4. ( ), spin-parity assignment still in some doubt.
5. ?, spin-parity assignment unknown.
6. p_{max}, momentum of each secondary in two-body decay, or maximum momentum of any secondary in three-body decay.

Particle	J^P	I^G	Mass MeV	Mean life sec	Mode	Decay Fraction	p_{max}(MeV/c)
Y							
p	1 $\frac{1}{2}^+$ $\frac{1}{2}$		938.259($\pm$5)	stable			
n	1 $\frac{1}{2}^+$ $\frac{1}{2}$		939.553($\pm$5)	932 $\pm$ 14	$pe^-\nu$	100%	1
Λ	0 $\frac{1}{2}^+$ 0		1115.60($\pm$8)	2.51($\pm$3) $\times$ 10^{-10}	$p\pi^-$	65%	100
					$n\pi^0$	35%	104
					$pe\nu$	0.85 $\times$ 10^{-3}	163
					$p\mu\nu$	1.3 $\times$ 10^{-4}	131
Σ^+	0 $\frac{1}{2}^+$ 1		1189.4($\pm$2)	0.802($\pm$7) $\times$ 10^{-10}	$p\pi^0$	52%	189
					$n\pi^+$	48%	185
					$\Lambda e^+\nu$	2 $\cdot$ 10^{-5}	72
					$p\gamma$	1.2 $\times$ 10^{-3}	225
					$n\pi^+\gamma$	1.3 $\times$ 10^{-4}	185
Σ^0	0 $\frac{1}{2}^+$ 1		1192.5($\pm$1)	<10^{-14}	$\Lambda\gamma$	100%	75
Σ^-	0 $\frac{1}{2}^+$ 1		1197.3($\pm$1)	1.49($\pm$3) $\times$ 10^{-10}	$n\pi^-$	100%	193
					$\Lambda e^-\nu$	6 $\cdot$ 10^{-5}	79
					$ne^-\nu$	1.1 $\times$ 10^{-3}	230
					$n\mu^-\nu$	0.5 $\times$ 10^{-3}	210
					$n\pi^-\gamma$	10^{-4}	193
Ξ^0	-1 $\frac{1}{2}^+$ $\frac{1}{2}$		1314.7($\pm$7)	3.0($\pm$2) $\times$ 10^{-10}	$\Lambda\pi^0$	100%	135
Ξ^-	-1 $\frac{1}{2}^+$ $\frac{1}{2}$		1321.3($\pm$2)	1.66($\pm$4) $\times$ 10^{-10}	$\Lambda\pi^-$	100%	139
Ω^-	-2 $\frac{3}{2}^+$ 0		1672.5($\pm$5)	1.3($\pm$3) $\times$ 10^{-10}	$\Xi^0\pi^-$	28 events	293
					$\Xi^-\pi^0$	world total	289
					ΛK^-		210

BARYONS

x

TABLE OF MESON RESONANCES

Meson resonances (nonstrange)

Particle	J^P	I^G	Mass MeV	Width Γ (MeV)	Mode	Decay Fraction
ρ	1^-	1^+	765 ± 10	125 ± 20	$\pi\pi$ e^+e^- $\mu^+\mu^-$	100% $6 \cdot 10^{-5}$ $6 \cdot 10^{-5}$
ω	1^-	0^-	$783.9(\pm 3)$	11.4 ± 0.9	$\pi^+\pi^-\pi^0$ $\pi^0\gamma$ $\pi^+\pi^-$ e^+e^-	90% 9% 0.9% $7 \cdot 10^{-5}$
δ $\pi(1016)$	— (0^+)	$\geqslant 1$ 1^-	962 ± 5 1016 ± 10	<5 ~ 25	$\eta\pi$ $K^{\pm}K^0$	$\left.\right\}$ same particle?
ϕ	1^-	0^-	$1019.5(\pm 6)$	3.9 ± 0.4	K^+K^- $K_L K_S$ $\pi^+\pi^-\pi^0$ e^+e^- $\mu^+\mu^-$	46% 36% 18% $4 \cdot 10^{-4}$ $4 \cdot 10^{-4}$
$A1$	(1^+)	1^-	1070 ± 20	95 ± 35	3π	100%
B	(1^+)	1^+	1235 ± 15	102 ± 20	$\omega\pi$	100%
f	2^+	0^+	1264 ± 10	151 ± 25	$\pi\pi$	$\sim 80\%$ + others
D	(1^+)	0^+	1288 ± 7	33 ± 5	$K\bar{K}\pi$	—
$A2$	2^+	1^-	~ 1300	~ 90	$\rho\pi$ $K\bar{K}$	$>50\%$ $\sim 5\%$
E	(0^-)	0^+	1422 ± 4	70 ± 8	$K\bar{K}\pi$	$\sim 100\%$
f'	2^+	0^+	1514 ± 5	73 ± 23	$K\bar{K}$ $K\bar{K}\pi$ $\eta\pi\pi$	$\sim 70\%$ $\sim 10\%$ $\sim 20\%$
g ρ'	— —	1^+ 1^+	1663 ± 20 1714 ± 20	~ 140 ~ 125	2π 4π	Dominant Dominant $\left.\right\}$ same particle?

Meson resonances (strange)

Particle	J^P	I^G	Mass MeV	Width Γ (MeV)	Mode	Decay Fraction
$K^*(890)$	1^-	$\frac{1}{2}$	$892.6(\pm 4)$	50 ± 1	$K\pi$ $K\pi\pi$	$\sim 100\%$ 0.2%
$K^*(1420)$	2^+	$\frac{1}{2}$	1409 ± 10	107 ± 15	$K\pi$ $K^*(890)\pi$ $K\rho$ $K\omega$ $K\eta$	57% 27% 8% 4% 2%
$K(1240 - 1420)$	1^+	$\frac{1}{2}$	Probably 2 resonances		$K\pi\pi$ $(K^*(890)\pi$ or $K\rho)$	

TABLE OF BARYON RESONANCES

Baryon resonances (nonstrange)

Resonance	J^P	I	Mass (MeV)	Width Γ (MeV)	Decay Mode	Fraction (%)
$N(1470)$	$\frac{1}{2}^+$	$\frac{1}{2}$	1435–1505	200 –400	$N\pi$ $N\pi\pi$	60 40
$N(1520)$	$\frac{3}{2}^-$	$\frac{1}{2}$	1510–1540	100–150	$N\pi$ $N\pi\pi$	50 50
$N(1535)$	$\frac{1}{2}^-$	$\frac{1}{2}$	1500–1600	50–160	$N\pi$ $N\eta$	34 66
$N(1670)$	$\frac{5}{2}^-$	$\frac{1}{2}$	1655–1680	105–175	$N\pi$ $N\pi\pi$	42 58
$N(1688)$	$\frac{5}{2}^+$	$\frac{1}{2}$	1680–1692	105–180	$N\pi$ $N\pi\pi$	60 40
$N(1700)$	$\frac{1}{2}^-$	$\frac{1}{2}$	1665–1765	100–400	$N\pi$ ΛK	70 5
$N(1780)$	$\frac{1}{2}^+$	$\frac{1}{2}$	1650–1860	50–450	$N\pi$ $N\eta$	34 ~ 10
$N(1860)$	$\frac{3}{2}^+$	$\frac{1}{2}$	1770–1900	180–330	$N\pi$ $N\pi\pi$	27
$N(2190)$	$\frac{7}{2}^-$	$\frac{1}{2}$	2000–2260	270–325	$N\pi, N\pi\pi$	
$N(2220)$	$\frac{9}{2}^+$	$\frac{1}{2}$	2200–2245	260–330	$N\pi, N\pi\pi$	
$N(2650)$	?	$\frac{1}{2}$	2650	360	$N\pi, N\pi\pi$	
$N(3030)$	?	$\frac{1}{2}$	3030	400	$N\pi, N\pi\pi$	
$\Delta(1236)$	$\frac{3}{2}^+$	$\frac{3}{2}$	1230–1236	115 ± 5	$N\pi$ $N\gamma$	99.4 0.6
$\Delta(1650)$	$\frac{1}{2}^-$	$\frac{3}{2}$	1620–1695	130–200	$N\pi$ $N\pi\pi$	27 73
$\Delta(1670)$	$\frac{3}{2}^-$	$\frac{3}{2}$	1650–1690	175–300	$N\pi, N\pi\pi$	
$\Delta(1890)$	$\frac{5}{2}^+$	$\frac{3}{2}$	1840–1920	135–380	$N\pi, N\pi\pi$	
$\Delta(1910)$	$\frac{1}{2}^+$	$\frac{3}{2}$	1780–1935	230–420	$N\pi, N\pi\pi$	
$\Delta(1950)$	$\frac{7}{2}^+$	$\frac{3}{2}$	1935–1980	140–220	$N\pi$ $\Delta(1236)\pi$ ΣK	45 50 2, + others
$\Delta(2420)$	$\frac{11}{2}^+$	$\frac{3}{2}$	~ 2420	~ 300	$N\pi, N\pi\pi$	
$\Delta(2850)$	?	$\frac{3}{2}$	2850	400	$N\pi, N\pi\pi$	
$\Delta(3230)$	?	$\frac{3}{2}$	3230	440	$N\pi, N\pi\pi$	

Baryon resonances ($S = -1$)

Resonance	J^P	I	Mass (MeV)	Width Γ (MeV)	Decay Mode	Fraction (%)
$\Lambda(1405)$	$\frac{1}{2}^-$	0	1405 ± 5	40 ± 10	$\Sigma\pi$	100
$\Lambda(1520)$	$\frac{3}{2}^-$	0	1518 ± 2	16 ± 2	$N\bar{K}$ $\Sigma\pi$ $\Lambda\pi\pi$	46 41 10 + other
$\Lambda(1670)$	$\frac{1}{2}^-$	0	1670	~ 30	$N\bar{K}$ $\Lambda\eta$ $\Sigma\pi$	15 35 50
$\Lambda(1690)$	$\frac{3}{2}^-$	0	1690	27–85	$N\bar{K}$ $\Sigma\pi$ $\Lambda\pi\pi$ $\Sigma\pi\pi$	20 55 15 10
$\Lambda(1815)$	$\frac{5}{2}^+$	0	1815 ± 5	75 ± 10	$N\bar{K}$ $\Sigma\pi$ $\Sigma(1385)\pi$	65 11 17

Baryon resonances ($S = -1$) (*continued*)

Resonance	J^P	I	Mass (MeV)	Width Γ (MeV)	Decay Mode	Decay Fraction (%)
$\Lambda(1830)$	$\frac{5}{2}^-$	0	1835	66–145	$N\bar{K}$ $\Sigma\pi$	10 30
$\Lambda(2100)$	$\frac{7}{2}^-$	0	2100	40 145	$N\bar{K}$	25 + other
$\Lambda(2350)$	?	0	2350	~ 150	$N\bar{K}$	
$\Sigma(1385)$	$\frac{3}{2}^+$	1	1385 ± 1	36 ± 3	$\Lambda\pi$ $\Sigma\pi$	90 10
$\Sigma(1670)$	$\frac{3}{2}^-$	1	1670	50	$\Sigma\pi$ $\Lambda\pi$	~ 50 ~ 30 + other
$\Sigma(1750)$	$\frac{1}{2}^-$	1	1750	80	$N\bar{K}$ $\Lambda\pi$	15
$\Sigma(1765)$	$\frac{5}{2}^-$	1	1765 ± 5	60–150	$N\bar{K}$ $\Lambda\pi$ $\Lambda(1520)\pi$ $\Sigma(1385)\pi$	45 15 15 13
$\Sigma(1915)$	$\frac{5}{2}^+$	1	1910	70	$N\bar{K}$ $\Lambda\pi$	~ 10 ~ 5
$\Sigma(2030)$	$\frac{7}{2}^+$	1	2030	80–170	$N\bar{K}$ $\Lambda\pi$ $\Sigma\pi$	~ 10 ~ 35 ~ 5
$\Sigma(2250)$	?	1	2250	~ 200	$N\bar{K}$	
$\Sigma(2455)$	?	1	2455	~ 100	$N\bar{K}$	
$\Sigma(2620)$	?	1	2620	~ 140	$N\bar{K}$	

Baryon resonances ($S = -2$)

Resonance	J^P	I	Mass (MeV)	Width Γ (MeV)	Decay Mode	Decay Fraction (%)
$\Xi(1530)$	$\frac{3}{2}^+$	$\frac{1}{2}$	1530 ± 1	7.3	$\Xi\pi$	100
$\Xi(1820)$	?	$\frac{1}{2}$	1820	~ 30	$\Lambda\bar{K}$ $\Xi\pi$ $\Xi(1530)\pi$ $\Sigma\bar{K}$	30 10 30 30
$\Xi(1930)$	?	$\frac{1}{2}$	1930	110	$\Xi\pi, \Lambda\bar{K}$	

Chapter 1

History and Basic Concepts

1.1 INTRODUCTION

High energy physics deals basically with the study of the innermost structure of matter, and as such is a logical development of the speculations which began with the early Greek philosophers. The high energies required for these investigations follow from the uncertainty principle: localization of the study of matter to within extremely small distances requires illumination with radiation of the highest possible frequency and energy. Experiments in this field are therefore carried out with giant particle accelerators and their associated detection equipment.

The subject matter of this book is concerned with the properties and interactions of the so-called elementary or fundamental particles—a term which arose in far-off days when the different types of particle were few and it was believed that perhaps they were themselves the ultimate, basic constituents of matter. Nowadays, it is customary to omit the misleading term "elementary" and refer to the subject as particle physics or high energy physics. Some forty years ago, only four basic particle states, the proton and the neutron, the electron and the neutrino, together with the photon or quantum of the electromagnetic field, were known. While these four or five states certainly constitute the vast bulk of all matter in our particular region of the universe, we now recognize them as merely the end products of a very large number of particle states, most of them extremely unstable and not commonly encountered, but nevertheless, we believe, quite as fundamental as the few common particles to our understanding of the basic constitution of matter. The multitude of particle states appears to fit into regular patterns or multiplets, rather akin to the spectroscopic levels of atoms and nuclei. The proton and neutron occupy undistinguished places in these patterns, and their predominance in our world

1

appears to arise solely from the ambient conditions and as an accident of the conservation laws.

Apart from the multiplicity of particle states, we are also faced with the existence of several distinct types of interaction. All physical phenomena can be described in terms of one or other of these interactions. The *strong* interactions are familiar as providing the binding forces between neutrons and protons in the nucleus, and through them are generated the sources of energy in the stars. The *electromagnetic* interactions account for extra-nuclear phenomena, and lead to the bound states of atoms and molecules. The *weak* interactions are exemplified by the extremely slow processes of nuclear β-decay. In addition, there is the *gravitational* interaction, to which all particles and radiation are subject, and which is extremely feeble in comparison with the other three interactions. At least two other types of interaction have been proposed, although the evidence for their existence is much less clear: a "superstrong" interaction, to account for certain regularities observed among the strongly interacting particles; and a "superweak" interaction, proposed to account for a very small but definite violation of the rule that the weak interactions are invariant under the simultaneous application of the operations of charge conjugation and space inversion.

Each interaction is distinguished not only by its inherent strength, but also by its own particular set of conservation laws and selection rules. Eventually, one hopes to arrive at a comprehensive picture of the particles and interactions, which would embrace all particle states and all forms of interaction in a coherent whole. Even the most sublime optimist would admit, however, that we are a very long way indeed from such a goal. In the meantime, our understanding of the subject rests on a framework of very general invariance principles and conservation laws, which allows us to obtain relations between the properties of the various particle states as well as the different interaction fields. This framework, which we hope to have a degree of permanency, is described in Chapter 3. Upon it detailed theoretical models describing particular types of interaction have been erected. Some, like quantum electrodynamics, describing the electromagnetic interactions, have been remarkably successful. Other models have enjoyed success in describing restricted classes of phenomena, but have failed completely when attempts have been made to generalize them. Some of the main features of the models of strong, electromagnetic, and weak interactions are introduced in Chapters 4, 5, 6, and 7.

At all stages in the book, the role played by experimental facilities and techniques, on which progress in this field of physics ultimately depends, is emphasized. The methods of particle detection and measurement are summarized in Chapter 2. Before considering experimental methods, however, it is useful first to trace briefly some of the historical developments of high energy physics—if only to explain the idiosyncrasies of the notation—and to describe the broad classification of particles and interactions.

1.2 HISTORICAL DEVELOPMENT

High energy physics proper had its foundation in experiments with cosmic rays, the only available source of very energetic particles up to the early 1950's. The majority of the relatively long-lived particle states—those unstable only against decay through the weak interactions—were discovered in experiments with Geiger counters, cloud chambers, and nuclear emulsions, at sea-level or at mountain altitudes, or carried by high-altitude balloons. The existence of a *meson*—a particle of mass intermediate between an electron and a proton— was deduced in cloud chamber experiments by Anderson and Neddermeyer (1938) and Street and Stevenson (1937). At that time, it was thought that the particle in question was to be identified with the strongly interacting quantum of the nuclear field, previously postulated by Yukawa in 1935. This quantum was visualized as the carrier of the nuclear force, just as the photon is the carrier of the electromagnetic field. Yukawa's proposed particle had a Compton wavelength $\hbar/mc$ equal to the range of nuclear forces (10^{-13} cm), and thus a mass m, of order 200 electron masses, approximately equal to that found experimentally.

Yukawa had been led to this conclusion by the following arguments. The relativistic relation between energy E, momentum p, and mass m for the meson will be

$$E^2 = p^2c^2 + m^2c^4. \tag{1.1}$$

The wave equation describing such free particles is obtained by substituting the appropriate quantum-mechanical operators for E and p:

$$E_{op} \rightarrow i\hbar \frac{\partial}{\partial t}, \qquad p_{op} \rightarrow -i\hbar\nabla = -i\hbar \frac{\partial}{\partial r}.$$

When these operate on the wave function ψ, Eq. (1.1) becomes

$$\nabla^2\psi - \frac{m^2c^2}{\hbar^2}\psi - \frac{1}{c^2}\frac{\partial^2\psi}{\partial t^2} = 0, \tag{1.2}$$

called the Klein-Gordon equation, which describes a scalar field ψ associated with particles of mass m. For $m = 0$, Eq. (1.2) reduces to the familiar wave equation of electromagnetism:

$$\nabla^2\psi - \frac{1}{c^2}\frac{\partial^2\psi}{\partial t^2} = 0,$$

where ψ is the electric potential in free space. For a static field this becomes Laplace's equation:

$$\nabla^2\psi = \frac{1}{r^2}\frac{\partial}{\partial r}\left(r^2\frac{\partial\psi}{\partial r}\right) = -4\pi\rho,$$

where ρ is the electric charge density in the medium; it has as a solution the familiar Coulomb potential due to a charge q:

$$\psi = \frac{q}{r},$$ (1.3)

with $r > R$, where $q = \int_0^R 4\pi\rho r^2 \, dr$. For $m \neq 0$, Eq. (1.2) has the time-independent form

$$\nabla^2\psi - \frac{m^2c^2}{\hbar^2}\psi = 0,$$

with a spherically symmetric solution

$$\psi = \frac{g}{r}e^{-r/R}, \qquad R = \hbar/mc,$$ (1.4)

as can be verified by substitution. Thus, the potential ψ associated with the "meson field" has a range of order $R = \hbar/mc$. This result also follows from an uncertainty principle argument; to create a meson of mass m implies a violation of energy conservation unless it is restricted within a time $\Delta t \leqslant \hbar/mc^2$. In this time, the meson could cover at most a distance $R \sim c\,\Delta t = \hbar/mc$, which is thereby the range of the interaction. To further reinforce the picture, the radioactive β-decay of the mesons was observed, with a mean lifetime of order 10^{-6} sec, in rough agreement with Yukawa's expectations.

However, experiments were then made with counters by Conversi et al. (1947), which showed that the interpretation of the penetrating cosmic ray mesons as Yukawa quanta was untenable. With the aid of electromagnets, they were able to observe separately the fate of the positive and negative mesons coming to rest in blocks of various absorbers. Negative mesons coming to rest in the absorber would be expected to fall into Bohr orbits about an atomic nucleus, and thence suffer rapid nuclear absorption (in 10^{-12} sec). On the contrary, negative mesons stopping in light materials like carbon were observed to decay. This was a surprising and unexpected result. If the nucleus is to capture the meson, then only those mesons which have reached the atomic S-state are important, since it is the only one with a wave function of appreciable amplitude at the nucleus. The formula for the "Bohr radius" r_0 of this orbit is given by

$$r_0 = \frac{137}{Z}\frac{\hbar}{mc},$$ (1.5)

where m is the meson mass, and Z the atomic number. Note that, because of the larger mass, this is two orders of magnitude smaller than for an electron. The meaning of r_0 is that the meson wave function has the form

$$\psi(r) = \psi(0)e^{-r/r_0},$$

where r measures radial distance from the nuclear center. On the other hand, the nuclear radius will be

$$R = R_0 A^{1/3} \simeq \frac{\hbar}{mc} A^{1/3},$$

where the unit radius R_0 is of the order of the force range, i.e. the Compton wavelength of the Yukawa quantum. The fraction of time spent by the meson inside the nucleus is thus, very roughly,

$$\left(\frac{R}{r_0}\right)^3 = \frac{AZ^3}{(137)^3}. \tag{1.6}$$

Since the velocity of the meson in the Bohr orbit is simply $Zc/137$, it follows that the distance traveled in nuclear matter in a mean lifetime, $\tau = 2 \times 10^{-6}$ sec, is

$$l = \frac{AZ^4}{(137)^4} c\tau = \frac{1.7}{10^4} AZ^4 \text{ cm}.$$

For carbon, for example, the meson will spend 10^{-3} of the time inside nuclear matter, in which it will travel a mean distance $l \approx 1$ cm, or 10^{13} times the internucleon distance R_0, without appreciable interaction, since according to the experiment such negative mesons nearly always decay. A strongly interacting particle should have an interaction mean free path of order R_0, so the mesons in the Conversi experiment must have very weak interaction. They could not be the Yukawa quanta. In heavy nuclei, however, the mesons *were* found to undergo nuclear absorption, rather than decay with emission of a fast electron. We see from Eq. (1.6) that, if we set $A \approx 2Z$, the absorption rate goes as Z^4. The absorption process is now known to be the *weak* interaction on one of the constituent protons:

$$\mu^- + p \rightarrow n + v,$$

with a coupling strength some 10^{-12} times less than that of the strong nuclear forces. The symbol v stands for a neutrino, a neutral particle of very small mass, and μ for the weakly interacting meson. This coupling happens to be just right to ensure that, in light nuclei, negative mesons nearly always undergo radioactive decay, whilst in heavy nuclei, with a large value of Z^4, they mostly suffer nuclear absorption.

After this setback, Marshak and Bethe (1947) suggested that there were *two* types of meson. The ones observed in cosmic rays were the decay products of heavier mesons, and these latter were to be identified with the Yukawa or nuclear-force quantum. Shortly afterwards, Lattes *et al.* (1947) found events in photographic emulsion demonstrating the existence of a pi-meson or pion, decaying at rest into a monoenergetic mu-meson or muon, and a neutral particle, later proved to be a neutrino. Thus,

$$\pi^+ \rightarrow \mu^+ + v.$$

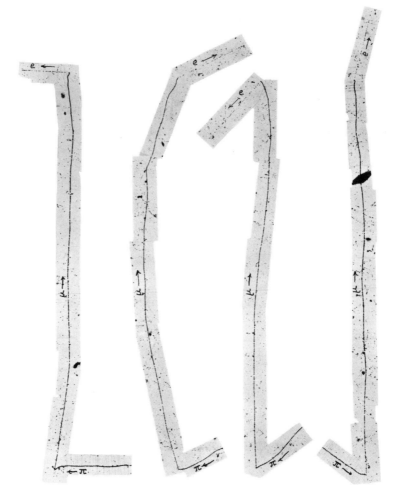

Fig. 1.1 Examples of the decay sequence $\pi \rightarrow \mu \rightarrow e$ in nuclear emulsion. Note that the constancy of range of the muon track implies that the pion decays at rest into two particles of unique masses: $\pi \rightarrow \mu + \nu$. The electron emitted in the β-decay of the muon, $\mu \rightarrow e + \nu + \bar{\nu}$, was not observed in the early experiments, which employed less sensitive emulsion. (Courtesy, University of Bristol.)

Typical decays are shown in Fig. 1.1. Events were also observed which occurred at about the same rate as the pi-mu decays and which were interpreted as nuclear capture of negative pions, leading to annihilation of the meson and disintegration of the nucleus. Thus, *pions* had strong interactions with nuclei, and were created by energetic cosmic ray primary particles (mostly protons) undergoing nuclear collisions in traversing the atmosphere. The pions usually decay in flight in the stratosphere, producing the daughter, noninteracting *muons* which constitute the bulk of the cosmic ray particles at sea-level and which had been observed in the previous experiments. Later work established that the muons in turn decay according to the scheme $\mu^+ \rightarrow e^+ + \nu + \bar{\nu}$. The muon decay is also shown in Fig. 1.1.

In these early experiments, it was established that the muon lifetime was about 2×10^{-6} sec, and that of the pion about 10^{-8} sec, and that they had masses of ~ 200 and ~ 270 electron masses, respectively. An impression of the scientific atmosphere of those days may be conveyed by remarking that the apparatus for the earliest pion lifetime estimate (correct to a factor 3) consisted of a few cocoa tins (containing photographic plates) tied to a vertical pole stuck in an Alpine glacier! Accurate determination of the masses and lifetimes of these mesons followed, with the production of beams of pions at the Berkeley synchro-cyclotron, from 1948 onwards. The evidence for the existence of the neutral pion, decaying to two gamma rays, $\pi^0 \rightarrow 2\gamma$, was obtained simultaneously, in 1950, by Carlson *et al.* in cosmic ray studies with nuclear emulsion, and by Bjorklund *et al.* at Berkeley. These showed the neutral and charged pions to have similar masses and to be produced with similar cross sections, as expected for charged and neutral counterparts of the same particle. Precise determination of the π^0-mass was made by Panofsky *et al.* (1951), by observations on the spectrum of γ-rays emitted when negative pions stopped in hydrogen. The γ-ray energies were measured by means of a pair spectrometer— see Fig. 1.2(a)—and could be attributed to two reactions:

$$\pi^- + p \rightarrow n + \gamma,$$
$$\rightarrow n + \pi^0$$
$$\quad\quad \hookrightarrow 2\gamma.$$

The first reaction yields a γ-ray of unique energy. In the second reaction, the π^0 has unique energy, with velocity βc. In the π^0 rest frame, the γ-rays have quantum energy $m_{\pi^0}/2$ and an isotropic angular distribution, so that the γ-spectrum in the laboratory is flat, extending from

$$(E_\gamma)_{\min} = m_{\pi^0}(1 - \beta)/2$$

to

$$(E_\gamma)_{\max} = m_{\pi^0}(1 + \beta)/2.$$

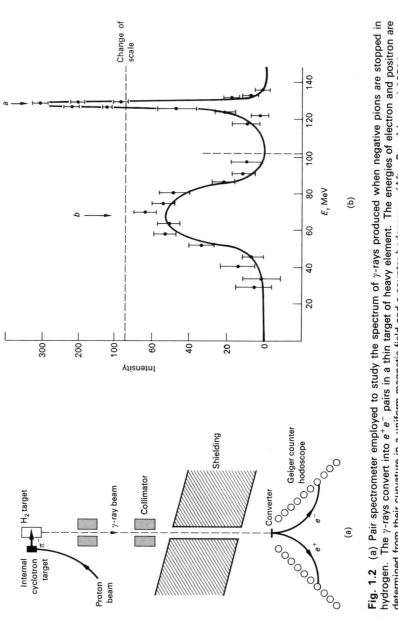

Fig. 1.2 (a) Pair spectrometer employed to study the spectrum of γ-rays produced when negative pions are stopped in hydrogen. The γ-rays convert into e^+e^- pairs in a thin target of heavy element. The energies of electron and positron are determined from their curvature in a uniform magnetic field and a counter hodoscope. (After Panofsky et al. 1951.)

(b) γ-ray spectrum from a similar, but improved, experiment by Crowe and Phillips (1954). Peak "a" is due to the reaction $\pi^- + p \to n + \gamma$, and the broad distribution centered about the point "b" is due to $\pi^- + p \to n + \pi^0$, $\pi^0 \to 2\gamma$.

The observed spectrum, shown in Fig. 1.2(b), thus allowed determination of β and an accurate measure of $(m_{\pi^-} - m_{\pi^0})$. The current value is 4.60 MeV.

The detection of neutral as well as charged pions was of great importance, since as Kemmer had emphasized some ten years previously, they were necessary in the Yukawa theory to account for the observed near equivalence of the proton-proton and neutron-proton potentials.

In the same year (1947) in which the pion was discovered, the cloud chamber observations of Rochester and Butler on penetrating cosmic ray showers revealed the existence of still heavier unstable particles. In one picture they observed a "V-event" ascribed to a neutral particle decaying into two charged particles; another picture showed a heavy charged particle decaying in flight in the gas of the chamber (Fig. 1.3). For almost two years, not a single further example was found, and then confirmation began to trickle in from many groups studying cosmic rays. At almost the same time, the Bristol group reported the decay at rest, in emulsion, of a heavy charged meson into three charged secondaries, which were soon to be identified as pions. By 1953, although the picture was still confused, the existence had been established of unstable particles heavier than nucleons, collectively called *hyperons:* notably the lambda-hyperon (at that time code-named V_1^0), decaying in the mode $\Lambda \to p + \pi^-$; the sigma-hyperon, decaying according to $\Sigma^+ \to p + \pi^0$ or $\Sigma^+ \to n + \pi^+$; the cascade or xi-hyperon, giving a "double V-event," $\Xi^- \to \pi^- + \Lambda$, $\Lambda \to p + \pi^-$. Furthermore, in the same year, artificial production of V-particles was observed in a hydrogen diffusion cloud chamber using a 1.4 GeV/c negative pion beam from the newly commissioned proton synchrotron (Cosmotron) at the Brookhaven National Laboratory.

Apart from the hyperons, a variety of decay modes had been established for particles, both neutral and charged, of mass between that of the pion and the proton—generically labeled *K*-mesons or *kaons*. Masses in the range 900 to $1150m_e$ were quoted, although the masses for the decays involving only charged particles, namely $\tau^+ \to \pi^+ + \pi^+ + \pi^-$ and $\theta^0 \to \pi^+ + \pi^-$, were quite well determined and closely similar. The confusion of 1953 arose from the fact that, although only one particle of unique mass was involved, it possessed many decay modes; $K^+ \to \mu^+ + \nu$, $\mu^+ + \pi^0 + \nu$, $\pi^+ + \pi^0$, $e^+ + \pi^0 + \nu$, $\pi^+ + \pi^+ + \pi^-$, $\pi^+ + 2\pi^0$, and $K^0 \to \pi^+ + \pi^-$ had been observed (though not all identified). This situation was not finally resolved until 1955, with the production from the 6 GeV Bevatron at Berkeley of an intense, momentum-analyzed beam of K-mesons, employing for the first time the newly discovered principle of strong-focusing. (Figure 2.16 shows examples of K^+ decay in a bubble chamber.)

However, the greatest problem posed by the V-particles was their relatively copious production (some 1% of the pion production) as compared with their long lifetimes, in the range 10^{-8} to 10^{-10} sec. Suppose a Λ-hyperon were produced in a reaction like

$$\pi^- + p \to \Lambda + \pi^0. \tag{1.7}$$

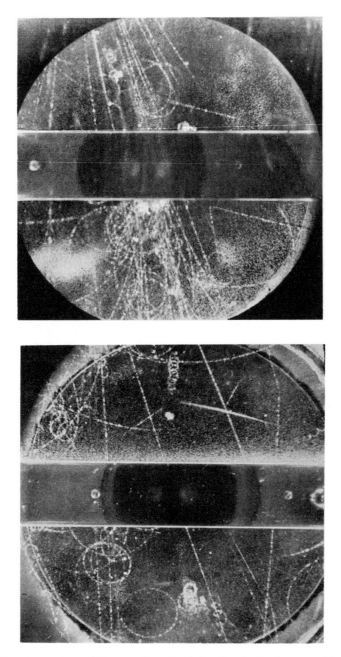

Fig. 1.3 First observations of *V*-events in a cloud chamber, by Rochester and Butler (1947). The upper picture is of a "neutral *V*-event," consisting of a wide-angle fork occurring in the gas a few millimeters below the horizontal plate. Subsequent analysis suggests that it was due to the decay $K^0 \rightarrow \pi^+ + \pi^-$. The lower picture is of a "charged *V*-event," seen as a fork near the right-hand top corner of the picture. The secondary traverses the 3 cm lead plate without interaction. The measured momenta are in fact consistent with the decay scheme $K^+ \rightarrow \mu^+ + \nu$, or what is now called the $K\mu2$ decay mode of the charged kaon. (Courtesy, Pergamon Press.)

Then, by the principle of microscopic reversibility, the inverse process,

$$\Lambda \to \pi^0 + \pi^- + p \to \pi^- + p, \tag{1.8}$$

can occur. The first stage in Eq. (1.8) violates conservation of energy; it is supposed that it can occur virtually, i.e. on such a short time scale as to be allowed by the uncertainty principle $\Delta E = \hbar/\Delta t$. The typical time involved in the production reaction is of the order of the range of the strong interaction (10^{-13} cm) divided by the velocity of light, i.e. $\sim 10^{-23}$ sec. Why then was the decay process not as rapid? The corresponding problem in the case of the slowly decaying, but apparently copiously produced muon, had been solved, as we saw, with the discovery of the pion.

The hypothesis of *associated production* introduced by Pais (1952) solved the problem elegantly; the kaons and hyperons must be created (and destroyed) in pairs. Then the production process would appear as

$$\pi^+ + n \to \Lambda + K^+. \tag{1.9}$$

The inverse strong decay process must then be

$$\Lambda \to K^- + n + \pi^+ \to K^- + p. \tag{1.10}$$

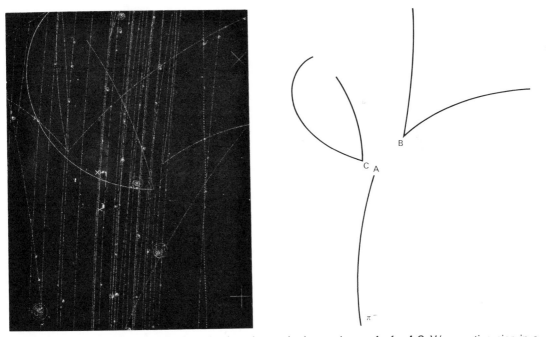

Fig. 1.4 An example of associated production, due to the interaction at A of a 4 GeV/c negative pion in a hydrogen bubble chamber: $\pi^- + p \to \Lambda + K^0$. The Λ-hyperon decays at B, $\Lambda \to p + \pi^-$, and the K^0-meson at C, according to $K^0 \to \pi^+ + \pi^-$. (Courtesy, CERN Information Service.)

Note that the principle of associated production compels us to keep a K^- on the right-hand side (i.e. we cannot write $\Lambda \to K^- + n + \pi^+ \to n + \pi^0$ as a fast process); and that $\Lambda \to K^- + p$ as a real decay process is forbidden by conservation of energy, since $m_\Lambda < m_K + m_p$ (see table on page ix). Notice also from this table that there *are* hyperon resonance states of higher mass, e.g. $\Lambda(1520)$, which *do* decay by strong interactions as in Eq. (1.10). Gell-Mann and Nishijima formalized the concept of associated production in 1955, by introducing an additive quantum number, called *strangeness* (S). The Λ-hyperon and K^- were assigned $S = -1$, and the K^+, $S = +1$. Conservation of strangeness in a strong interaction then led to the associated production hypothesis, Eq. (1.9). Violation of strangeness conservation, as in Eq. (1.8), implies that the decay process must have the long time scale typical of weak interactions. Indeed, the so-called "τ-θ paradox"—the observation that the K-meson could decay to either two or three pions—led directly to the suggestion by Lee and Yang (1956) that the weak interactions were not invariant under the operation of space inversion (parity violation). This is discussed in Chapter 4. An example of the process of associated production is shown in Fig. 1.4.

These early cosmic ray experiments have been described at some length, in order to explain how some of the nomenclature originated, and because they were of the greatest importance in stimulating the construction of high energy, high intensity accelerators, which alone made possible the precision measurements required to put the subject on a quantitative footing.

1.3 BROAD CLASSIFICATION OF PARTICLE STATES

1.3.1 Bosons and Fermions

On the basis of rather general arguments, it can be proved that there is a correspondence between the spin angular momentum and the statistics obeyed by a particle. Particles with half-integral spin ($\frac{1}{2}\hbar$, $\frac{3}{2}\hbar$, $\frac{5}{2}\hbar$, etc.) obey Fermi-Dirac statistics and are thus called *fermions*. Those with integral spin (0, $\hbar$, $2\hbar$, etc.) obey Bose-Einstein statistics and are called *bosons*. Some examples are:

	Spin J (units of $\hbar$)	Particle
Bosons	0	π, K-meson
	1	γ (photon); ρ, ω-mesons
	2	f-meson
Fermions	$\frac{1}{2}$	μ, e, p, n, Λ, Σ
	$\frac{3}{2}$	$N^*(1238)$, $\Omega^-(1675)$
	$\frac{5}{2}$	$N^*(1688)$, $\Lambda(1815)$

The statistics obeyed by a particle are intimately connected with the symmetry of the wave function ψ, describing a pair of identical particles, under interchange. If the particles are identical, the square of the wave function, giving the probability of particle 1 at one coordinate and particle 2 at another, cannot be altered by the interchange $1 \leftrightarrow 2$. Thus

$$|\psi|^2 \xrightarrow{1 \leftrightarrow 2} |\psi|^2,$$

so that

$$\psi \xrightarrow{1 \leftrightarrow 2} \pm \psi.$$

The following rule holds:

$$\left. \begin{array}{l} \text{Identical bosons:} \quad \psi_{\text{total}} \xrightarrow{1 \leftrightarrow 2} + \psi_{\text{total}}; \quad \text{symmetric } (S) \\[4pt] \text{(e.g. } 2\pi^0) \\[4pt] \text{Identical fermions:} \quad \psi_{\text{total}} \xrightarrow{1 \leftrightarrow 2} - \psi_{\text{total}}; \quad \text{antisymmetric } (A) \\[4pt] \text{(e.g. } 2p) \end{array} \right\} \quad (1.11)$$

The wave function ψ describes various properties of the pair of particles; for example, their relative spatial separation, charge, spin, and so on. We may express the total wave function as the product of functions depending on the individual variables. For the moment, consider the space and spin parts, so that

$$\psi_{\text{total}} = \Phi(\text{space}) \times \alpha(\text{spin}).$$

The space wave function Φ will contain a part describing the orbital motion of one particle about the other, which can be represented by a spherical harmonic $Y_l^m(\theta, \phi)$, as described in Section 3.3. Interchange of spatial coordinates of particles 1 and 2 (leaving spin alone) is then equivalent to the replacement $\theta \to \pi - \theta$, $\phi \to \pi + \phi$, which introduces a factor $(-1)^l$ multiplying the function Φ. Thus, if the relative orbital angular momentum, l, of the particles is even (odd), Φ is symmetric (antisymmetric) under the interchange.

The rules embodied in Eq. (1.11) imply that, for identical bosons, Φ and α must be both symmetric or both antisymmetric; while for fermions, a symmetric Φ function implies an antisymmetric α function, and vice versa. As an example, consider the case of two neutral pions. Since these particles have zero spin, the function describing the total spin is necessarily symmetric, so that the space function Φ is also symmetric, with l even. Thus the decay $\rho^0 \to 2\pi^0$, where ρ^0 is a neutral rho-meson with spin $J = 1$, is forbidden by such symmetry requirements.

The phrase "identical particles" needs amplification. If we specify only the spin and orbital angular momentum quantum numbers of the system, "identical" means that the particles must have all other properties (mass, charge, lifetime) the same. We may, however, extend the meaning of identical by including a charge coordinate; for strongly interacting particles this takes the form of an isotopic spin wave function:

$$\psi_{\text{total}} = \Phi(\text{space}) \times \alpha(\text{spin}) \times \chi(\text{isopin}). \tag{1.12}$$

Then "identical" implies any pair of pions, nucleons, and so on, regardless of charge. Since two particles with *all* quantum numbers identical must have ψ symmetric, it follows that two fermions cannot exist in the same quantum state. This is the famous Pauli Principle.

In the following pages we shall appeal time and again to the symmetry of pairs of particles under interchange. The spin-statistics relation has been well checked for photons, electrons, neutrons, protons and, very recently, for muons (Russell *et al.* 1971). No direct checks with other types of particle have been possible, largely because it is difficult to produce them in sufficient intensity; however, there are no reasons whatever to doubt the general validity of the relations (1.11).

1.3.2 Baryons, Leptons, Photons, and Mesons

Fermions and bosons can be further subdivided:

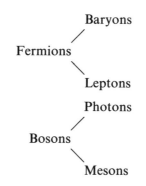

The relativistic wave equation proposed in 1928 by Dirac was able to account for the intrinsic angular momentum, or spin quantum number, of the electron, which had previously been postulated by Uhlenbeck and Goudsmit in order to account for the Zeeman and other effects in atomic physics. A consequence of the Dirac theory was that free electrons were described by 4-component wave functions, which, in the particle rest frame, corresponded to two spin states each of positive and negative energy. The negative energy states were interpreted in terms of an antiparticle, the positron. (See Appendix B.) The existence of positive electrons was demonstrated in cosmic ray experiments by Anderson in 1932 and Blackett and Occhialini a year later. The existence of antiparticles is a general property of fermions, the antiparticle having the same mass as the particle, but opposite charge and magnetic moment. The antiproton was discovered in experiments at Berkeley by Chamberlain *et al.* in 1955 (see Section 2.8). Antideuterons and various antihyperons have since been observed.

Fermions and antifermions can only be created or destroyed in pairs. For example, a γ-ray, in the presence of a nucleus (to conserve momentum), can

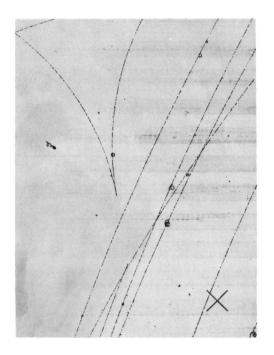
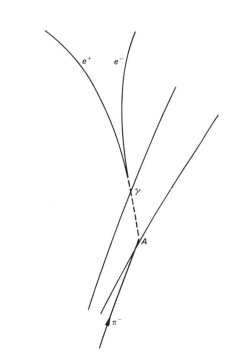

Fig. 1.5 Conversion of a photon into an electron-positron pair in a bubble chamber. The photon originated from the decay $\pi^0 \rightarrow 2\gamma$. Since the neutral pion has a short lifetime (10^{-16} sec), the pair appears to point straight at the interaction vertex, A, corresponding to the charge exchange reaction $\pi^- + p \rightarrow \pi^0 + n$.

"materialize" into an electron-positron pair—see Fig. 1.5; and an e^+e^- bound state, called positronium, annihilates into two or three γ-rays (Section 3.14). *Baryons*, or heavy fermions, comprise all fermions of mass greater than the proton mass; a baryon state is always characterized by the appearance of a neutron or proton as one of the final decay products, as indicated in the table on page x. *Leptons*, or light fermions, include all fermions of mass less than the proton, and do not include neutrons or protons as decay products. The pair creation of fermions is formally expressed by assigning an additive quantum number, called the *baryon number B* or *lepton number L*, which is absolutely conserved.

The known leptons consist of the electron e^-, muon μ^-, and neutrino ν. Each is assigned a lepton number $L = +1$; the antileptons, consisting of the positron e^+, positive muon μ^+, and antineutrino $\bar{\nu}$, have $L = -1$. All other particles have $L = 0$. Lepton conservation means: total lepton number $L =$ number of leptons $-$ number of antileptons $=$ a constant.

A similar situation exists for baryons and antibaryons. The experimental evidence for conservation of L and B is very strong. Thus, the lifetime for the decay of the proton into electrons or γ-rays is greater than 10^{21} years. The best evidence for conservation of L comes from the absence of double β-decay, as discussed in Section 4.9. Examples of these rules are the following processes:

$$\pi^+ \rightarrow \mu^+ + v_\mu; \qquad\qquad\qquad \gamma \rightarrow e^+ + e^-;$$

$$L(= L_\mu): \quad 0 \quad -1 \quad +1 \qquad\qquad L(= L_e): \quad 0 \quad -1 \quad +1$$

$$v_e + n \rightarrow e^- + p; \qquad\qquad p + n \rightarrow \bar{p} + n + p + p;$$

$$L(= L_e): \quad +1 \quad 0 \quad +1 \quad 0 \qquad\qquad B: \quad +1 \quad +1 \quad -1 \quad +1 \quad +1 \quad +1$$

$$\mu^+ \rightarrow e^+ + v_e + \bar{v}_\mu; \qquad\qquad p + \bar{p} \rightarrow 5\pi.$$

$$L_e: \quad 0 \quad -1 \quad +1 \quad 0 \qquad\qquad B: \quad +1 \quad -1 \quad 0$$

$$L_\mu: \quad -1 \quad 0 \quad 0 \quad -1$$

$$L: \quad -1 \quad -1 \quad +1 \quad -1$$

Experiments in Brookhaven and CERN in 1962/3 showed the existence of two types of neutrino; one (labeled v_e) associated with electrons (as in β-decay), and the other (labeled v_μ) associated with muons (as in $\pi \rightarrow \mu$ decay). To account for these observations, we write $L = L_e + L_\mu$, where L_e is the electron number and L_μ the muon number, these two quantities being separately conserved. The present experimental upper limits on possible nonconservation of L_e and L_μ are of order one part per million.

In contrast to the situation for fermions, there are no absolute rules on number conservation for bosons. The bosons, or integral spin objects, can be divided into *photons* (of zero rest mass) and *mesons*. Antiparticles, as usual, carry the opposite charge and magnetic moment to the particles. The term "meson," meaning particle of mass intermediate between electron and proton, was originally applied to the muon, now classified as a lepton. "Meson" has now lost its original meaning, and refers to any boson of finite rest mass.

A word may be added here about the definition of spin angular momentum of particles. An atom is said to have an intrinsic angular momentum J (in units of $\hbar$) if it can exhibit $(2J + 1)$ components J_z, of angular momentum along the quantization axis (z-axis). Thus, if a magnetic field is applied along the z-axis, one can observe $(2J + 1)$ separate energy levels, owing to interaction of the atomic moment and the field. The x-, y-, and z-components of J are here defined in the rest frame of the atom or particle (only J_z has nonzero eigenvalues). For particles of zero rest mass, traveling with light velocity c, like photons or neutrinos, one cannot make observations in the rest frame; in this case, there are only two measurable components of the spin, $J_z = \pm J$, the z-axis being in the direction of motion of the particle. Thus, massless particles are completely spin-polarized either along or against the direction of motion, this being the only possible description of the spin which is independent of the reference frame.

1.3.3 Hadrons and Strangeness

Mesons and baryons, which undergo strong interactions, are sometimes collectively called *hadrons*; in contrast, *photons* and *leptons* undergo only electromagnetic and/or weak interactions. As we have mentioned above, it has been found necessary to introduce two types of hadrons—*strange* and *nonstrange*. The allocations of the strangeness quantum number to the kaons and hyperons are indicated in Table 3.2, p. 91. They follow from the assignment of the isospin quantum numbers to the various particle states, as discussed in Chapter 3; however, the introduction of strangeness is more elegant than a description in terms of isospin alone. Conservation of strangeness incorporates simultaneous conservation of charge, baryon number, and the third component of isospin, as in Eq. (3.36).

   In place of the strangeness quantum number S, an alternative quantity, called the *hypercharge*, Y, is frequently employed. This is the sum of the strangeness and the baryon number

$$Y = B + S.$$

1.4 TYPES OF INTERACTION

1.4.1 Electromagnetic Interactions

As already noted, each type of interaction is typified by an inherent strength or coupling constant. This coupling constant enters in the matrix element for the process under consideration, which when squared gives the decay probability or cross section. In electromagnetic phenomena involving the interactions between charged particles and photons, the characteristic coupling constant is

$$\frac{e^2}{\hbar c} = \alpha \simeq \tfrac{1}{137}. \tag{1.13}$$

The quantity α is called the fine structure constant, because it determines the magnitude of the spin-orbit splitting in atomic spectra. As an example, the cross section for Compton scattering of a photon by an electron is of order $(\hbar/mc)^2(e^2/\hbar c)^2$. Here the magnitude of length involved is the electron Compton wavelength $\hbar/mc$. We can visualize the process as the absorption of a photon by an electron followed by its reemission as in Fig. 1.6(a). At each vertex we put a factor e, so that when the amplitude is squared we obtain a factor e^4 or α^2. For Compton scattering by a proton, the cross section would be $(\hbar/Mc)^2\alpha^2$. For pair production in the field of a nucleus Z, as in Fig. 1.6(b), photons are emitted or absorbed at three vertices, so that the cross section is of order $(\hbar/mc)^2\alpha^3$. Relatively few particle states decay principally via the electromagnetic interaction; it occurs where decay by strong interactions is forbidden by the conservation laws. Examples are

$$\pi^0 \rightarrow 2\gamma,$$
$$\Sigma^0 \rightarrow \gamma + \Lambda.$$

In the π^0-decay one of the γ's may undergo "internal conversion" to give a so-called Dalitz pair, $\pi^0 \rightarrow e^+e^-\gamma$; the branching ratio for this process is of

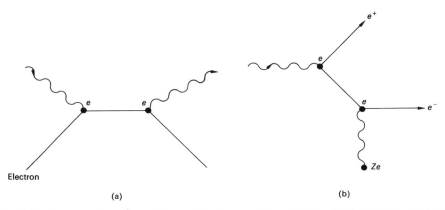

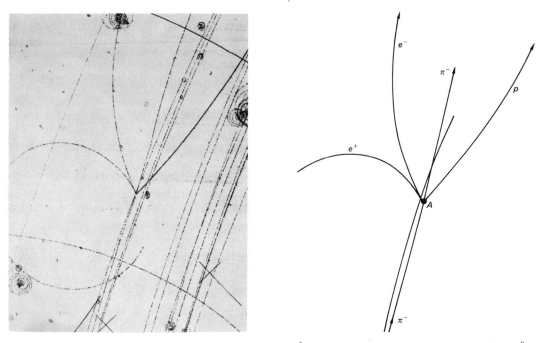

Fig. 1.6 Diagrammatic representation of electromagnetic processes: (a) Compton scattering; (b) pair production in the field of a nucleus, *Z*.

Fig. 1.7 Example of the Dalitz decay of a neutral pion, $\pi^0 \to \gamma + e^+e^-$, in a bubble chamber. The π^0 is created at the point *A*, in a π^-p inelastic collision. The π^0 travels a distance of $<1\ \mu$ in a mean lifetime, so that the e^+e^- pair appears to come directly from the interaction.

order α. (See Fig. 1.7.) The lifetimes involved in these processes are of the order of 10^{-16} sec, i.e. long compared with the characteristic nuclear time, 10^{-23} sec. Electromagnetic decays do not necessarily have to include real photons as end products. For example, the decay of the eta-meson,

$$\eta \to \pi^+ + \pi^- + \pi^0,$$

is electromagnetic, since it is forbidden by the selection rules for strong interactions (Section 3.16).

1.4.2 Strong Interactions

Although the strong interactions between the hadrons cannot be characterized by a unique coupling constant, it is clear that the strong coupling, as judged from observed cross sections, is much larger than the electromagnetic coupling. For example, at about 1 GeV incident energy, the total pion-nucleon cross section is about 10^{-26} cm^2, compared with 10^{-29} cm^2 for the electromagnetic process of pion photo-production, $\gamma + p \to \pi^0 + p$. It is possible to analyze the low energy pion-nucleon elastic scattering in terms of a dimensionless coupling constant. If this cross section is measured in terms of the nucleon Compton wavelength, one obtains for this particular strong coupling

$$\frac{g^2}{\hbar c} \simeq 15. \tag{1.14}$$

The lifetimes of states decaying by strong interactions are of the order of the "characteristic nuclear time." This is equal to the range R_0 of the strong forces, of order $\hbar/m_\pi c = 1.4 \times 10^{-13}$ cm, divided by c, and is thus $\sim 10^{-23}$ sec. Because of the very short lifetime τ, such states are generally rather broad, the width being given by

$$\Gamma = \frac{\hbar}{\tau},$$

i.e. of order 10 to 100 MeV. Such states are called *resonances*, since they can be formed by collision of the particles into which they decay, and are then signified by a bump or resonance in the collision cross section at the appropriate center-of-mass energy (equal to the mass of the resonance). An example is the enhancement or resonance peak appearing in the $\pi^+ p$ elastic scattering cross section, for a pion-proton invariant mass of 1236 MeV, and called the $N^*(1236)$ state:

$$\pi^+ + p \to N^*(1236) \to \pi^+ + p.$$

The width of the resonance in this case is 120 MeV. Although it does not have a unique mass and has an extremely transient existence, traveling only 10^{-13} cm before decay, it has definite quantum numbers (spin-parity $\frac{3}{2}^+$, isospin $\frac{3}{2}$), and is thus a well-defined state. Its lifetime cannot, of course, be measured, but is inferred from the observed width.

The conventional notation is that states decaying by strong interactions are called resonances, and those decaying by weak or electromagnetic interactions are called particles. Generally, one can measure the lifetime, but not the width, for particles. Thus, a pion decays weakly, with a measured lifetime $\tau = 2.5 \times 10^{-8}$ sec; the characteristic decay length $c\tau$ is therefore quite large—7.5 meters. Its width, according to the above formula, is 2×10^{-8} eV and thus unmeasurable.

1.4.3 Weak Interactions

The weak interactions, as exemplified by nuclear β-decay, are described in terms of the Fermi constant G. In cgs units, $G = 1.4 \times 10^{-49}$ erg cm^3, and the rate of β-decay is proportional to G^2 and to kinematic factors (phase space). A dimensionless measure of the weak interaction can only be obtained if one defines a length, for example, the pion Compton wavelength. This fact arises because the proposed carrier of the weak field, the intermediate vector boson W—analogous to the photon of the electromagnetic field—has not yet been observed, and its mass is unknown. So, the weak interaction could be inherently weak, or rather strong but of extremely short range. If we take the pion Compton wavelength as the dimension of length we obtain

$$G = 1.4 \times 10^{-49} \text{ erg cm}^3 = 2.3 \times 10^{-7} \hbar c \left(\frac{\hbar}{m_\pi c}\right)^2 = 10^{-5} \hbar c \left(\frac{\hbar}{M_p c}\right)^2, \quad (1.15)$$

or

$$\frac{(\sqrt{G})^2}{\hbar c} \simeq 10^{-7} \left(\frac{\hbar}{m_\pi c}\right)^2,$$

where $\sqrt{G}$ can be thought of as the "weak charge" carried by a particle analogous to the electromagnetic charge, e. It will be noted that this value is very small compared with $e^2/\hbar c \sim 10^{-2}$ or $g^2/\hbar c \sim 10$ for the other interactions.

As one might expect, the cross sections for weak reactions are extremely feeble. For example, at 1 GeV bombarding energy, the cross section for weak pion production by neutrinos,

$$v + p \rightarrow \pi^+ + p + \mu^-,$$

is only about 10^{-38} cm^2, i.e. some 10^{-12} times that for typical hadron cross sections. The strong interaction cross sections correspond to mean free paths in solid materials measured in terms of centimeters, whilst the neutrino mean free path is of the order of the earth-sun distance! The lifetimes for decay via the weak interactions are correspondingly long. Nuclear β-decay lifetimes range from 10^{-6} sec to thousands of years. For the decays of pions, kaons, and hyperons, the lifetimes are in the range 10^{-10} to 10^{-8} sec, i.e. of order 10^{14} times longer than the characteristic nuclear time, and thus typical of the weak interactions.

TABLE 1.1

Interaction	Typical coupling constant	Typical cross sections	Typical lifetimes
Strong	$g^2/\hbar c \sim 10$	10^{-26} cm^2 = $10^4\,\mu$b	10^{-23} sec
Electromagnetic	$e^2/\hbar c = \frac{1}{137}$	10^{-29} cm^2 = $10\,\mu$b	10^{-16} sec
Weak	$\sim 10^{-7}$	10^{-38} cm^2 = $10^{-8}\,\mu$b	10^{-8} sec
Gravitational	$\sim 10^{-45}$	—	—

A summary of the couplings in the various interactions is given in Table 1.1.

1.4.4 Gravitational Interactions

These do not concern us here, but we mention them for completeness. If K is the Newtonian constant of gravitation ($K = 6.7 \times 10^{-8}$ cgs units), Km^2 has the same dimensions as e^2. If we take for m the electron mass, then the dimensionless gravitational constant becomes

$$Km^2/\hbar c = 2 \times 10^{-45}, \tag{1.16}$$

and is thus enormously weaker than the other three interactions.

A quantum, called the *graviton*, can be postulated for the gravitational field, just as for the other interactions. Since the gravitational field has infinite range, the graviton mass, like that of the photon, is zero (see Eq. 1.4). The graviton has spin $J = 2$. This is a consequence of the fact that, whereas positive and negative electric charges—and thus electric dipole oscillators—exist, negative mass does not. The simplest radiator of gravitons is therefore an oscillating mass quadrupole. Intensive searches for cosmic pulses of gravitational radiation (gravitons) are currently under way (Weber, 1971).

1.5 CONSERVATION RULES

From the foregoing discussion it will be clear that the numerical strengths of the various interactions are not too well defined. However, they are also quite clearly differentiated in terms of the type of conservation rules which they are found to obey. These are listed in Table 1.2. Some of the symbols have been explained above and the rest will be discussed in Chapter 3.

Just like charge and energy, momentum and angular momentum, baryon and lepton number are absolutely conserved in all interactions. Other quantities, conserved in the strong interactions, are violated in weak and electromagnetic interactions. These violations usually occur in a regular way, however. Thus, the weak interactions appear to obey the rule $\Delta I = 1$ for nonstrange decays, and $\Delta I = \frac{1}{2}$ for $\Delta S = 1$ decays, while $\Delta S = 2$ transitions are forbidden.

TABLE 1.2

Conserved quantity	Interaction		
	Strong	Electromagnetic	Weak
B (baryon no.)	Yes	Yes	Yes
L (lepton no.)	Yes	Yes	Yes
I (isospin)	Yes	No	No ($\Delta I = 1$ or $\frac{1}{2}$)
G (G-parity)	Yes	No	No
S (strangeness)	Yes	Yes	No ($\Delta S = 1$)
P (parity)	Yes	Yes	No
C (charge conjugation parity)	Yes	Yes	No
CP	Yes	Yes	Yes (but 10^{-3} violation in K^0 decay)
CPT	Yes	Yes	Yes

1.6 UNITS

The familiar systems of units (cgs, mks) employed in physics are not very appropriate in nuclear physics. Lengths are normally quoted in terms of the *fermi* (1 fermi $= 10^{-13}$ cm), and cross sections in *barns* (1 barn $= 10^{-24}$ cm^2), millibarns (10^{-27} cm^2), or microbarns (10^{-30} cm^2). The unit of energy is basically the *electron volt* (1 eV $= 1.6 \times 10^{-12}$ ergs), with the larger units MeV (10^6 eV) and GeV (10^9 eV). Masses of particles are usually measured in MeV/c^2, meaning that if the mass is M the rest energy is Mc^2 MeV. For brevity, the mass is frequently given as the value of Mc^2 in MeV. Thus, the proton mass is 938.256 MeV, or 0.938 GeV. Similarly, the momentum p multiplied by c has the dimensions of energy, so that the momentum is quoted in MeV/c or GeV/c. It is in practice important to distinguish between the momentum and energy of a beam of particles, so that although the terminology "MeV" is applied indiscriminately to both mass and energy, momenta are invariably quoted in MeV/c.

In actual theoretical calculations, the quantities $\hbar = h/2\pi$ and c occur repeatedly. This is a nuisance which can be avoided by choosing as the unit of mass that of some standard particle, m_0 (for example, the proton mass). Thus we set

$$m_0 = 1.$$

The natural unit of length is then taken as the Compton wavelength of the standard particle:

$$\lambda_0 = \frac{\hbar}{m_0 c} = 1;$$

that of time as:

$$t_0 = \frac{\lambda_0}{c} = \frac{\hbar}{m_0 c^2} = 1;$$

and that of energy as:

$$E_0 = m_0 c^2 = 1.$$

In these "natural" units, we see that $\hbar = c = 1$. The standard particle may be chosen at will, a usual choice being the proton. Then the unit of energy is equivalent to 938 MeV, of length 0.21 fermi $= 2.1 \times 10^{-14}$ cm, etc.

As an example of the use of these "natural units," let us calculate the cross section for elastic scattering of neutrinos on electrons at high energy. The cross section is then simply the product of the weak interaction constant, squared, times a phase-space factor, equal to the square of the center-of-mass momentum, p, of either particle:

$$\sigma = \frac{4G^2}{\pi} p^2.$$

From Eq. (1.15), we have, with $\hbar = c = 1$, and in terms of the proton mass as the unit,

$$G = 10^{-5}/M_p^2.$$

Then,

$$\sigma \approx \frac{10^{-10}}{M_p^2} \left(\frac{p^2}{M_p^2} \right).$$

This is the square of a length. Taking the proton mass as the standard, the unit of length is the proton Compton wavelength, so that

$$\sigma \approx 10^{-10} \left(\frac{2}{10^{14}} \right)^2 p^2 \approx 10^{-38} p^2 \text{ cm}^2,$$

where p is in units of $M_p c = 0.94$ GeV/c, or 1 GeV/c approximately. If the same calculation is carried out using cgs units, for example, many powers of h and c are involved, and the possibility of arithmetic errors is much greater.

BIBLIOGRAPHY

Gell-Mann, M., and A. H. Rosenfeld, "Hyperons and heavy mesons (systematics and decay)," *Ann. Rev. Nucl. Science* **7**, 407 (1957).

Jackson, J. D., *The Physics of Elementary Particles*, Princeton University Press, Princeton, 1958.

Nishijima, K., *Fundamental Particles*, Benjamin, New York, 1963.

Segre, E., *Nuclei and Particles*, Benjamin, New York, 1964, Chs. 13, 14.

Swartz, C. E., *The Fundamental Particles*, Addison-Wesley, Reading, Mass., 1965.

Yang, C. N., *Elementary Particles*, Princeton University Press, Princeton, 1961.

Older reviews of historic interest:

Leprince-Ringuet, L., "Mesons and heavy unstable particles in cosmic rays," *Ann. Rev. Nucl. Science* **3**, 39 (1953).

Powell, C. F., "Mesons," *Rep. Prog. Phys.* **13**, 350 (1950).

PROBLEMS

1.1 The law of radioactive decay of free muons may be written $dN/dt = -\lambda_d N$, where $\lambda_d = 1/\tau$ is the decay constant and $\tau = 2.16$ μsec. For a negative muon captured in an atom of atomic number Z, the decay constant is $\lambda = \lambda_d + \lambda_c$, where λ_c is the probability per unit time of nuclear capture. For aluminum ($Z = 13, A = 27$) the mean lifetime for decay of negative muons has the value 0.88 μsec. Calculate λ_c, and assuming Eq. (1.6) to be valid, compute the interaction mean-free-path Λ of a muon in nuclear matter. From this quantity, estimate the magnitude of the coupling constant in the reaction $\mu^- + p \to n + \nu$.

1.2 In calcium ($Z = 20, A = 40$) the negative muon lifetime is 0.33 μsec. As in Problem 1.1, calculate Λ for this case. How do you account for the difference as compared with aluminum? [*Hint:* consider effects of nuclear size (Chapter 5) and of the Pauli principle.]

1.3 The cross section for the reaction $\pi^- + p \to \Lambda + K^0$ at 1 GeV/c incident momentum is approximately 1 mb (10^{-27} cm²). Both Λ- and K^0-particles decay with a mean lifetime of about 10^{-10} sec. From this information, estimate the relative magnitude of the couplings responsible for the production and decay, respectively, of the Λ- and K^0-particles.

1.4 Compute the maximum and minimum laboratory energies of the γ-rays produced in the decay of neutral pions, of mass m and velocity βc. Show that, for $\beta \sim 1$, $E_{\max}$ is equal to the total pion energy. Verify the formula on p. 7 for nonrelativistic neutral pions. (For relativistic transformations, see Appendix A.)

1.5 State which of the following reactions are allowed by the conservation laws and which are forbidden, and give the reasons in either case:

$$\pi^0 \to e^+ + e^-, \tag{i}$$

$$p \to n + e^+ + \nu_e, \tag{ii}$$

$$\mu^+ \to e^+ + e^- + e^+, \tag{iii}$$

$$K^+ + n \to \Sigma^+ + \pi^0. \tag{iv}$$

1.6 It was once suggested that the numerical values of the charge of electron and proton might differ by a small amount $|\Delta e|$, so that expansion of the universe could be attributed to an electrostatic repulsion between hydrogen atoms in space. Estimate the minimum value of $|\Delta e/e|$ required for this hypothesis. [References: Theory; H. Bondi, and R. A. Lyttleton, *Nature* **184**, 974 (1959); *Proc. Roy. Soc.* **A252**, 313 (1959). Experimental disproof; A. M. Hillas, and T. E. Cranshaw, *Nature* **184**, 892 (1959).]

Chapter 2

Experimental Methods

The first sections of this chapter on experimental methods deal briefly with the interaction of charged particles and radiation with matter in bulk. All methods of detecting particles depend ultimately on this interaction. The derivation of the formulas quoted can be found in the bibliography at the end of the chapter.

2.1 IONIZATION LOSS OF CHARGED PARTICLES IN MATTER

An energetic charged particle traversing a medium loses energy continuously in collisions with atomic electrons. These lead to excitation and ionization of the atoms. The number of collisions in traversing an interval dx g cm^{-2} of the medium, resulting in energy transfers $E' \rightarrow E' + dE'$, is

$$f(E') \, dE' \, dx = \frac{2\pi z^2 e^4 N_0 Z}{mv^2 A} \frac{dE'}{E'^2} \left(1 - \frac{v^2}{c^2} \frac{E'}{E'_{max}} \right) dx, \qquad (2.1)$$

where v and ze are the velocity and charge of the incident particle (assumed spinless), N_0 is Avogadro's number, Z is the atomic number and A the mass number of the medium, and m the electron mass. Equation (2.1) results from a quantum-mechanical calculation; for $v^2 \ll c^2$, a classical impact-parameter approach gives the same result. The formula is valid provided the particle mass, M, is large compared with m. If the total energy of the incident particle $E \ll M^2 c^2/m^2$—a condition valid in practice for all incident particles other than electrons—then the maximum kinematically allowed energy transfer is

$$E'_{max} = 2mv^2/(1 - \beta^2), \qquad (2.2)$$

where $\beta = v/c$. The Bethe-Bloch formula for the total rate of ionization energy loss per g cm^{-2} traversed is given by

$$\frac{dE}{dx} = \frac{2\pi N_0 Z z^2 e^4}{mv^2 A}\left[\ln\left(\frac{2mv^2 E'_{max}}{I^2(1-\beta^2)}\right) - 2\beta^2\right],$$

$$= \frac{4\pi N_0 Z z^2 e^4}{mv^2 A}\left[\ln\left(\frac{2mv^2}{I(1-\beta^2)}\right) - \beta^2\right], \tag{2.3}$$

where the quantity I is an effective ionization potential averaged over all electrons. An approximate expression for I suggested by the Thomas-Fermi model of the atom is

$$I = 10Z \text{ eV}. \tag{2.4}$$

The important features of the ionization loss formula are that:

i)   The dependence on the properties of the incident particle is expressed through the relation

$$\frac{dE}{dx} = z^2 f(v),$$

independent of the mass, M.

ii)   For a given value of z, dE/dx varies as $1/v^2$ at nonrelativistic velocities. After passing through a minimum value, dE/dx increases very slowly with increasing value of $\gamma = E/Mc^2 = (1 - \beta^2)^{-1/2}$. This variation is shown in Fig. 2.1.

iii)   Apart from a weak dependence through I in the logarithmic term, dE/dx depends on the medium chiefly through variations in Z/A. Numerically, for singly charged particles of near minimum ionization, dE/dx varies from approximately 2 MeV g^{-1} cm^2 for light elements to 1 MeV g^{-1} cm^2 for heavy elements.

The formulas above refer strictly to heavy, spinless particles. The effect of *spin* is to increase the probability of high energy transfers, in the form of extra terms in (2.1); the effect on the mean energy loss, (2.3), is negligible, however. For incident *electrons*, the ionization loss is typically a few per cent less than that given by Eq. (2.3); what is far more important in the case of electrons is that additional radiation losses are severe (see Section 2.5.2).

The increase in dE/dx at relativistic velocities, $\gamma \gg 1$, predicted by Eq. (2.3) arises from two causes. One is that the maximum transferable energy E'_{max} in Eq. (2.2) increases as γ^2. These high energy electrons, or δ-rays as they are called, themselves ionize the medium and produce characteristic tracks branching off the main trajectory. The other cause is that the transverse component of the electric field due to the incident particle, when transformed into the laboratory frame, acquires a relativistic factor γ. Thus the field due to the

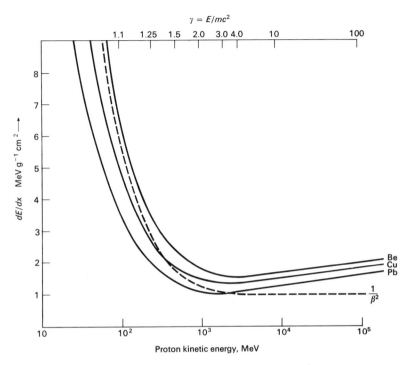

Fig. 2.1 Value of the ionization energy loss, dE/dx, in MeV g^{-1} cm^2, as a function of kinetic energy for protons, for beryllium, copper, and lead. The dashed curve corresponds to a $1/\beta^2$ dependence. (After Sternheimer, 1961.)

charged particle extends to greater distances from the trajectory, and more "distant" collisions become important. The effect of the high energy transfers can be excluded if we consider only the local ionization loss (say $E' < 10$ keV), such as is relevant to formation of tracks in a bubble chamber or cloud chamber.

The logarithmic rise in ionization from the second cause, the effect of distant collisions, does not continue indefinitely, however, because the electric field of the incident particle polarizes the medium and this produces a shielding effect, so that the ionization rate flattens off to a constant value. This factor, which is not included in Eq. (2.3), is important whenever the impact parameter becomes comparable with interatomic distances, so that the dielectic properties of the medium have to be considered. The effect is often called the density effect, because it is proportional to the electron concentration in the medium, and is therefore more important for solids and liquids than for gases. The result is that, for $\gamma > 10$, the local ionization reaches a plateau value of up to 10% above the minimum value in most solids and liquids. For gases, as shown in Fig. 2.2, the plateau is much higher and not attained until correspondingly higher energies.

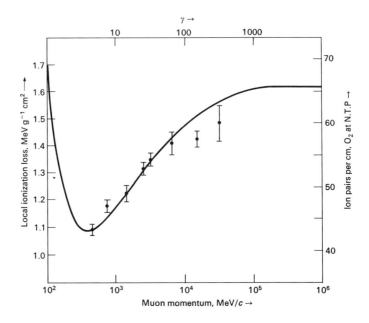

Fig. 2.2 Measurements of the local ionization loss (energy transfers < 1 keV) of cosmic ray muons in oxygen, as determined by counting droplets along the tracks in a cloud chamber (Ghosh *et al.*, 1954). The curve shows the predicted variation, taking into account the density effect (Sternheimer, 1961). Note that the plateau, at about 1.5 times the minimum value, is reached for $\gamma > 100$.

2.2 RANGE-ENERGY RELATION

An important practical result of the ionization energy loss of charged particles is that there is a relation between the energy or momentum of a particle and its *range* in an homogeneous medium. A measurement of range frequently affords the most accurate method of determining the energy of a particle. If the initial momentum is known, for example from the radius of curvature in a magnetic field, the range allows one to determine the mass, in terms of that of a particle for which the range-momentum relation is already known. The first accurate determinations of the masses of the muons and charged pions and kaons were made in this way.

For a particle of initial kinetic energy T, the range is

$$R = \int_0^T \frac{dE}{(dE/dx)} . \tag{2.5}$$

For $v^2 \ll c^2$, Eq. (2.3) suggests $(dE/dx)^{-1} \propto v^2 \propto E$. Thus, one might expect $R \propto T^2$. This is not true in practice, because of the velocity term in the logarithm of Eq. (2.3) and because I itself is weakly velocity-dependent.

Empirically, the power law relation,

$$R = \text{const. } T^{1.75},$$

holds for medium weight elements and $0.1 < \beta < 0.7$.
From the expressions

$$\frac{dE}{dx} = z^2 f(v),$$

and

$$T = Mg(v),$$

it follows that

$$R = \frac{M}{z^2} \phi(v) = \frac{M}{z^2} F\left(\frac{T}{M}\right). \qquad (2.6)$$

From this scaling law, for example, the range R_π of a pion of kinetic energy T_π can be simply obtained if we know the range-energy relation for protons:

$$R_\pi = \left(\frac{M_\pi}{M_p}\right) R_p,$$

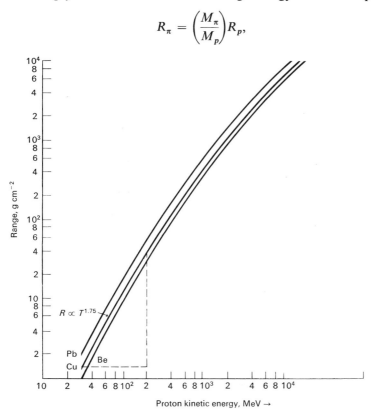

Fig. 2.3 Range-energy relation for protons in different materials, as calculated by Sternheimer (1961). Note that, in the low energy region, there is an approximate power-law relation $R \propto T^{1.75}$.

where R_p is the range of a proton of energy

$$T_p = (M_p/M_\pi)T_\pi.$$

The range-energy relation for protons in various elements is given in Fig. 2.3.

The range does not provide a unique determination of initial energy and the above relations refer to the mean value, $\langle R \rangle$. Fluctuations about the mean value, called *straggling*, result from the statistical fluctuations in the number of ionizing collisions along the particle trajectory. (See Problem 2.1). As an example, the fractional rms fluctuation in range, $\sigma_R/\langle R \rangle$, is of order 4% for protons of $\beta = 0.1$ in medium weight elements; it is a slowly varying function of initial velocity v, and, for given v, varies as $M^{-1/2}$.

2.3 CERENKOV RADIATION

We have seen that, when high energy charged particles traverse dielectric media, the relativistic increase in the local ionization loss predicted by the Bethe-Bloch formula is halted by the polarization or density effect. A phenomenon closely related to the polarization effect is the production of *Cerenkov radiation*, which accounts for a part of the relativistic increase in energy loss by fast charged particles. The Cerenkov effect, so named after its discoverer, provides an important experimental method of measuring particle velocity and of differentiating particles of different velocities, in the extreme relativistic region (see Section 2.7.5). Cerenkov radiation is produced whenever the velocity βc of the charged particle exceeds c/n, where n is the refractive index of the medium. The Huyghens construction of Fig. 2.4 shows that radiation from

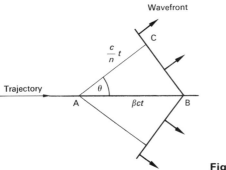

Figure 2.4

excited atoms along the path AB of the particle can form a coherent wavefront BC at one particular angle. The wavefront forms the surface of a cone with axis along the trajectory, the normals AC to the wavefront making an angle θ to AB given by

$$\cos\theta = \frac{c}{n}\, t/\beta ct = \frac{1}{\beta n}; \qquad \beta > \frac{1}{n}. \tag{2.7}$$

Cerenkov radiation appears as a continuous spectrum, and, in a dispersive medium, n and θ will be functions of the particular frequency considered. For water, $n = 1.33$ in the visible region of the spectrum, and charged particles of $\beta > 0.75$ will therefore produce visible Cerenkov light in water. This is beautifully demonstrated by the "swimming pool" nuclear reactor; the fast electrons from β-decay of fission fragments produce intense blue light as they pass through the water moderator.

The total energy appearing in Cerenkov radiation per unit length of track is

$$\frac{dE}{dx} = \frac{4\pi^2 z^2 e^2}{c^2}\int\left(1 - \frac{1}{\beta^2 n^2}\right)v\,dv, \tag{2.8}$$

where the integration over frequency v extends over the range for which $\beta n > 1$. The dependence on the density of the medium is only through the refractive index n. Over a small range of frequencies, we can neglect the dependence of n on v and, with some rearrangement, Eq. (2.8) becomes

$$\frac{dE}{dx} = \frac{z^2}{2}\left(\frac{e^2}{\hbar c}\right)^2\left(\frac{mc^2}{e^2}\right)\left[\frac{(hv_1)^2 - (hv_2)^2}{mc^2}\right]\left(1 - \frac{1}{\beta^2 n^2}\right)_{\text{average}}. \tag{2.9}$$

For a singly charged particle of $\beta \sim 1$ in water, this expression gives a value for dE/dx of 400 eV/cm for visible light ($\lambda = 4000$ to 7000 Å), and thus about 200 photons/cm. Note that this is very small compared with the total ionization energy loss, of order 2 MeV/cm.

2.4 COULOMB SCATTERING

In traversing a medium, a charged particle suffers electromagnetic interactions with both electrons and nuclei. As Eq. (2.3) indicates, the energy transfer is inversely proportional to the mass of the target particle, and in comparison with electrons, that lost in Coulomb collisions with nuclei is negligibly small. On the other hand, because of the greater mass of the target, the particle is *scattered* appreciably in these collisions.

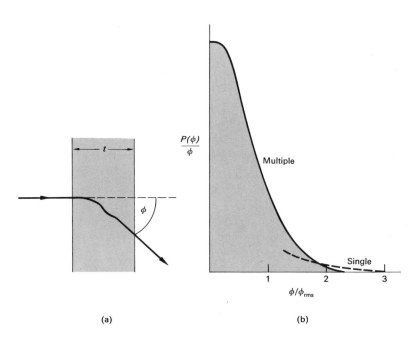

(a)   (b)

Fig. 2.5 (a) Multiple scattering of a charged particle in traversing a layer of material of thickness t. (b) If the number of particles scattered through angle $\phi \rightarrow \phi + d\phi$ is $P(\phi)\, d\phi$, the distribution of $P(\phi)/\phi$ is approximately Gaussian. For very large deflections, however, the main contribution comes from single scattering (shown dotted).

The differential cross section per nucleus for scattering the particle into the element of solid angle $d\Omega$ at angle θ is given by the familiar Rutherford formula

$$\frac{d\sigma(\theta)}{d\Omega} = \frac{1}{4}\left(\frac{Zze^2}{pv}\right)^2 \frac{1}{\sin^4 \dfrac{\theta}{2}}, \tag{2.10}$$

where p, ze, and v denote the momentum, charge, and velocity of the particle, Z is the atomic number of the scattering medium, and the nucleus is assumed to act like a point charge. This formula refers strictly to the center-of-mass frame of the colliding particles. It is valid also in the laboratory, provided θ is small (i.e. $\cos \theta \approx 1$), and in that approximation also holds relativistically. Equation (2.10) refers to a *single* act of scattering. It can be seen that, if θ is small enough, and the particle traverses a thick layer of material, the probability of scattering will become very large—in other words, one obtains the phenomenon of *multiple* scattering (Fig. 2.5, see also Fig. 1.1). The resultant space angle ϕ, compounded of a large number of small deviations in a thick layer t of material, follows a roughly Gaussian distribution:

$$P(\phi)\, d\phi = \frac{2\phi}{\langle\phi^2\rangle} \exp\left[\frac{-\phi^2}{\langle\phi^2\rangle}\right] \times d\phi, \tag{2.11}$$

where the mean square deflection $\langle \phi^2 \rangle$ is given approximately by

$$\langle \phi^2 \rangle = z^2 \left(\frac{E_s}{\rho \beta c} \right)^2 \frac{t}{X_0},$$ (2.12)

where

$$E_s = \sqrt{4\pi \times 137} \times mc^2 = 21 \text{ MeV},$$

and

$$\frac{1}{X_0} = \frac{4Z(Z+1)r_e^2 N_0}{137A} \ln \left(\frac{183}{Z^{1/3}} \right).$$ (2.13)

The quantity X_0, for reasons that will appear below, is called the *radiation length* of the medium. In Eq. (2.12) all the properties of the medium are incorporated in X_0. Values of X_0 for different elements are given in Table 2.1.

TABLE 2.1 Radiation lengths in various elements (after Bethe and Ashkin, 1953)

Element	Z	E_c (MeV)	X_0 (g/cm^2)
Hydrogen	1	340	58
Helium	2	220	85
Carbon	6	103	42.5
Aluminum	11	47	23.9
Iron	26	24	13.8
Lead	82	6.9	5.8

Multiple Coulomb scattering is of considerable importance in many high energy physics experiments. For example, it often limits the accuracy with which one can determine the momentum of a particle from the magnetic curvature of the track in a bubble chamber. For simplicity, let us suppose that the momentum p of a particle of unit charge is normal to the field B. The particle then travels in a circular path, of radius ρ, where $pc = Be\rho$. After traveling a distance s, the magnetic deflection will be

$$\phi_B = \frac{s}{\rho} = \frac{Bes}{pc} = \frac{0.3Bs}{pc},$$ (2.14)

where B is in kG, s in cm, and pc in MeV. On the other hand, the root mean square angle of scattering in the plane of the track will be, from Eq. (2.12),

$$\phi_{sc} = \frac{21}{\sqrt{2}} \frac{1}{p\beta c} \sqrt{\frac{s}{X_0}},$$ (2.15)

where the factor $1/\sqrt{2}$ is included because we are considering the projection of the space angle in a plane. Thus, the fractional uncertainty in the magnetic deflection has an rms value

$$\frac{\phi_{sc}}{\phi_B} = \frac{50}{\beta B \sqrt{sX_0}} \, . \tag{2.16}$$

The error can thus be made small by employing a high field and long track lengths. For a hydrogen bubble chamber, we have typically $B = 20$ kG, $s = 25$ cm, and $X_0 = 1000$ cm, so that for relativistic particles ($\beta = 1$), $\phi_{sc}/\phi_B \approx 0.016$. For heavier liquids like propane or freon, X_0 is much less than in hydrogen and the errors on momentum measurement are correspondingly larger (see Table 2.1).

2.5 ABSORPTION OF γ-RAYS IN MATTER, AND RADIATION LOSS BY FAST ELECTRONS

There are three main types of phenomenon responsible for the absorption of energetic γ-rays in matter. They are:

  i) photoelectric absorption;

  ii) Compton scattering;

  iii) pair production.

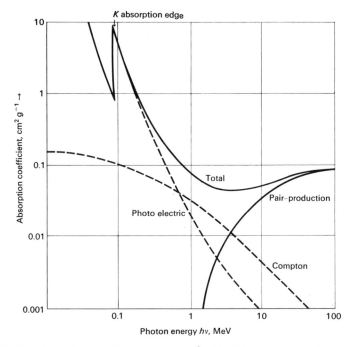

Fig. 2.6 The absorption coefficient per g cm^{-2} of lead for γ-rays as a function of quantum energy.

Figure 2.6 indicates the contributions to the absorption coefficient from these processes, for lead. Very roughly, for a quantum energy hv and an absorber of atomic number Z, the photoelectric cross section varies as $Z^4/(hv)^3$, the Compton cross section as Z/hv, and the pair production cross section as Z^2 times a function of hv which rises rapidly from threshold ($2mc^2 \simeq 1$ MeV) and tends to a constant for $hv > 1$ GeV. Both the photoelectric and pair production processes result in catastrophic absorption of the γ-ray, while in the Compton process the photon is scattered with loss of energy; the absorption coefficient refers to attenuation in the number of photons of the initial energy. Note also that the photoelectric cross section exhibits discontinuities at the absorption edges corresponding to the binding energies of the $K, L \cdots$ electron shells of the particular element.

As the graph indicates, the photoelectric effect is of minor importance for $hv > 1$ MeV and we shall not discuss it further. The phenomenon of production of an electron-positron pair by a γ-ray in the electromagnetic field of a nucleus or electron is closely related to the radiation of γ-rays or *bremsstrahlung* by fast electrons in a Coulomb field, and both processes are therefore discussed in this section.

2.5.1 Compton Effect

Application of relativistic kinematics to the elastic collision of a photon with a stationary, free electron gives the Compton formula for the change in wavelength as a function of scattering angle:

$$\lambda' - \lambda = \frac{h}{mc}(1 - \cos\theta). \tag{2.17}$$

Because the unit of length, h/mc, appears in this formula, it is called the Compton wavelength (of the electron).

For low-frequency photons, such that $hv \ll mc^2$, the scattering cross section per electron is given by the classical Thomson formula

$$\sigma_T = \frac{8\pi}{3}\left(\frac{e^2}{mc^2}\right)^2 = \frac{8\pi}{3}r_e^2, \tag{2.18}$$

where $r_e = 2.8 \times 10^{-13}$ cm is the classical electron radius. In this non-relativistic region, σ_T is independent of frequency v. In the relativistic case, when the recoil velocity of the electron becomes large, the cross section is given by the

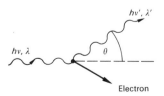

Fig. 2.7 Compton effect.

quantum-mechanical Klein-Nishina formula

$$\sigma_c = 2\pi r_e^2 \left\{ \frac{(1+\alpha)}{\alpha^2} \left[\frac{2(1+\alpha)}{(1+2\alpha)} - \frac{1}{\alpha} \ln(1+2\alpha) \right] \right.$$

$$\left. + \frac{1}{2\alpha} \ln(1+2\alpha) - \frac{(1+3\alpha)}{(1+2\alpha)^2} \right\}, \tag{2.19}$$

where $\alpha = h\nu/mc^2$. This can be expanded to give at low frequency ($\alpha \ll 1$):

$$\sigma_c \approx \sigma_T(1 - 2\alpha), \tag{2.20}$$

and at high frequency ($\alpha \gg 1$):

$$\sigma_c \approx \tfrac{3}{8}\sigma_T \left(\frac{2 + \ln 2\alpha}{\alpha} \right). \tag{2.21}$$

Thus, the high energy cross section varies approximately as v^{-1}; per atom, σ_c will be proportional to Z, the number of electrons.

2.5.2 Radiation Loss by Fast Electrons

For energies above a few MeV, the radiation loss or *bremsstrahlung* becomes the principal form of energy loss in the medium and dominates over the ioniza-tion loss. The radiative collisions take place with the nuclei, and to a lesser extent the electrons, of the medium. A physical picture of what happens can be obtained by viewing the process from the rest frame of the incident electron (Fig. 2.8). Then, the charged nucleus approaches the stationary electron with a velocity $\beta = v/c \approx 1$ and the transverse component $E_\perp$ of the electric field of the nucleus is increased by a relativistic factor $\gamma = (1 - \beta^2)^{-1/2}$. Further-more, there will be a transverse magnetic field $H_\perp = \beta E_\perp \approx E_\perp$ perpendicular to $E_\perp$. Thus, the electron "sees" a pulse of transverse electromagnetic radiation, or in other words, a stream of virtual photons emitted by the nucleus. The electron may then scatter one of these photons into a real state by the Compton

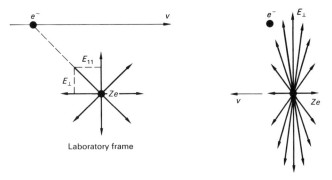

Laboratory frame

Figure 2.8   Electron frame

effect, itself losing energy in the process. It is in fact possible to carry out a semiclassical calculation of the bremsstrahlung process along these lines.

A quantum-mechanical treatment of the radiation loss gives the Bethe-Heitler formula for the cross section per nucleus,

$$\frac{d\sigma(E, E')}{dE'} = \frac{4}{E'} Z^2 \alpha r_e^2 \ln\left(\frac{183}{Z^{1/3}}\right) g(w), \tag{2.22}$$

where E is the electron energy, E' the photon energy, and $w = E'/E$. The function $g(w)$ depends on a screening parameter, defined as

$$\Gamma = \frac{100mc^2}{E}\left(\frac{w}{1 - w}\right) Z^{-1/3}. \tag{2.23}$$

When $\Gamma \ll 1$, i.e. for very high electron energies, the maximum "impact parameter" for the collision is determined by the screening of the nuclear charge by the atomic electrons. In this case of so-called "complete screening," the function $g(w)$ in Eq. (2.22) has the form

$$g(w) = [1 + (1 - w)^2 - \tfrac{2}{3}(1 - w)] + \frac{(1 - w)}{9 \ln\left(\dfrac{183}{Z^{1/3}}\right)}. \tag{2.24}$$

This function is displayed in Fig. 2.9 for lead. We may remark that Eq. (2.22) predicts that the number of photons radiated should tend to infinity as

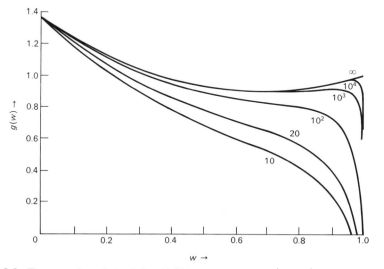

Fig. 2.9 The quantity $g(w)$ of Eq. (2.22), which equals $E'\phi(E, E')$ where $\phi(E, E')$ is the probability in unit energy interval per radiation length for emission of a photon of energy $E' = wE$ by an electron of energy E. The curves apply to lead, the numbers indicating electron energies in MeV. (After Rossi and Greisen, 1941.)

$E' \to 0$. This so-called "infra-red divergence" is a fiction of the theory, since it does not appear when one asks sensible questions about the real, physical world. Firstly, we note that if the emitted photon is very "soft," one cannot distinguish experimentally between such a bremsstrahlung process and elastic scattering, since the equality in energy between incident and scattered electrons can only be established within a finite error. In quantum electrodynamics, the divergence in fact disappears as a radiative correction to the Coulomb scattering.* Secondly, we may remark that the Bethe-Heitler formula refers to collisions with a *single* nucleus. In practice, one necessarily deals with the result of successive collisions with many nuclei of the medium. When $E' \to 0$, i.e. the photon wavelength becomes large and comparable with the distance between collisions, it turns out that successive photons interfere destructively and the low-frequency bremsstrahlung is suppressed (Landau and Pomerancuk, 1953; Migdal, 1956; Bell, 1958). This suppression effect has been detected in cosmic-ray experiments with nuclear emulsion, for electrons in the 1000 GeV energy region (Fowler *et al.*, 1959).

Regardless of these effects, however, integration of Eq. (2.22) to obtain the total energy radiated, $\int E' \, d\sigma(E, E')$, will yield a finite result. The average radiation energy loss for an electron of initial energy E, in traversing dx g cm^{-2} of the medium, for the case of complete screening, is

$$\frac{1}{E}\frac{dE}{dx} = \frac{4N_0 Z^2 r_e^2}{137A}\left[\ln\left(\frac{183}{Z^{1/3}}\right) + \frac{1}{18}\right], \tag{2.25}$$

where N_0 is the number of nuclei per g of medium. Note that in Eqs. (2.22) and (2.25) the quantity $r_e^2 = e^4/m^2c^4$ enters. Classically, this dependence on m, the mass of the radiating particle, arises because the energy radiated is proportional to the square of the acceleration, which varies inversely as m. Thus particles of mass M, and a given velocity, will radiate at about $(m/M)^2$ times the rate of electrons of the same velocity. Thus, for a muon—the next lightest particle after the electron—the radiation loss is extremely small, and only becomes comparable with the ionization loss in the ultra high energy region (of order 5000 GeV).

If we omit the last term in Eq. (2.25), making a contribution of order 1% only, we obtain

$$\frac{dE}{E} = -\frac{dx}{X_0}, \tag{2.26}$$

where

$$\frac{1}{X_0} = \frac{4N_0 Z^2 r_e^2}{137A}\left(\ln\frac{183}{Z^{1/3}}\right).$$

---

* See, for example, S. Gasiorowicz, "Elementary Particle Physics," Chapter 13 (John Wiley, New York, 1966).

If ionization losses can be neglected in comparison with the radiation loss, Eq. (2.26) gives the mean energy $\langle E \rangle$ of an electron, initial energy E_0, after having traversed a thickness x g cm^{-2}:

$$\langle E \rangle = E_0 e^{-x/X_0}. \tag{2.27}$$

The radiation length X_0 may therefore be defined as that thickness of material required to reduce the mean energy of an electron beam by a factor e.

In this treatment, we have so far neglected radiation losses in collision with atomic electrons. Since, in bremsstrahlung, the momentum transfers involved are small ($<mc$), the mass of the target is not important and we can therefore include the effect of atomic electrons simply by adding to the (nuclear charge)^2, the sum of the squares of the electron charges. This means replacing Z^2 in Eq. (2.25) by $Z^2 + Z = Z(Z + 1)$, when the definition of X_0 becomes identical with that in Eq. (2.13).

The angular distribution of the bremsstrahlung emission is peaked around a value

$$\theta = \frac{1}{\gamma} = \frac{mc^2}{E}, \tag{2.28}$$

independent of the photon energy. This result may be obtained qualitatively from the classical picture of Fig. 2.8. In the electron rest frame, the cross section for scattering the virtual photons from the nucleus is largest for frequencies in the electron rest frame $h\nu \ll mc^2$; the Thomson scattering, Eq. (2.18), is isotropic in this frame. In the laboratory frame (see Appendix A), the angle of emission is related to that in the electron frame by

$$\gamma \tan \theta = \frac{\sin \theta^*}{(\cos \theta^* + \beta)}.$$

Since the right-hand side of this equation is of magnitude unity for isotropic scattering, Eq. (2.28) follows, when $\gamma \gg 1$.

2.5.3 Pair Production

The cross section for a γ-ray to convert into an electron-positron pair in the field of a nucleus is closely related to that for bremsstrahlung of a fast electron. This is illustrated by the diagrams of Fig. 2.10. We see that the diagram representing bremsstrahlung (b) differs only from that for pair creation (a) in

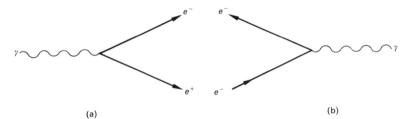

(a)   (b)

Fig. 2.10 Diagrams for the processes of (a) pair creation, and (b) bremsstrahlung.

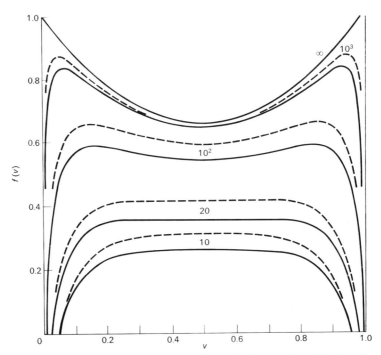

Fig. 2.11 The quantity $f(v)$ of Eq. (2.29), is plotted against v where E is the photon energy, and $E' = vE$ is the total energy of the positron $f(v) = E\chi(E, E')$ where $\chi(E, E')$ is the probability per radiation length of production of a positron in unit energy interval at E'. Full line, air; dotted line, lead. Numbers are photon energies in MeV. (After Rossi and Greisen, 1941.)

reversing the sign of charge and momentum of an electron. Thus, the pair creation cross section differs from the bremsstrahlung formula (2.22) only in kinematic factors. It is given for the case of complete screening by

$$\frac{d\sigma_{\text{pair}}(E, E')}{dE'} = \frac{4}{E} Z^2 \alpha r_e^2 f(v) \ln\left(\frac{183}{Z^{1/3}}\right), \qquad (2.29)$$

where

$$f(v) = [v^2 + (1 - v)^2 + \tfrac{2}{3}v(1 - v)] - \frac{v(1 - v)}{9 \ln\left(\dfrac{183}{Z^{1/3}}\right)}.$$

Here, E is the photon energy, E' is the total energy of the positron, and $v = E'/E$. "Complete screening" corresponds to the case of high photon energies, when

$$\Gamma = \frac{100mc^2}{Ev(1 - v)Z^{1/3}} \ll 1.$$

The variations of the bremsstrahlung and pair production cross sections with the energy partition parameters w and v are shown in Figs. 2.9 and 2.11, for lead. When integrated over positron energy, Eq. (2.29) gives, for the total probability of pair production per nucleus,

$$\sigma_{\text{pair}} = 4Z^2\alpha r_e^2 \left[\frac{7}{9} \ln \left(\frac{183}{Z^{1/3}} \right) - \frac{1}{54} \right]. \tag{2.30}$$

Again, replacing Z^2 by $Z(Z + 1)$ accounts for pair production in the field of atomic electrons as well as nuclei. Thus, the probability of pair production per radiation length is approximately 7/9—this asymptotic value being reached only for photon energies of order 1 GeV, as seen in Fig. 2.6. Again, as in Eq. (2.28), the average angle between the direction of emission of the created electron or positron and the gamma ray is $\theta \approx mc^2/E$. Thus, electron pairs of energy 100 MeV or more appear as a "V" track of extremely small divergence (see Figs. 1.5 and 6.6 for some examples).

2.6 ELECTRON-PHOTON SHOWERS

For electrons and photons of high energy, a dramatic result of the combined phenomena of bremsstrahlung and pair production is the occurrence of *cascade showers*. An example of a shower is shown in Fig. 4.25. A parent electron radiates in traversing the medium, and the photons convert to electron-positron pairs, which radiate in turn. The number of electrons and photons will therefore increase exponentially with depth of material traversed. This increase is halted, however, by the ionization energy loss of the electrons and positrons, which gradually saps the energy of the shower. From Eq. (2.25) we have for the radiation loss of the electronic component

$$-\left(\frac{dE}{dx} \right)_{\text{rad.}} = E \frac{4N_0 Z^2 r_e^2}{137A} \ln \left(\frac{183}{Z^{1/3}} \right),$$

where N_0 is Avogadro's number, and A is the mass number of the nuclei of the medium. The log term has typically a magnitude ≈ 4. For the ionization loss we have, since $\beta = 1$,

$$-\left(\frac{dE}{dx} \right)_{\text{ion.}} = 4\pi N_0 \frac{Z}{A} r_e^2 mc^2 \left[\ln \left(\frac{2mv^2\gamma^2}{I} \right) - 1 \right],$$

and where, if we include polarization effects, the term in square brackets has typically a magnitude ≈ 11. Thus the ratio

$$\frac{\left(\dfrac{dE}{dx} \right)_{\text{rad.}}}{\left(\dfrac{dE}{dx} \right)_{\text{ion.}}} = \frac{ZE}{137\pi mc^2} \times \frac{4}{11} \approx \frac{ZE}{600},$$

where E is the electron energy in MeV. Thus, there is a *critical energy*

$$E_c \approx \frac{600}{Z} \, \text{MeV}, \qquad (2.31)$$

such that, for $E > E_c$, radiation losses dominate, while for $E < E_c$, ionization loss is more important (see Table 2.1).

The development of a shower can now be discussed according to the following very simplified model. Starting off with a primary electron of energy E_0, suppose that, in traversing one radiation length, it radiates half its energy, $E_0/2$, as one photon. Assume that, in the next radiation length, the photon converts to a pair, the electron and positron each receiving half the energy (that is, $E_0/4$), and that the original electron radiates a further photon carrying half the remaining energy, $E_0/4$. Thus, after two radiation lengths, we will have a photon of energy $E_0/4$ and two electrons and one positron, each of $E_0/4$. By proceeding in this way, it is easily seen that, after t radiation lengths, there will be $N = 2^t$ particles, with photons, electrons, and positrons approximately equal in number. We have here neglected ionization loss and dependence of radiation and pair production cross sections on energy. The energy per particle at depth t will then be $E(t) = E_0/2^t$. This process continues until $E(t) = E_c$, when we suppose that ionization loss suddenly becomes important, and no further radiation is possible. The shower will thus reach a maximum and then cease abruptly. The maximum will occur at

$$t = t_{\text{max}} = \frac{\ln (E_0/E_c)}{\ln 2}, \qquad (2.32)$$

the number of particles at the maximum being

$$N_{\text{max}} = \exp \left[t_{\text{max}} \ln 2 \right] = \frac{E_0}{E_c}. \qquad (2.33)$$

The number of particles of energy exceeding E will be

$$N(>E) = \int_0^{t(E)} N \, dt = \int_0^{t(E)} e^{t \ln 2} \, dt$$

$$\approx \frac{e^{t(E) \ln 2}}{\ln 2} = \frac{E_0/E}{\ln 2},$$

where $t(E)$ is the depth at which the particle energy has fallen to E. Thus, the differential energy spectrum of particles $dN/dE \propto 1/E^2$. The total integral track length of *charged* particles (in radiation lengths) in the whole shower will be

$$L = \tfrac{2}{3} \int_0^{t_{\text{max}}} N \, dt = \frac{2}{3 \ln 2} \frac{E_0}{E_c} \approx \frac{E_0}{E_c}. \qquad (2.34)$$

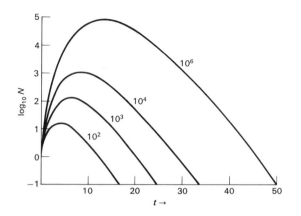

Fig. 2.12 Average number, N, of electrons in a shower, plotted as a function of thickness of medium traversed in radiation lengths. The shower is assumed initiated by an electron of energy E_0. The numbers on the curves give E_0 as a multiple of the critical energy E_c. (After Rossi and Greisen, 1941.)

The last result also follows from the definition of E_c and conservation of energy; nearly all the energy of the shower must eventually appear in the form of ionization loss of charged particles in the medium.

In practice, the development of a shower consists of an initial exponential rise, a broad maximum, and a gradual decline. Figure 2.12 shows the results of a more refined analytical calculation, due to Rossi and Greisen. Nevertheless, the above Eqs. (2.32 to 2.34) indicate correctly the main qualitative features, which are:

a) a maximum at a depth increasing logarithmically with primary energy E_0;

b) the number of shower particles at the maximum proportional to E_0; and

c) a total track length integral proportional to E_0.

These features are of importance in the measurement of γ-ray and electron energies in high energy experiments. For precise work, the analytical solution of the cascade problem is inadequate, especially at low energy, and recourse must be made to shower tables computed using the Monte-Carlo technique. This procedure allows the use of exact (energy-dependent) cross sections and also takes account of the lateral spread of the particles through Coulomb scattering. It may also be remarked that the development of showers is a statistical phenomenon. For fixed E_0, the fluctuation in position of the maximum is of order one radiation unit, being determined by the fate of the primary particle in the early stages. At the maximum, the rms fluctuation in the number of shower particles is approximately $\sqrt{N_{max}}$. Thus for a shower of energy 1.4 GeV in iron, Eqs. (2.33), (2.34), and Table 2.1 suggest a statistical error in N_{max}, or the track length integral L, of order $\pm 10\%$. In practice, the errors are in fact somewhat greater. Nevertheless, a shower count provides a useful method of estimating γ-ray or electron energies in the GeV region.

2.7 METHODS OF ACCELERATING AND DETECTING HIGH ENERGY PARTICLES

The early experimental work in particle physics—up to the early 1950's—was carried out mostly with cloud chambers and nuclear emulsions as visual detectors, and Geiger counters, proportional counters, and ionization chambers as electrical detectors. Although well suited to cosmic ray observations, these instruments have been superseded in work with accelerators. The most important visual detectors are, at present, the bubble chamber and the spark chamber, while the principal electronic instruments are the scintillation counter, Cerenkov counter, sonic spark chamber, and wire chamber. By the term "visual detector" we mean any system depending on a photographic image of the tracks or event as an essential stage in the analysis; "electronic instruments" are those in which the record consists only of electrical impulses, usually stored in some suitable form, such as on magnetic tape. All records, photographic or otherwise, usually depend heavily on digital computers, off-line or on-line, for their analysis and interpretation.

Our purpose in this section is not to give a comprehensive account of all detection techniques, which the interested reader can find in the bibliography at the end of the chapter, but rather to outline the essential features of the different types of detector commonly employed in this field, in order to illustrate how they are suited to particular types of experiment. The use of the various detectors will become apparent in the description of the salient experiments throughout the text of later chapters. The techniques employed depend not only on the intrinsic characteristics of the physical processes being investigated, but are also closely linked with the properties of the beams available from high energy accelerators. Therefore, an extremely brief discussion of accelerators is a first essential.

2.7.1 Accelerators

Accelerating machines, designed to produce high energy beams of protons or electrons, are of two main types. In cyclic accelerators, called synchrotrons, the particles are accelerated by radio frequency (RF) fields and constrained in a circular path of constant radius by a ring of electromagnets. In linear accelerators, the particles are simply accelerated in a straight line by a series of RF cavities. Four main types of machine are currently in use for experimentation in the GeV region:

  i) proton synchrotrons;
 ii) electron synchrotrons;
iii) electron linear accelerators;
 iv) clashing beam machines.

TABLE 2.2 Some accelerators at present operating or under construction*

Peak energy (GeV)	Laboratory	Date of first operation	Strong or weak focusing	Radius (m)
Proton synchrotrons:				
200–500	NAL, Batavia, Illinois	—	S	1000
70	Serpukhov, USSR	1967	S	235
33	Brookhaven, Long Island	1960	S	129
28	CERN, Geneva	1960	S	100
12.5	Argonne, Illinois	1963	W	27
7	Rutherford Lab., England	1963	W	24
6.2	Lawrence Rad. Lab., Berkeley	1954	W	18
Electron synchrotrons:				
6.5	DESY, Hamburg	1964	S	50
6	CEA, Cambridge, Mass.	1962	S	36
Electron linacs:				
21	SLAC, Stanford	1966	—	—
Storage rings:				
25 p–p	CERN, Geneva	1971	S	150
0.7 e^+e^-	Novosibirsk, USSR	1966	W	5
0.55 e^+e^-	Orsay, Paris	1967	S	3.5
1.5 e^+e^-	Frascati, Rome	1970	S	17

* N.B. This is not a complete list.

A list of some accelerators is given in Table 2.2. In the proton synchrotron, the circulating proton beam is accelerated once (or more) per turn by an RF field. As the momentum increases, the field in the magnet ring must be increased to keep the orbit radius constant. The last feature is desirable as it obviously minimizes the volume of field and weight of steel required. Furthermore, the RF frequency must be increased as the speed of the protons builds up, reaching a constant value as the velocity approaches c. It is usual to inject the protons from a linear accelerator with energies of order 10 to 50 MeV (depending on the peak energy), so that the initial velocity is of order $0.2\ c$. As an example, in the CERN 28 GeV proton synchrotron (Fig. 2.13) the increase of the beam energy by the RF acceleration is about 100 keV per turn, so that 3×10^5 turns are required to achieve full energy. The "ring" is of radius 100 m and contains 240 magnet sectors.

After reaching peak energy, the pulse of $\sim 10^{12}$ protons is steered to an internal target, or ejected onto an external one. All or part of the pulse can be ejected in one turn (i.e. in 2 μsec) or, as is more useful in counter experiments, spilled out more slowly over a period of, say, 0.1 sec. The whole acceleration

Fig. 2.13 View of an internal target area of the CERN 28 GeV proton synchrotron. The magnets of the main accelerator ring curl away on the left. Towards the bottom and right, secondary beam lines diverge through vacuum pipes. In the exact center of the picture is a window-frame bending magnet, with (diamond-shape) quadrupole focusing magnets upstream and downstream. At extreme right center is the tank of an electrostatic separator. (Courtesy, CERN Information Service.)

cycle lasts 2 to 3 sec, and is then repeated with another pulse of protons. The magnetic field is used to focus as well as guide the beam, this being achieved by shaping the magnet pole faces to produce vertical and horizontal gradients. The older synchrotrons are of the so-called weak-focusing type. The gradient is small, and the cross section of the beam is large (several centimeters across), and hence vacuum pipe and magnet apertures are large. Strong-focusing machines use large field gradients, alternating in sign from one magnet to the next. The result is that the beam is successively focused (defocused) horizontally and defocused (focused) vertically; it is possible to get a net focusing effect (as can be seen for a light beam by alternating convergent and divergent lenses of equal power). This design leads to a much reduced beam cross section (millimeters rather than centimeters), and therefore smaller and cheaper magnets.

In electron synchrotrons, the principles are quite similar, except that the electrons are always relativistic so that the circulation time (RF frequency) is constant, and only the magnetic field has to be varied. The problem with such accelerators is that, because of its small mass, the electron radiates under the circular acceleration (synchrotron radiation). The energy lost in this way varies as E^4/ρ where E is the electron energy and ρ the orbit radius. Acceleration to very high energies would therefore involve prohibitive demands on RF power, to replace this loss, or enormously large magnet rings. Thus for energies above 10 GeV the linear electron accelerator is the only practicable device. The principle of the electron linac is to accelerate electrons down a long vacuum tube by means of a traveling electromagnetic wave, of phase velocity equal to the particle velocity (c, for electrons above a few MeV). The RF power is generated by klystron oscillators, and the phase velocity of the wave in the guide is reduced to c by interposing suitable irises in the tube. The largest existing electron linac is that at Stanford, 2 miles long, producing beams of 15 to 20 GeV electrons. This machine produces extremely intense pulses of electrons of 2 μsec duration, 360 times per second. The high intensity and energy are therefore to some extent offset by the poor duty cycle, which makes counter experiments relying on coincidence methods extremely difficult (if not impossible), since the length of each electron pulse is comparable with the counter resolving times.

Finally we mention clashing beam machines. The disadvantage of experiments using, for example, protons of energy E on a *stationary* proton target, is that the available energy in the center of momentum of the collision is approximately $\sqrt{2ME}$ (for $E \gg M$, the proton mass)—see Appendix A. So, if the bombarding proton energy is increased by a factor x, the available energy for production of new particles increases by a factor of $x^{1/2}$ only. On the other hand, if *both* protons have energy E and oppositely directed velocities, the available energy in a head-on collision will be about $2E$, i.e. much greater than for the first case. For proton–proton colliding beam experiments, protons from a synchrotron are injected in turn into two concentric magnet rings, in

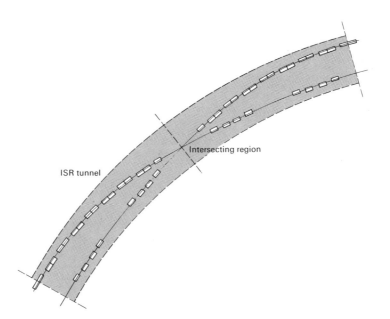

Fig. 2.14 Diagram of one octant of the CERN intersecting storage rings. Both rings of magnets are accommodated in one tunnel, of circumference 943 m. The intersection angle is 14.8°.

which they circulate in opposite directions. In the CERN intersecting storage rings (ISR) design, the rings cross over in eight regions, so that the beam–beam interactions can be observed and measured at several stations (Fig. 2.14). Note that for 25 GeV colliding protons, the available energy is the same as that for a 1300 GeV proton on a stationary nucleon target. Electron-positron storage rings have also been built; in this case, because of the opposite charges, one ring will accommodate both beams. Experiments with such rings are described in Chapter 5.

2.7.2 Beams

A full discussion of the methods of selecting, focusing, and separating secondary beams from targets around an accelerator would be inappropriate in this book, and we can again only mention the bare essentials and refer the interested reader to the bibliography.

   A target irradiated by, for example, a high energy proton beam will emit all sorts of secondary particles, having a wide range of momentum and angles. As one might expect, the most energetic particles are emitted in the "forward" direction. The desirability of an external target, in a region free of magnetic field, is therefore apparent; it is usually difficult to get such charged particles from an internal target, located in the vacuum pipe inside a synchrotron ring, because of the effects of the field. The direction of emission of the secondaries from the target, and their momenta and sign of charge, can next be chosen by using suitably placed collimating slits and bending and focusing magnets. For example, one could thus define a narrow beam of positive particles of well defined momentum. For some experiments, this might be enough, the discrimination between the different possible constituents (proton, π^+, K^+) being provided by the detection system (see, for example, the antiproton experiment described in Section 2.8.1). For other experiments, particularly in bubble chambers, pure beams of one type of particle are desirable. These can be obtained by using electrostatic separators (consisting of a pair of long parallel plates maintained at a high potential difference) which, for a particular momentum, deflect particles of different masses by different amounts. Such a system, in one or two stages, will separate π^+, K^+, and p (or π^-, K^-, and $\bar{p}$) cleanly for momenta below about 5 GeV/c. For higher momenta, radio-frequency separators may be used. These consist of two (or more) RF cavities, spaced apart along the beam direction, with a suitable relative phase relation. It can then be arranged that the velocity of the wanted particle is such that the sideways deflection produced by the oscillating field in the first cavity is exactly counterbalanced by that in the second, i.e. the wanted particle is undeflected. On the other hand, unwanted particles of the same momentum have slightly different velocity, so the two kicks will not cancel and there is a resultant deflection. In either type of separator, suitably placed collimating slits transmit the wanted particles and absorb the rest.

2.7.3 Bubble Chambers

The bubble chamber relies for its operation on the fact that in a superheated liquid boiling will start with the formation of gas bubbles at nucleation centers in the liquid, and particularly along the trail of ions left by the passage of a charged particle. The superheating can be produced by a sudden expansion of the liquid. The first bubble chamber was conceived and built by Glaser in 1952, and consisted of a 3×1 cm cylinder of glass, containing diethyl-ether. Like all the early chambers, it was "clean" in the sense that the superheated state could be maintained for a long time, the only nucleation centers being charged ions. A year later, Hildebrand and Nagle demonstrated that one could obtain bubble tracks in liquid hydrogen, giving the possibility of observing interactions of incident particles with free protons, rather than those bound in the complex nuclei of heavy liquids. The third step in bubble chamber technology took place shortly afterwards, when it was demonstrated that small, clean chambers were unnecessary. Gasketed chambers, with metal bodies and glass windows, although "dirty" in the sense that boiling would occur at the seals and other discontinuities, could operate successfully if the bubble tracks were photographed before the onset of such spontaneous boiling. The final step which ensured the success of the instrument as an important tool in particle physics was taken at Berkeley by Alvarez, who, in the middle

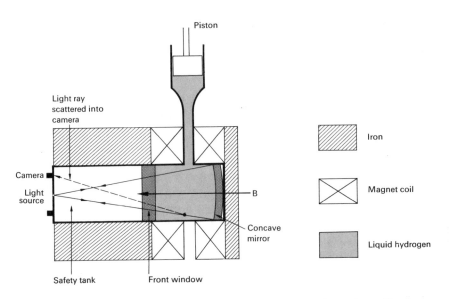

Fig. 2.15 Sketch of principal components of a bubble chamber, end-on view. The beam enters normal to the plane of the paper.

1950's, when little was known of the technique, pioneered the design and construction of a large (500-liter) hydrogen chamber.

Different types of expansion systems have been used, employing either a gas phase or pistons, diaphragms or bellows directly in contact with the liquid. The principal components of a bubble chamber are sketched in Fig. 2.15. The chamber is immersed in the field of a large electromagnet to provide a means of measuring momentum from track curvature. After the liquid has been made sensitive by the expansion, the bubbles grow to a diameter of order $10\,\mu$ in a period of 10 to 50 msec, and are then photographed using flash illumination. The bubbles then collapse under a recompression stroke. The cycle time is typically 1 sec, and thus is well matched to cyclic pulsed accelerators, with repetition rates of the same order. In this respect, the bubble chamber is far superior to its forerunner, the cloud chamber, with a recovery time of several minutes.

The typical time required to expand a bubble chamber is some 10 msec, being determined by the transmission of the expansion wave through a large liquid volume. On the other hand, the lifetime of the nucleation centers is at least an order of magnitude smaller, which means that it is impossible to operate the chamber expansion in time to record particles which have previously traversed an external counter system, acting as the trigger. Since the spill time of a secondary beam from an accelerator is accurately predictable, however, this is no great disadvantage.

TABLE 2.3 Characteristics of bubble chamber liquids

Liquid	Operating temperature and pressure	Geometrical interaction mean free path, cm	Radiation length, cm	Coulomb error on momentum ($\beta = 1$) for 25 cm track, $H = 20$ kG	Density, g/cm^3
Hydrogen	27°K, 4.7 atm	440	970	1.8%	0.06
Deuterium	30°K, 5.2 atm	240	890	1.9%	0.12
Propane (C$_3$H$_8$)	58°C, 20 atm	120	109	5.5%	0.44
Freon (CF$_3$Br)	30°C, 18 atm	60	11	17%	1.5

Bubble chambers employing a wide variety of liquids have been built. Table 2.3 shows properties of some common liquid fillings. For given track length, Coulomb scattering errors, which determine the precision in momentum measurement (Section 2.4), are smallest for hydrogen, and large for a heavy liquid like freon. By the same token, heavy liquids have short radiation and interaction lengths, and good stopping power, and thus are suitable for experiments requiring efficient detection of γ-rays and electrons. Such chambers have also been used in the study of K^+ decay modes, when the charged secondary is brought to rest inside the chamber, making possible an accurate measure of momentum from the range—see Fig. 2.16. However, the bulk of bubble chamber experiments have been carried out in hydrogen/deuterium chambers.

A brief mention should be made of the analysis of the bubble chamber pictures. The chamber volume is viewed by three (or more) cameras, permitting the reconstruction of tracks in space from the film image, using fiducial marks inscribed on the front window. The light rays from track to camera may traverse the liquid, front window (several inches thick), a vacuum tank, and several intermediate glasses (heat shields) before they reach the lens. So, optical reconstruction is a complex procedure, and the film measurements are processed by computer, using a "geometry program" which includes a list of optical constants, magnetic field calibration, and so forth. After the geometrical reconstruction of tracks and interaction vertices, a "kinematics program" is used to fit the angles and momenta of the various tracks from an interaction to a particular hypothesis. For example, for a 4-prong event produced by an incident K^- beam of known momentum, one might try the hypothesis $K^- p \rightarrow K^- \pi^+ \pi^- p$, or $p\pi^+ \pi^- \pi^- \overline{K}^0$, or $\Lambda \pi^+ \pi^+ \pi^- \pi^-$ and so on, to discover the appropriate fit.

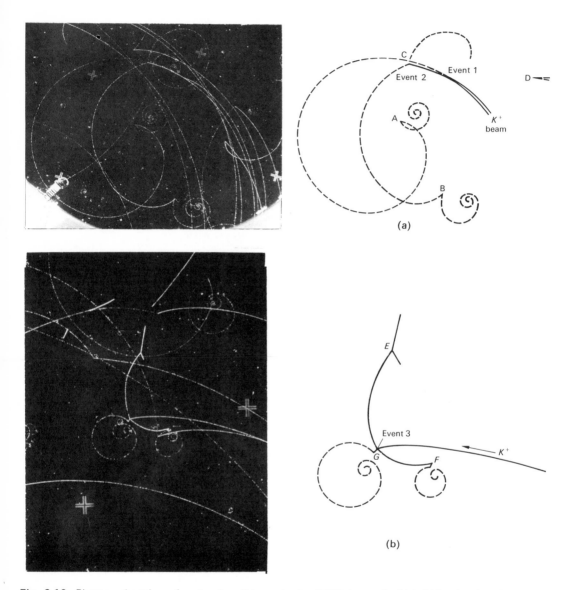

Fig. 2.16 Photographs taken of a stopping K-beam in the CERN heavy liquid bubble chamber, filled with freon (CF_3Br). (a) Event 1; $K_{\mu 2}$-decay, $K^+ \rightarrow \mu^+ + \nu$. The positive muon comes to rest and decays at A, according to the scheme $\mu^+ \rightarrow e^+ + \nu + \bar{\nu}$. Event 2; $K_{\pi 2}$-decay, $K^+ \rightarrow \pi^+ + \pi^0$. The positive pion comes to rest at B, showing the decay sequence $\pi^+ \rightarrow \mu^+ \rightarrow e^+$. The 4 MeV μ^+ track is very short. The electron at C and electron pair at D are from conversion of the γ-rays from the π^0-decay. They point back to the K^+ decay vertex. (b) Event 3; $K_{\pi 3}$-decay, $K^+ \rightarrow \pi^+ + \pi^+ + \pi^-$. The negative pion undergoes capture at the end of the range, giving a nuclear disintegration "star" at E. The $\pi^+ \rightarrow \mu^+ \rightarrow e^+$ decay sequence of the positive pions appears at F and G.

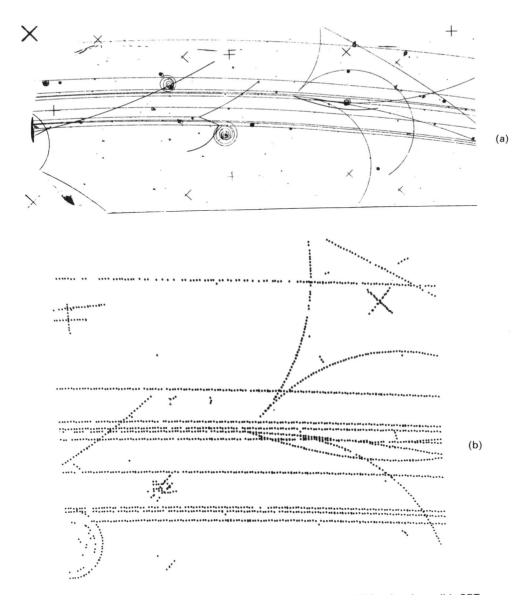

Fig. 2.17 (a) Original picture from the CERN 2 m hydrogen bubble chamber. (b) CRT display of scan output from an automatic scanning and measuring system (University of Oxford PEPR system). The film is scanned by a computer-controlled electron beam with cross section in the form of a short, fine line. Events of simple topology can be automatically scanned and measured at the rate of up to 400/h.

Since a bubble chamber records faithfully all interactions, and not simply those of particular interest, it is necessary to scan and measure up a large number of events. Various automatic or semiautomatic devices have been developed to speed this process. These usually incorporate a cathode ray tube, the electron beam scanning the tube face, near which the film negative is located. "Hits" occur when the spot or line locates a track in the negative and is registered by a photomultiplier, which then feeds back to an on-line computer controlling the deflection of the electron beam. A CRT display from one such device is shown in Fig. 2.17.

Recent developments in bubble chamber technique have been in the direction of very large chambers of volume 10 m³ or more. These make possible a more complete record of an interaction, being more efficient at detecting neutrons and γ-rays (from π^0 and Σ^0 decay), from the primary vertex. At high energies particularly, the average number of secondaries from an interaction is large and will generally include several neutral particles. A large mass of sensitive liquid is also desirable for the study of interactions of energetic neutrino beams, for which the cross section per nucleon is small. For such large volumes, superconducting magnets, with their lower power consumption and larger attainable field strengths, are preferable to conventional electromagnets. Because of the great volume of liquid to be photographed, it is necessary to employ cameras with wide angle lenses close to the liquid surface, and special illumination systems. Other developments have been in the direction of hydrogen chambers designed to combine the virtues of both hydrogen and heavy liquid. Several experiments have been carried out with metal plates immersed in the hydrogen, in order to convert γ-rays and at the same time have good momentum resolution on tracks of charged particles and to retain the kinematic constraints and simplicity of interactions on free protons. Another innovation is to fill the chamber with a rich neon-hydrogen mixture (radiation length of order 50 cm), in which is immersed an expandible plastic target filled with hydrogen (or deuterium). Both target and heavy liquid can be made simultaneously sensitive. Measurements on charged particles from target interactions are made in the hydrogen "bag," while γ-rays are converted in the heavy liquid.

2.7.4 Scintillation Counters

The scintillation counter is one of the most widely used instruments in high energy physics. It is really a development of an old device used by Rutherford and his colleagues over fifty years ago in the study of α-particle scattering by nuclei. They used a specially activated zinc sulfide screen which, upon being hit by an α-particle, produced a scintillation which could be seen in a microscope by an observer whose eyes had previously been dark-adapted. The scintillation materials used nowadays consist of plastics such as polystyrene; organic liquids (for example, toluene); inorganic crystals, principally of sodium or

TABLE 2.4 Characteristics of typical scintillators

	Pulse height (relative to anthracene)	Decay time (nsec)	λ_{max} (Å)	Density (g/cm^3)
Polystyrene + p-terphenyl	0.28	3	3550	0.9
+ tetraphenyl-butadiene	0.38	4.6	4800	
Sodium iodide (+ thallium)	2.1	250	4100	3.7
Anthracene	1.0	32	4100	3.7
Toluene	0.7	<3	4300	0.9

cesium iodide; and organic crystals like anthracene. The passage of an ionizing particle through such a medium results in emission of fluorescent light over a broad wavelength range. In organic scintillators, the emission is principally in the short wavelength region, below 4000 Å, and a wavelength-shifting dye is employed, which will reemit in the visible region. An example is shown in Table 2.4 for polystyrene.

The light from the scintillating medium is arranged to fall onto the cathode of a photomultiplier, either placed in direct contact with the block of scintillator or optically connected to it via a lucite "light pipe" (see Fig. 2.18). Most photomultipliers employed have a spectral response peaking in the region of 4500 Å. For each photoelectron ejected from the cathode, the successive secondary-emission electrodes (dynodes) will generate some 10^8 electrons at the anode. The output pulse from the photomultiplier is fed into suitable amplifiers, discriminators, and scalers, the essential function of which is to store the number of photomultiplier pulses of a particular magnitude.

The performance of a scintillation counter can be illustrated with a few numbers. In traversing 1 cm of plastic scintillator, a minimum-ionizing particle will lose about 1.5 MeV in ionization, liberating on average 10^4 photons of mean quantum energy $h\nu = 3$ eV (thus converting 2% of the energy loss to fluorescence). Assuming 10% of the light is collected, and that the photo-multiplier cathode efficiency (number of photoelectrons per incident photon) is also 10%, approximately 2000 photoelectrons will be liberated. Clearly therefore, even with very thin scintillators, the efficiency for recording particles is essentially 100%. For inorganic crystals (NaI/Tl) the light output is very closely proportional to the ionization energy loss. Organic scintillators are not nearly so linear; for example, the pulse height per unit energy loss of a 5 MeV α-particle is only 10% of that for a singly charged particle at minimum ionization.

The response time of a scintillator, as measured by the width of the output pulse, is determined by a number of factors. One must consider the time re-quired for the light to travel through the material by various routes, and the

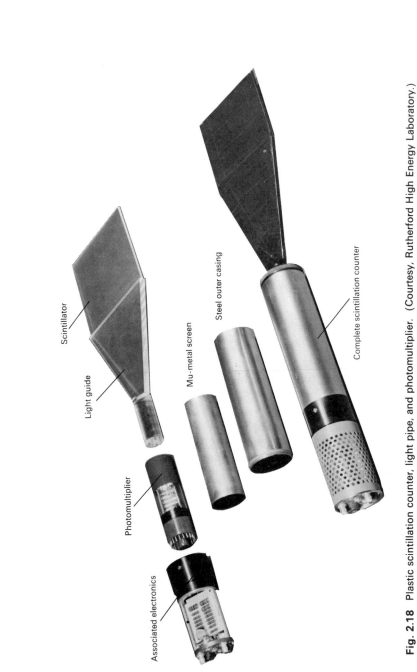

Fig. 2.18 Plastic scintillation counter, light pipe, and photomultiplier. (Courtesy, Rutherford High Energy Laboratory.)

fact that the light emitted from the fluorescing material decays exponentially with a characteristic decay constant. This is 3 to 30 nsec for organic materials and some 250 nsec for sodium iodide (1 nsec $= 10^{-9}$ sec). There are also fluctuations in transit time through the photomultiplier, typically a few nanoseconds. Under favorable conditions, therefore, one can obtain final electrical pulses of width down to about 10 nsec (for plastics). The great advantages of the scintillation counter are that it is robust, simple, and efficient, giving large, sharp output pulses. Its spatial resolution is poor, since the pulse is not related in an obvious way to the location of the trajectory of the charged particle through the counter. Thus, if spatial information is required, it is necessary to use a large hodoscope array of very small counters. Because all photomultipliers generate random noise, it is usual, when counting particles by this method, to place two or more scintillators in coincidence. The accidental coincidence rate is then small because of the good resolution time.

2.7.5 Cerenkov Counters

As discussed in Section 2.3, the Cerenkov radiation emitted by a medium traversed by a relativistic particle provides a method of measuring the particle velocity. The intensity of Cerenkov light is given by Eq. (2.9). For $\beta \approx 1$, giving the maximum signal, one finds for water ($n = 1.33$) some 200 photons/cm, in the visible region. The photons are recorded by a photomultiplier; assuming most of the light is collected and a photocathode efficiency of 10%, this gives 20 photoelectrons/cm of water radiator, to be compared with about 500 photoelectrons for a particle of $\beta \approx 1$ traversing 1 cm of plastic scintillator. Thus, the Cerenkov light output is meager, and one must aim at good efficiency of light collection—already made easier by the fact that it is restricted to a cone at the Cerenkov angle, Eq. (2.7).

A Cerenkov counter consists basically of a radiator (the Cerenkov medium), a light focusing system, and one or more photomultipliers. The particle velocity is given by the angle of the Cerenkov cone, and this characteristic angle must be preserved by the transmission system if it is required to measure it. A radiator is therefore usually in the form of a cylinder of transparent material, with the particle beam parallel to the axis, as shown in Fig. 2.19. Since

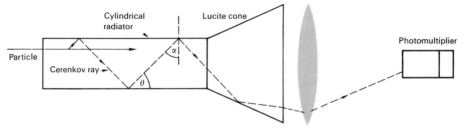

Fig. 2.19 Basic design of a Cerenkov counter.

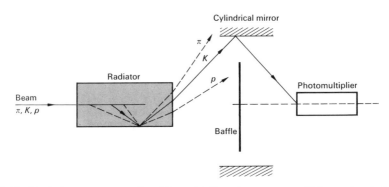

Fig. 2.20 Typical arrangement of a differential Cerenkov counter. This can be designed to select light from only one of the three components of the beam (K-mesons in the case shown).

$\cos \theta = 1/\beta n$, $\sin \alpha = \cos \theta > 1/n$ and the light is propagated to the end of the cylinder by total internal reflection. If the refractive index n is large, such that $n^2 > 1 + 1/\beta^2$, the rays will be internally reflected at the end surface, so that optical coupling is necessary to the photomultiplier. This may be in the form of a lucite cone as shown, followed by a focusing lens.

For selecting particles of different mass from a mixed high momentum beam, where all particles have $\beta \approx 1$, it is necessary to use gases such as ethylene or carbon dioxide at a high and variable pressure as the radiators in order to obtain an n value close to unity. Gas counters operating at pressures up to about 100 atm can cover the range $n = 1.0$ to 1.2.

A Cerenkov counter can be used as a threshold device, responding only to particles above a given velocity, and thus, for a beam of fixed momentum, to particles of mass below some chosen value; or it may be a differential counter, which depends for its mass resolution on the angular resolution of the Cerenkov light. These are related by the equation

$$\tan \theta \, d\theta = \frac{m \, dm}{E^2},$$

where E is the particle energy. A typical system which would accept light at a particular angle is shown in Fig. 2.20.

Another type of counter is the total absorption Cerenkov counter, used for measuring the total shower energy produced by an energetic photon or electron. The radiator is made of a block of lead glass, of dimensions several radiation lengths, so that the entire shower is dissipated in the block. The relativistic electrons in the shower produce Cerenkov radiation, the total light output being proportional to the track length integral since all the electrons have $\beta \approx 1$. The energy resolution of such a device is of order 20% in the GeV region.

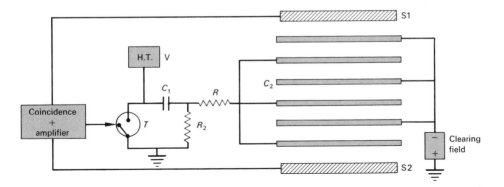

Fig. 2.21 Sketch of the main components of a parallel plate spark chamber. $S1$, $S2$: scintillation counters; T: thyratron or spark gap.

2.7.6 Spark Chambers

The essential components of a parallel plate spark chamber are shown in Fig. 2.21. It consists of a number of parallel conducting plates or foils, of separation of order 1 cm, immersed in an inert gas such as neon or argon at a pressure of 1 to 1.5 atm. The passage of an ionizing particle through the chamber is signaled by the scintillation counters $S1$ and $S2$, the resulting coincidence pulse being used to trigger a small spark gap or hydrogen thyratron, T. The condenser C_1, connected to a high voltage supply V of around 10 kV, is discharged to earth, thus producing a pulse of magnitude $VC_1/(C_1 + C_2)$ across the chamber, capacity C_2. The rise time of the voltage pulse $\tau = C_2R$ (for

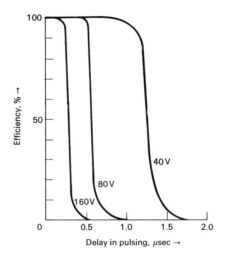

Fig. 2.22 Spark chamber efficiency as a function of delay in H.T. pulse and value of clearing field, for argon-alcohol filling.

$C_1 \gg C_2$). An electrical breakdown occurs in the chamber gas, along the trail of ions left by the charged particle. The sparks rapidly reduce the voltage across the chamber plates, and the current decays with a time constant of order RC_1. Typically, the rise time $\tau \approx 10^{-8}$ sec, i.e. short compared with the formative time of the electron avalanche (a fraction of a microsecond).

The efficiency of the chamber, i.e. the probability of any gap being discharged, depends on the delay involved in applying the voltage pulse, and on the magnitude of the clearing field, usually of order 100 V.

Figure 2.22 shows some typical results. The important point is that if the clearing field is reasonably large, essentially all ions are removed from the gas within 0.5 μsec of the passage of the charged particle, so that if the high voltage is delayed by more than this, no spark will occur. Thus a spark chamber has a good time resolution, of order 1 μsec. This means it can operate in a flux of 10^5 particles/sec, and can select and register only one track of interest. This ability can be contrasted with that of a bubble chamber, which has no selectivity. Once discharged, the recovery time of a spark chamber (the so-called dead time) is rather long, about 10 msec. This is not usually a great handicap; in an optical spark chamber, winding-on of the cameras takes much longer than this. The other important feature of the spark chamber is its spatial resolution as compared with a scintillation counter, for which the resolution is of the same magnitude as the size of the counter. Sparks can be located within an accuracy of about 0.3 mm.

Most commonly, sparks are recorded photographically, using two cameras at 90° stereo-angle. Examples of such pictures are given in Figs. 3.15 and 4.26. An alternative method is to use transducers (for example, a piezoelectric crystal) to provide an electrical record from the sonic pulse following the spark. Two small sonic probes per gap, placed along the edges of the chamber, are sufficient to determine the x-, y-coordinates of the position of a single spark, using the time-of-flight of the sound wave. If several sparks occur in the one gap (multiparticle event) a large number of probes is required and the technique becomes unmanageable. The obvious advantage of sonic detection is that one obtains a digitized record direct, without the tedium of film analysis.

Spark chambers in which the electrodes consist of arrays of parallel wires have also been used. One set of wires is connected via resistors in parallel to the high tension supply. This gives good multispark efficiency, since all the energy available for discharge cannot then be transmitted through a single spark. The earthed set of wires may each be equipped with a small ferrite magnetic core, which is switched in magnetization sense if a spark occurs on that wire. Sensing wires thread the cores, and the voltage pulses appearing on these when a core switches can be read off in sequence. Two chambers with wires at right angles, one above the other, then determine the x-, y-coordinates of the sparks.

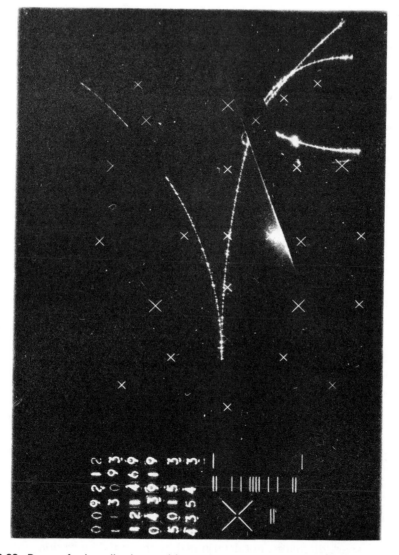

Fig. 2.23 Decay of a long-lived neutral kaon observed in a streamer chamber operated at SLAC, and attributed to the mode $K_L \rightarrow \mu^+ \pi^- \nu$. The left-hand track is probably due to the μ^+, the gap being where it traverses 2 radiation lengths of lead without interaction. The right-hand track interacts in a similar plate and is ascribed to a π^-. The streamers are normal to the plane of the picture, as is the applied magnetic field. (Courtesy, Stanford Linear Accelerator Laboratory.)

2.7.7 Streamer Chambers

A development of the narrow gap spark chamber described above, is the wide gap or streamer chamber, intended to record a three-dimensional image of the particle trajectory. An extremely short, high voltage pulse, of order 10 nsec duration and peak magnitude 10 to 20 kV/cm, is applied between plates at top and bottom of the chamber, which may have a volume of order 1 m^3. An electron avalanche develops along the trail of ions, as before, but because of the short pulse duration, this is arrested at an early stage, so that the "track" appears as a row of short streamers extending typically a few millimeters in the direction of the electric field. If the plate electrodes are transparent, the streamers may be viewed along the field and then appear as a series of dots similar to a track in a bubble chamber (Fig. 2.23).

Since no spark is allowed to develop, the amount of light output is small, but can be recorded on special fast film. Streamer chambers operate with inert gas fillings (neon or helium, for example). Cylindrical targets of hydrogen gas, inserted in the chamber, have been used in conjunction with pencil photon beams at the Stanford Laboratory. In such an application, the actual inter-action vertex in the hydrogen is not, of course, visible. An obvious advantage of the streamer chamber over the bubble chamber is the fact that it can be triggered, and thus made highly selective.

2.7.8 Proportional Counters

As indicated above, the disadvantage of the scintillation counter is that it has essentially no spatial resolution; and of the spark chamber, that, once discharged, the recovery time is long. Modern proportional wire chambers suffer from neither of these drawbacks. The proportional counter as such is of course a very old device, which works as follows. A fine wire anode is located along the axis of a sealed glass or metal tube, serving as the cathode. The voltage and the pressure of the gas in the tube can be adjusted so that ions, liberated by passage of a charged particle, are accelerated to sufficient velocity to produce fresh ionization by collision—a phenomenon called gas multiplication. The output voltage pulse generated when the electron avalanche reaches the anode is then proportional to the initial ionization. (If the electric field is further increased, actual breakdown of the gas will eventually occur, and one obtains a spark discharge, not determined by the initial ionization and limited only by the external circuit parameters. In this case the instrument acts as a Geiger counter.) Because breakdown, as in the spark chamber, does not take place, the recovery of a proportional counter is rapid, and typically the dead-time is of order 10^{-6} sec.

In the modern proportional counter, as developed by Charpak et al. (1968), spatial information is obtained by having many fine parallel wires in a plane as one set of electrodes, between planar steel meshes serving as the other. The applied voltage is typically 2000 V, with wires of diameter of order 50 μ and separations 0.1 to 1 cm. The gas filling is usually an argon/alcohol mixture, at atmospheric pressure. Each wire has to have its individual transistor amplifier, and acts like a cylindrical proportional chamber. Amplification factors are of order 5000 times, and this is sufficient for single minimum-ionizing particles to be detected with 100% efficiency. Two counters, with wires at right angles, one above the other, can be used to give the spatial position of a particle within about $\frac{1}{4}$ mm.

2.8 EXAMPLES OF THE APPLICATION OF DETECTION TECHNIQUES TO EXPERIMENTS

In Chapter 1, some of the early discoveries in the field of particle physics were outlined. Progress, as in any field of science, has depended on the close interplay between the discovery of experimental phenomena, usually resulting from application of new techniques, and the introduction of fresh ideas and principles which they stimulate. In the early days, as we have seen, the introduction of the counter-controlled cloud chamber led to the discovery of mesons. The observation of the vital link of the $\pi \to \mu$ decay chain, proving the essential correctness of Yukawa's ideas, had to await the introduction of a different technique—in this case, special photographic emulsions, with a stopping power and spatial resolution superior to that of the cloud chamber. In more recent times, the invention of the hydrogen bubble chamber, ideally matched to pulsed cyclic accelerators, and with a momentum resolution for charged particles of order 1%, has led to the discovery of a host of new baryon and boson resonant states.

The more precise evaluation of particular physical quantities is as important as the discovery of completely new phenomena. The agreement or disagreement between experiment and theory is always a function of the experimental precision. As an example, Table 2.5 lists some determinations of the muon mass, illustrating the improvement in accuracy over a 25-year period, with the introduction of new techniques. At the present time, the accuracy of 1 in 10^5 is of no intrinsic interest, as there is no theory which can predict the muon mass. But one day there will be.

Discussion of the application of the detection methods outlined above to particular experiments appears whenever appropriate throughout the text of this book. At this point, however, it is perhaps worth singling out a couple of experiments, because they illustrate two quite different aspects of the experimental method.

The first experiment deals with the discovery of the antiproton. Before it was undertaken, there was no evidence for its existence; two cosmic ray events in nuclear emulsion had previously been ascribed to antiprotons, although it is now virtually certain that they were wrongly interpreted. If such a particle did exist, however, it was expected to have well-defined properties of charge and mass, and to undergo annihilation in ordinary matter. The problem was then to observe a very specific signal due to antiprotons, from a very large background due to more prolifically produced particles.

TABLE 2.5 Measurement of muon mass

Year	Authors	Method	Mass m_μ/m_e
1937	Street and Stevenson (*Phys. Rev.* **52**, 1003)	Momentum/ionization of cosmic ray mesons in cloud chamber	175 ± 50
1949	Brode and Fretter (*Rev. Mod. Phys.* **21**, 37)	Momentum/range of mesons in cloud chamber with plates	215 ± 3
1956	Barkas *et al.* (*Phys. Rev.* **101**, 778)	Momentum/range in emulsion of muons from synchrocyclotron	206.9 ± 0.2
1960	Devons *et al.* (*Phys. Rev. Lett.* **5**, 330) Lathrop *et al.* (*Il Nuov. Cim.* **17**, 109)	X-rays from $3D_{5/2} \rightarrow 2P_{3/2}$ transition in mesic phosphorus	206.78 ± 0.03
1963	Hutchinson *et al.* (*Phys. Rev.* **131**, 1351)	Gyromagnetic ratio of muon and precession frequency in magnetic field	206.767 ± 0.003

The other experiment we describe is the determination of the neutral pion lifetime. In this case, numerous attempts had been made to determine this quantity directly, based on small numbers of events, using the nuclear emulsion technique stretched to the very limit of its capability. Successive estimates had given smaller and smaller values, ranging from 10^{-14} sec in 1950 to $\approx 2 \times 10^{-16}$ sec in 1961. There was no unambiguous theoretical prediction of the lifetime, although 10^{-16} sec was thought to be an upper limit. The experiment described, using a counter technique and exploiting the very large intensities available at an accelerator, under carefully controlled conditions, was able to detect an extremely small effect (1.5% in counting rate) and measure it with precision.

2.8.1 Discovery of the Antiproton

The experiment was performed at the Berkeley Bevatron in 1955 by Chamberlain, Segre, Wiegand, and Ypsilantis. The Bevatron had been expressly designed to accelerate protons of momentum up to 6.3 GeV/c, approximately the threshold for the production of a nucleon-antinucleon pair in a proton–proton collision. The value of the total energy for an incident proton in collision with a stationary proton is $E/Mc^2 = 7$, or $p_{\text{threshold}} = 6.5$ GeV/c. (See Problem 2.6, and Appendix A.) In the actual experiment, the circulating protons collided with a copper target. In the copper nucleus, a nucleon has Fermi momentum $\boldsymbol{p}_f$, so that if the latter is in the opposite sense to that of the incident proton, the

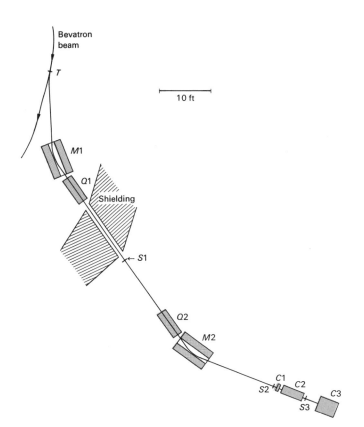

Fig. 2.24 Experimental set up used in the discovery of the antiproton. Q, quadrupole magnet; M, bending magnet; $S1$, $S2$, $S3$, scintillators; $C1$, $C2$, Cerenkov counters; $C3$, total absorption Cerenkov counter. (After Chamberlain *et al.*, 1955.)

available center-of-mass energy of the collision is increased, and the incident threshold momentum is reduced by a factor of approximately $(1 - p_f/Mc)$, i.e. to about 4.8 GeV/c, if we take $p_f/Mc \approx \frac{1}{4}$ as a typical value.

Figure 2.24 shows the experimental set-up employed. Secondary negative particles of 1.2 GeV/c momentum, from an internal copper target T, are bent out in the fringe field of the Bevatron magnet ring, further deflected by a bending magnet, and focused by means of a quadrupole magnet onto a scintillation counter $S1$. The negative beam is further deflected by a second bending magnet and refocused by another quadrupole onto scintillator $S2$. $C1$ is a threshold Cerenkov counter, recording particles of $\beta > 0.79$, and $C2$ is a differential counter, of the type shown in Fig. 2.20, recording particles of $0.78 > \beta > 0.75$. When account is taken of energy loss in the counters and other material traversed, it turns out that negative pions of the beam momentum have $\beta = 0.99$, and particles of protonic mass have $\beta = 0.76$. Thus, the coincidence

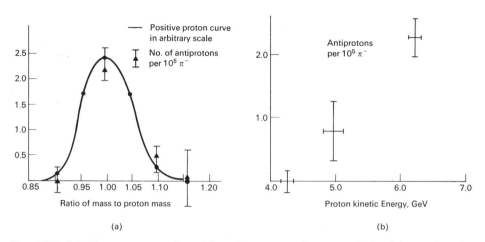

Fig. 2.25 (a) Mass spectrum of particles of near-protonic mass, obtained by varying the momentum of the beam in the antiproton experiment: ▲, negative particles; ●, positive particles. (b) Rate of production of antiprotons relative to negative pions, as a function of the kinetic energy of the incident proton beam on the target.

$S1 + S2 + C1 + S3 - C2$ (meaning a signal from the first four, not accompanied by a signal from $C2$) corresponds to the passage of a negative pion, while $S1 + S2 + C2 + S3 - C1$ corresponds to an antiproton. In this experiment, the ratio of negative pions to antiprotons in the beam was of order 100,000, so that there are severe background problems. Thus, an independent check of the events was made by measuring the time of flight between counters $S1$ and $S2$, separated by some 40 ft, which was expected to be 40 nsec for a negative pion ($\beta = 0.99$) and 51 nsec for an antiproton ($\beta = 0.76$). Time of flight could be determined by displaying the $S1 - S2$ pulses on an oscilloscope. Only those antiproton candidates (from the Cerenkov counter signals) with correct flight times were accepted as bona fide antiprotons.

Several further checks were then made. First, the beam was tuned to slightly different momenta, and the antiproton yield plotted as a function of momentum. Figure 2.25 shows the result. The abscissae plotted are particle mass, as deduced from momentum and velocity. Because the mass resolution of the apparatus is finite, one obtains a mass spectrum of finite width. By changing the position of the target T and reversing the current in the bending magnets, so that π^+ and positive protons are transported by the beam, a similar mass spectrum was deduced for protons. The agreement between the distributions for proton and antiproton events confirms that the negative heavy particles have the same mass as the proton within about 1%.

Secondly, a search was made by Brabant and others for large pulses from the final 14 in. lead glass Cerenkov counter $C3$, contemporaneous with antiprotons from the previous selection system. In order to exclude Cerenkov light from the enormous flux of fast negative pions, only Cerenkov radiation emitted in the backward direction was recorded, by placing the photomultipliers at the front of the radiator. Pulses corresponding to energy release > 1 GeV were indeed observed, thus proving that some of the antiprotons annihilated in flight in the lead glass counter, giving charged and neutral pions. Neutral pions from annihilation would decay to γ-rays, generating a cascade shower in the lead glass, and hence a large Cerenkov pulse. Since the energy released in neutral pions alone exceeded the kinetic energy of the negative protons, this proved that the latter were annihilating with nucleons, as expected. (Such annihilations were shortly afterwards observed in stacks of nuclear emulsion exposed to an antiproton beam.)

Finally, the yield of antiprotons relative to pions in the beam was measured as a function of the momentum of the circulating proton beam in the Bevatron. The observed excitation curve (Fig. 2.25) corresponds closely to what would be expected for the production of a nucleon-antinucleon pair in a nucleon-nucleon collision, taking account of Fermi motion of the target nucleon.

2.8.2 Lifetime of the Neutral Pion

The experiment was carried out at the CERN proton synchrotron in 1963 by Von Dardel, Dekkers, Mermod, Van Putten, Vivargent, Weber, and Winter. The neutral pion decays into γ-rays, $\pi^0 \rightarrow 2\gamma$, with a lifetime now known to be $\approx 10^{-16}$ sec. To see what this implies for the experimenter, we note that the mean distance traveled in the laboratory system, for a neutral pion of momentum p, is $\lambda = \gamma\beta c\tau = c\tau p/m_\pi c$. Thus, even for very energetic pions, with $p = 5$ GeV/c, for example, $\lambda \sim 10^{-4}$ cm or 1μ only.

The principle of the experiment is first to select neutral pions of a nearly unique momentum. Suppose for the moment one is able to do this. In Fig. 2.26, t represents the thickness of a thin metal foil bombarded by 18 GeV protons; neutral pions are produced uniformly through the foil, so that the rate of production in dx is $K \times dx$. The probability that a pion survives to the layer dy and there decays is $\exp(-y/\lambda) \times dy/\lambda$, and the probability that either photon should then convert inside the foil is $(t - y - x)/X$, where X

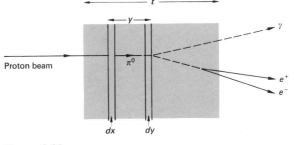

Figure 2.26

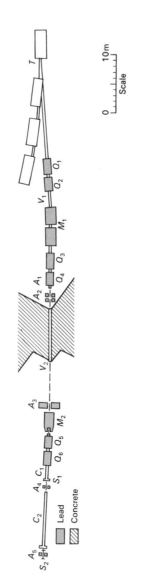

Fig. 2.27 Experimental arrangement for measurement of the π^0 lifetime. T, target; Q_1 to Q_6, quadrupole focusing magnets; $M_{1,2}$, bending magnets; A_1 to A_5, collimators; $C_{1,2}$, hydrogen Cerenkov counters; $S_{1,2}$, scintillators. (After Von Dardel *et al.*, 1963.)

is the γ-ray conversion length $(X \gg t)$. These considerations lead to a pair-production rate, as a function of thickness t, given by

$$R(t) = Kt\left\{B + \frac{1}{X}\left[\frac{t}{2} - \lambda + \frac{\lambda^2}{t}(1 - e^{-t/\lambda})\right]\right\}, \qquad (2.35)$$

where the first term arises from the internal conversion contribution i.e. to pairs produced in the Dalitz decay mode $\pi^0 \to e^+e^-\gamma$. If we neglect a small correction term due to energy degradation in the foil, Eq. (2.35) gives the t-dependence of the number of positrons emerging from the foil, assuming they originate from π^0-decay. We note that, for large t, the effect of a finite lifetime λ is to reduce the average path length for conversion, $t/2$, by the quantity λ. In the experiment, foils of platinum were used, for which $X \sim 1$ cm, $\lambda/X \sim 10^{-4}$, whilst the Dalitz pair branching ratio $B \sim 10^{-2}$. Thus $\lambda/X \ll B$. In principle, it might be possible to measure λ by using thick targets of different Z and hence X-values, but this would require absolute measurements of the constant term in the curly brackets $(B - \lambda/X)$ to a precision better than 1% in different materials. On the other hand, if one expands Eq. (2.35) for small t/λ one obtains

$$R(t) = Kt\left\{B + \frac{t^2}{6\lambda X} + \cdots\right\}, \qquad (2.36)$$

and one is faced with the somewhat easier problem of detecting a quadratic term in t using very thin targets $(t \sim \lambda)$ of a given material, but of a different and accurately measurable thickness. This was the method employed by Von Dardel *et al.*

The experimental set-up is shown in Fig. 2.27. The target system T consisted of four strips of platinum, thicknesses 3, 4, 18, and 58 μ, mounted on a rotatable head, so that different foils could be flipped in turn into identical radial and azimuthal positions relative to the circulating proton beam. The bending magnets, focusing magnets, and collimators serve to define a beam of positive particles of 5 GeV/c momentum at 6° angle, and they are recorded by scintillators $S1$ and $S2$ in coincidence. Positrons among the heavy particles (π^+, protons) were detected by the Cerenkov counters $C1$ and $C2$. These were threshold counters, 10 m long, filled with hydrogen gas, and with a 100% efficiency for positrons and an acceptance for heavier particles of less than 5×10^{-6}. The coincidence ratio $(S_1 + S_2 + C_1 + C_2)/(S_1 + S_2)$ measures the fraction of positrons, i.e. the quantity within the curly brackets in Eq. (2.35). Five Gev/c positrons must come from neutral pions of higher momentum. On the other hand, the flux of pions of momentum p falls off with p in a known way (roughly, exponentially) and if one folds into this spectrum the kinematics of the decay $\pi^0 \to 2\gamma$ and the conversion $\gamma \to e^+e^-$, it is found that the effective π^0-spectrum (contributing 5 GeV/c positrons) is narrow and centered around

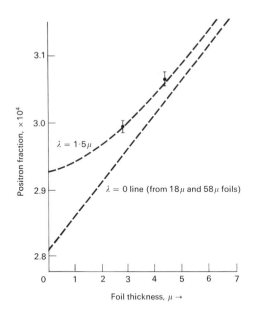

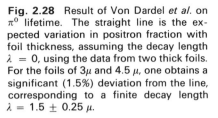

Fig. 2.28 Result of Von Dardel *et al.* on π^0 lifetime. The straight line is the expected variation in positron fraction with foil thickness, assuming the decay length $\lambda = 0$, using the data from two thick foils. For the foils of 3μ and $4.5\ \mu$, one obtains a significant (1.5%) deviation from the line, corresponding to a finite decay length $\lambda = 1.5 \pm 0.25\ \mu$.

7.1 GeV/c. It is, near enough, a unique energy, so that a given value of τ, the proper lifetime, corresponds to a unique decay path λ.

Special precautions were necessary to ensure the measured coincidence rates were not influenced by counting losses dependent on the length or intensity of the proton burst on the target, and that the accidental coincidence rates were subtracted in a rate-independent manner. These are technical details which we do not discuss. The results on the positron fraction of the beam, for different foil thicknesses, were then fitted to the parameters of formula (2.35). Figure 2.28 shows the result, and proves that the neutral pion has a finite lifetime, with $\lambda = 1.5 \pm 0.25\ \mu$, and thence

$$\tau_{\pi^0} = (1.05 \pm 0.18) \times 10^{-16} \text{ sec.} \tag{2.37}$$

In this experiment, the value of B deduced from the fit was reasonably consistent with the Dalitz decay rate found from independent measurements. Furthermore, from the rates it could be shown that the effect measured was due principally to neutral pions and not to other processes (for example, $\eta \rightarrow 2\gamma$).

We should emphasize that it is possible to determine the lifetime *indirectly* by other methods. In 1951, Primakoff had shown that the cross section for photoproduction of a neutral pion in the Coulomb field of a heavy nucleus would be inversely proportional to the lifetime. An experiment by Bellatini and others in 1965 at Frascati (Rome) using this principle, gave

$$\tau_{\pi^0} = (0.73 \pm 0.11) \times 10^{-16} \text{ sec.} \tag{2.38}$$

The Coulomb cross section has to be extracted from the total cross section, which includes nuclear effects, and corrections must be made for pion absorption in the nucleus. It is believed that such corrections can be made accurately and have a small effect on τ. Obviously, the values (2.37) and (2.38) are consistent.

BIBLIOGRAPHY

Bethe, A. A., and J. Ashkin, "Passage of radiations through matter" in Segre (ed.), *Experimental Nuclear Physics*, John Wiley, New York, 1953, Vol. 1, p. 166.

Blewett, M. H., "Characteristics of typical accelerators," *Ann. Rev. Nucl. Science* **17**, 427 (1967).

Bradner, H., "Bubble chambers," *Ann. Rev. Nucl. Science* **10**, 109 (1960).

Charpak, G., "Evolution of Automatic Spark Chambers," *Ann. Rev. Nucl. Sci.* **20**, 195 (1970).

Hutchinson, G., "Cerenkov detectors," *Prog. Nucl. Phys.* **8**, 195 (1960).

Livingstone, M. S., and J. P. Blewett, *Particle Accelerators*, McGraw-Hill, New York, 1962.

McMillan, E. M., "Particle accelerators," in Segre (ed.), *Experimental Nuclear Physics*, John Wiley, New York, 1959, Vol. 3, p. 639.

Price, W. J., *Nuclear Radiation Detection*, McGraw-Hill, New York, 1964.

Rossi, B., *High Energy Particles*, Prentice-Hall, New York, 1952.

Rutherglen, J. G., "Spark chambers," *Prog. Nucl. Phys.* **9**, 1 (1964).

Segre, E., *Nuclei and Particles*, Benjamin, New York, 1964, Chs. 2–4.

Sternheimer, R. M., "Interaction of radiation with matter," in Yuan and Wu (eds.), *Methods of Experimental Physics*, Academic Press, 1961, Vol. 5A, p. 1. (This volume also contains articles on the different types of particle detector.)

Wenzel, W. A., "Spark chambers," *Ann. Rev. Nucl. Science* **14**, 205 (1964).

"Bubble and spark chambers," in Shutt (ed.), *Pure and Applied Physics* **27** (1967).

"Instrumentation for high energy physics," in Farley (ed.), *Proc. 1962 CERN Conf.*, North-Holland, 1963.

PROBLEMS

2.1 Equation (2.1) gives the average number of collisions, say $\bar{n}$, resulting in energy loss $E' - E' + dE'$ when a particle of velocity β traverses a layer of thickness Δx g cm^{-2}. For individual particles, the distribution in n follows the Poisson law, so that the mean square fluctuation in n about $\bar{n}$ is $[(n - \bar{n})^2]_{\text{average}} = \bar{n}$. If we multiply Eq. (2.1) by E^2 and integrate over energy loss, we obtain the mean square deviation in energy loss $\varepsilon^2 = [(\Delta E - \overline{\Delta E})^2]_{\text{average}}$ about the mean $\overline{\Delta E}$. Show, using Eq. (2.2), that this has the form

$$\varepsilon^2 = 0.6\frac{Z}{A}(mc^2)^2\gamma\left(1 - \frac{\beta^2}{2}\right) \cdot \Delta x.$$

Calculate the fractional rms deviation, $\varepsilon/\overline{\Delta E}$, for protons of kinetic energy 500 MeV, traversing (i) 0.1, (ii) 1, (iii) 10 g cm^{-2} of plastic scintillator ($Z/A \sim \frac{1}{2}$). Take dE/dx as 3 MeV g^{-1} cm^2.

2.2 A narrow pencil beam of singly charged particles of momentum p, traveling along the x-axis, traverses a slab of material, s radiation lengths in thickness. If ionization loss in the slab may be neglected, calculate the rms lateral spread of the beam in the y-direction, as it emerges from the slab. [*Hint:* consider an element of slab of thickness dx at depth x, and find the contribution $(dy)^2$ which this element makes to the mean square lateral deflection; then integrate over the slab thickness.] Use the formula you derive to compute the rms lateral spread of a beam of 10 GeV/c muons in traversing a 100 m pipe filled with (i) air, (ii) helium, at NTP.

2.3 Extensive air showers in cosmic rays contain a "soft" component of electrons and photons, and a "hard" component of muons. Suppose the central core of a shower, at sea-level, contains a narrow, vertical, parallel beam of muons of energy 1000 GeV, which penetrate underground. Assume the ionization loss in rock is constant at 2 MeV g^{-1} cm^2. Find the depth in rock at which the muons come to rest, assuming the rock density to be 3.0. Using the formula of the preceding problem estimate their radial spread in meters, taking account of the change in energy of the muons as they traverse the rock. (Radiation length in rock = 25 g cm^{-2}.)

2.4 Verify the approximations (2.20) and (2.21) to the Klein-Nishina formula.

2.5 Show that the available kinetic energy in the head-on collision of two 25 GeV protons is equivalent to that in the collision of a 1300 GeV proton with a stationary nucleon.

2.6 A high energy electron collides with an atomic electron. What is the threshold energy for production of an e^+e^- pair?

2.7 A proton of momentum p, large compared with its rest mass M, collides with a proton inside a target nucleus, with Fermi momentum p_f. Find the available kinetic energy in the collision, as compared with that for a free nucleon target, when p and p_f are (i) parallel, (ii) antiparallel, (iii) orthogonal.

2.8 Verify Eq. (2.35).

2.9 It is sometimes possible to differentiate between the tracks due to relativistic pions, protons, and kaons in a bubble chamber by virtue of the high energy δ-rays which are produced. For a pion of momentum 5 GeV/c, what is the minimum energy of a δ-ray which must be observed, to prove it is not produced by a kaon or proton? What is the probability of observing such a knock-on electron in 1 m of liquid hydrogen (density 0.06)?

Chapter 3

Conservation Laws and Invariance Principles

As emphasized in Chapter 1, the different types of interaction between particles are characterized by specific sets of conservation laws. These rules can usually be interpreted in terms of invariance of the equations describing the interaction under a particular mathematical operation. This often allows one to go further than the conservation laws themselves, since combinations of several such operations may be taken together, to predict further and more restrictive selection rules, and to relate the amplitudes for different processes.

3.1 INVARIANCE IN CLASSICAL MECHANICS

In order to illustrate the relation between a conserved quantity and invariance under a mathematical operation, let us consider conservation of momentum and energy in classical mechanics. These are embodied in Newton's Laws of Motion, but a more general and elegant formulation is given by the Lagrange or Hamiltonian equations. These equations express the energy of a mechanical system, and its derivatives, in terms of generalized coordinates. They have the advantage of avoiding the often complicated problems of resolution of forces and transformation of coordinates involved in the Newtonian approach. The generalized coordinates of an isolated system of n particles are denoted by p_i and q_i where $i = 1, 2, \ldots, 3n$, and $3n$ is the number of degrees of freedom (i.e. there are six coordinates per particle). p_i and q_i can be taken as the momentum and space coordinates, for example, $(p_i = m \, dq_i/dt = m\dot{q}_i)$, or as the angular momentum and the azimuthal angular coordinates about a given axis. In quantum mechanics, a pair of such coordinates (or the uncertainties therein) are always related by the uncertainty principle, so that the product of a pair of generalized coordinates has the dimensions of erg sec. The Hamiltonian equations of motion describing the system are

$$\dot{p}_i = \frac{dp_i}{dt} = -\frac{\partial H}{\partial q_i},$$ (3.1)

and

$$\dot{q}_i = \frac{\partial H}{\partial p_i},$$

where the Hamiltonian function H equals the total energy (kinetic plus potential) of the system. If we now make an infinitesimally small translation,

$$q_i \rightarrow q_i + \delta q,$$

then

$$\delta H = \delta q \sum \frac{\partial H}{\partial q_i} = -\delta q \sum \dot{p}_i.$$ (3.2)

If the momentum is a conserved quantity, it is a constant of the motion, so $\sum \dot{p}_i = 0$ and $\delta H = 0$. Therefore, conservation of momentum may equally be described as invariance of the Hamiltonian under translation of the space coordinates (a finite translation can always be compounded of a succession of infinitesimal ones). This result is intuitively obvious: if there are no external forces (or fields) on the system, the momentum must be internally conserved and the total energy H cannot be changed by a bodily movement of the centre-of-mass in space. Similarly, taking the generalized coordinates as the energy and time, one can show that conservation of energy corresponds to invariance of H with respect to time translations.

3.2 INVARIANCE IN QUANTUM MECHANICS

Turning now to the situation in quantum mechanics, the system can be represented by a state function or state vector, ψ. The result of a physical measurement corresponds to some operator D acting on the state function, and is given by the expectation value $\int \psi^* D \psi \, dV$, or, in the more compact Dirac notation, $\langle \psi | D | \psi \rangle$. The variation of this quantity with time describes the temporal development of the system. It can be attributed to a time dependence of the state function ψ, and a time-independent operator D_0, as in the Schrödinger representation, where the equation of motion has the familiar form

$$i\hbar \frac{\partial \psi}{\partial t} = H\psi,$$ (3.3)

where H is the Hamiltonian (or energy) operator.

An alternative and completely equivalent viewpoint is to regard the state function ψ_0 as time independent, and the time dependence to apply to the operator D. This latter is the Heisenberg representation, which has the advantage for the present discussion that the operators are then closely related to the corresponding time-dependent classical variables (whereas ψ itself has no classical analog). The Heisenberg equation of motion for the operator D has the form*

$$\frac{dD}{dt} = \frac{i}{\hbar}[H, D],\qquad(3.4)$$

where the commutator

$$[H, D] = HD - DH.$$

Thus, the expectation value of the operator, $\langle\psi| D |\psi\rangle$, will generally depend on time, unless D *commutes* with the Hamiltonian, when

$$\frac{dD}{dt} = [H, D] = 0.\qquad(3.5)$$

If H and D commute, D is then a constant of the motion, and the eigenvalue of the operator D is a conserved quantity. Then H is invariant under the D operator, so that the order of the operators does not matter, i.e. $HD = DH$. As an example, consider the parity operator P which inverts all space coordinates:

$$P\langle r | \psi\rangle = \langle -r | \psi\rangle.$$

Then

$$PH(r)\langle r | \psi\rangle = H(-r)\langle -r | \psi\rangle$$
$$= H(-r)P\langle r | \psi\rangle.$$

If P is a constant of the motion, then from (3.5)

$$PH(r) = H(r)P,$$

so that

$$H(-r) = H(r),$$

i.e. the Hamiltonian is invariant under space inversions brought about by the P operator.

To each operator commuting with the Hamiltonian we can ascribe a physically measurable property of the system, the eigenvalue of the operator. We now take D to be the generator of an infinitesimal *space* translation:

$$D = 1 + \delta r \frac{\partial}{\partial r},\qquad(3.6)$$

---

* See, for example, "Quantum Mechanics" by L. I. Schiff, p. 132 (McGraw-Hill, New York, 1955); or "Quantum Mechanics" by F. Mandl, p. 110 (Butterworths, London, 1957).

so that

$$\psi(r + \delta r) = D\psi(r).$$

If (3.5) holds, then the eigenvalue of D, or equivalently the gradient operator $\partial/\partial r$, must be conserved. This quantity is the momentum, through the familiar relation

$$-i\hbar \frac{\partial}{\partial r} \psi = p\psi. \tag{3.7}$$

To summarize, the following statements are all equivalent:

momentum is conserved in an isolated system;
the Hamiltonian is invariant under space translations;
the momentum operator, which is the generator of space translations, commutes with the Hamiltonian.

In a similar manner, the generator of infinitesimal *rotations* about some axis is

$$D = 1 + \delta\phi \frac{\partial}{\partial \phi},$$

i.e.

$$\psi(\phi + \delta\phi) = D\psi(\phi).$$

The operator of the z-component of angular momentum is

$$L_z = -i\hbar\left(x\frac{\partial}{\partial y} - y\frac{\partial}{\partial x}\right) = i\hbar\frac{\partial}{\partial \phi}, \tag{3.8}$$

where ϕ is the azimuthal angle about the z-axis.

Thus

$$D = 1 - \frac{i}{\hbar} L_z \,\delta\phi. \tag{3.9}$$

A finite rotation is accomplished by repeating this process n times; $\phi = n\,\delta\phi$, where $n \to \infty$ as $\delta\phi \to 0$. Then

$$D = \left(1 - \frac{i}{\hbar}L_z\,d\phi\right)^n = 1 + n\left(-\frac{i}{\hbar}L_z\,d\phi\right) + \frac{n(n-1)}{2!}\left(-\frac{i}{\hbar}L_z\,d\phi\right)^2 + \cdots.$$

As $n \to \infty$,

$$D = 1 + \left(-\frac{i}{\hbar}L_z n\,d\phi\right) + \frac{1}{2!}\left(-\frac{i}{\hbar}L_z n\,d\phi\right)^2 + \cdots$$

$$= \exp\left(-\frac{i}{\hbar}L_z\phi\right). \tag{3.10}$$

Conservation of angular momentum about an axis then corresponds to invariance of the Hamiltonian under rotations about that axis.

TABLE 3.1

Transformation	Conserved quantity, or eigenvalue
Space displacement	Momentum
Time displacement	Energy
Spatial rotation	Angular momentum
Space inversion	Parity, ± 1
Rotation in isospin space	Isospin
Charge conjugation	C-parity, ± 1
G-conjugation	G-parity, ± 1
Gauge transformation	Electric charge, baryon number, lepton number

Some of the common transformations, together with the appropriate conserved quantities, are given in Table 3.1.

A word may be said about the class of transformations called gauge transformations—although we shall not have call to use them as such. In classical electromagnetic theory, the fields E and H can be represented as derivatives of the vector and scalar potentials A, ϕ. The *scale* of A and ϕ is arbitrary and we can add terms to both

$$A' = A + \boldsymbol{grad}\ \chi; \qquad \phi' = \phi - \frac{1}{c}\frac{\partial \chi}{\partial t}, \tag{3.11}$$

where χ is any scalar, without altering E and H. This is called a gauge transformation. Invariance of the electromagnetic interaction under a gauge transformation corresponds to the fact that the scale of an electrostatic potential is entirely arbitrary; and this in turn is consistent with energy conservation *only* if electric charge is conserved. That this is so may be shown by the following argument, due to Wigner: if the potential scale is arbitrary, then any physical process must be independent of its absolute value on this scale. Thus, if charge is not conserved, suppose a charge Q is created in a region where the electrostatic potential is V. An amount of work, W say, will be performed to create the charge Q, and this will be independent of V. If the charge Q is now moved to a point where the potential on our chosen scale is V', energy $Q(V - V')$ will be gained. If the charge is then destroyed, the energy W will be recovered, so that the net energy gain is $Q(V - V')$. Thus an arbitrary scale of potential implies that charge has to be conserved if conservation of energy is not to be violated. In field theory, charged particles are described by complex field functions, Φ, and a gauge transformation merely alters the phase of such a field, i.e. $\Phi \rightarrow \Phi \exp(i\lambda)$, where λ is arbitrary. The interaction energy depends on $\Phi\Phi^*$ and is clearly invariant under this transformation. On the other hand, if charge were *not* conserved, the phase (say, between a state of charge $Q = 1$ and one of $Q = 0$) should be a physically measurable quantity. Formally it can be shown that invariance of an interaction under a gauge transformation leads to a conserved current, and thus conservation of charge (and similarly, baryon number and lepton number).

3.3 PARITY

An important concept in quantum mechanics—and one which has no strict classical analog—is the behavior of the wave function representing a system, under inversion of all space coordinates (reflection through the origin). Many physical systems, but not all, can be described by wave functions which are eigenfunctions of the inversion or parity operator, P. In such a case we can write

$$P\psi(r) \rightarrow \psi(-r) = \xi\psi(r),$$

where ξ = a constant. A further operation P must return us to the original system:

$$P^2\psi(r) \rightarrow \xi P\psi(r) \rightarrow \xi^2\psi(r) \equiv \psi(r),$$

so that $\xi^2 = 1$ and $\xi = \pm 1$. $\xi = +1$ corresponds to a state of even parity, $\xi = -1$ to a state of odd parity. As an example, a system with a central potential, like a hydrogen atom, must have a well-defined parity, since the Hamiltonian $H(r)$ is clearly invariant under inversion. The actual angular dependence of the wave function for the hydrogen atom is given by the spherical harmonics,* specified by the orbital angular momentum l and its projection m on the z-axis:

$$Y_l^m(\theta, \phi) = \sqrt{\frac{(2l + 1)(l - m)!}{4\pi(l + m)!}} \times P_l^m(\cos\theta) \times e^{im\phi}; \qquad m \leqslant l, \quad (3.12)$$

where ϕ and θ are azimuthal and polar angles (see Fig. 3.1). The quantity $P_l^m(\cos\theta)$ is an associated Legendre polynomial:

$$P_l^m(\cos\theta) = (-1)^m \sin^m\theta \left[\left(\frac{d}{d(\cos\theta)}\right)^m P_l(\cos\theta)\right]$$

where the Legendre polynomial

$$P_l(\cos\theta) = \frac{1}{2^l l!}\left[\left(\frac{d}{d(\cos\theta)}\right)^l (-\sin^2\theta)^l\right]$$

for m positive. The phase factor $(-1)^m$ in the expression for P_l^m corresponds to the commonly used convention (Condon and Shortley 1951).

---

* See p. 329 for a table of spherical harmonics. For a discussion of spherical harmonic functions, see the texts "Quantum Mechanics" by L. I. Schiff (McGraw-Hill, New York, 1955) p. 71 or F. Mandl, (Butterworths, London, 1957) p. 51.

Under a space inversion P we have, from the figure

$$\theta \xrightarrow{\ P\ } \pi - \theta,$$
$$\phi \longrightarrow \pi + \phi.$$

Using the following relations

$$\cos(\pi - \theta) = -\cos\theta,$$
$$\sin(\pi - \theta) = \sin\theta,$$
$$e^{im(\pi + \phi)} = (-1)^m e^{im\phi},$$
$$\left[\frac{d}{d(\cos(\pi - \theta))}\right]^m = (-1)^m \left[\frac{d}{d(\cos\theta)}\right]^m,$$

it is seen that

$$Y_l^m(\theta, \phi) \xrightarrow{\ P\ } Y_l^m(\pi - \theta, \pi + \phi) = (-1)^l(-1)^{2m} Y_l^m(\theta, \phi)$$
$$= (-1)^l Y_l^m(\theta, \phi). \tag{3.13}$$

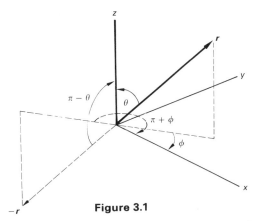

Figure 3.1

Thus the spherical harmonic functions have parity $(-1)^l$; an $s, d, g \ldots$ wave state has even parity, while a $p, f, h \ldots$ wave state has odd parity. Electric dipole transitions between atomic states are characterized by the selection rule $\Delta l = \pm 1$, so that, as a result of the transition, the parity of the atomic state must change. At the same time, a photon is emitted. As indicated below, bosons have been assigned an *intrinsic parity*. The parity of the electromagnetic radiation (photon) emitted in this case is -1. Thus, the parity of the whole system (atom + photon) is conserved.

3.4 SPIN OF THE PION

Before discussing the parity of the pion, we must first describe how its spin has been shown to be zero. For the positive pion, this has been done by applying the principle of detailed balance to the reversible reaction

$$p + p \rightleftarrows \pi^+ + d. \tag{3.14}$$

If the initial state is denoted by i and the final state by f, this principle states that the matrix elements squared, for the forward and backward reactions, are equal, i.e. $M_{if}^2 = M_{fi}^2$ at the same center-of-mass energy. This allows one to make statements about their relative cross sections in terms of the spins of the particles concerned.

The cross section for any reaction of the form

$$a + b \rightarrow c + d, \tag{3.15}$$

will be proportional to the product of M_{if}^2 and a density of final states or phase-space factor ρ_f. Here M_{if} contains all the dynamical features of the interaction, such as coupling strength, energy dependence, angular distribution, and so forth. The rate at which transitions occur from state i to state f is given by*

$$W = \frac{2\pi}{\hbar} |M_{if}|^2 \rho_f. \tag{3.16}$$

In first-order perturbation theory only, M_{if} can be simply interpreted as the overlap integral over volume, $\int \psi_f^* H' \psi_i \, d\tau$, where H' is the interaction potential, and ψ_i and ψ_f are initial and final state wave functions. When the interaction H' is strong, M_{if} cannot be calculated explicitly, and (3.16) can be regarded as a definition of M_{if}. To calculate W, we restrict the particles to some arbitrary volume (which cancels in the calculation) and which we shall take as unity. Then

$$\rho_f = \frac{dn}{dE_0} = \frac{d\Omega}{h^3} p_f^2 \frac{dp_f}{dE_0} g_f, \tag{3.17}$$

where E_0 is the total energy in the center-of-mass frame, p_f is the final state momentum in this frame ($|\boldsymbol{p}_c| = |\boldsymbol{p}_d| = p_f$), and $d\Omega$ is the solid angle containing the final state particles; g_f is a spin multiplicity factor. The cross section is, by definition, related to the reaction rate by

$$\sigma = \frac{W}{\phi} = \frac{W}{v_i n_a}, \tag{3.18}$$

where n_a is the density of incident particles of type a per unit volume, and v_i is the relative velocity of a and b. Thus $\phi = n_a v_i$ is the flux per square centimeter per second. Multiplying this by the cross section per target particle b gives the reaction rate W. Equation (3.15) refers to one particle of each type in the normalization volume (taken as unity). Therefore, $n_a = 1$ and, integrating over all

---

* See, for example, "Quantum Mechanics", by L. I. Schiff, p. 197 (McGraw-Hill, New York, 1955); "Quantum Mechanics" by F. Mandl, p. 161 (Butterworths, London, 1957); or "Basic Quantum Mechanics" by J. M. Cassels, p. 126 (McGraw-Hill, London, 1970).

angles of the products c and d ($\int d\Omega = 4\pi$), one obtains from (3.16), (3.17), and (3.18),

$$\sigma = \frac{W}{v_i} = |M_{if}|^2 \frac{p_f^2}{v_i} \frac{dp_f}{dE_0} g_f, \tag{3.19}$$

where numerical constants are absorbed in $|M_{if}|^2$. Conservation of energy gives

$$\sqrt{p_f^2 + m_c^2} + \sqrt{p_f^2 + m_d^2} = E_0,$$

and thus $\tag{3.20}$

$$\frac{dp_f}{dE_0} = \frac{E_c E_d}{E_0 p_f} = \frac{1}{v_f},$$

where E_c, E_d are total energies, and $v_f = v_c + v_d$ is the relative velocity of c and d. If s_c and s_d are the spins of particles c and d, the number of possible substates for each are $(2s_c + 1)$ and $(2s_d + 1)$, and thus $g_f = (2s_c + 1)(2s_d + 1)$. Equation (3.19) then becomes

$$\sigma_{a+b \rightarrow c+d} = |M_{if}|^2 \frac{(2s_c + 1)(2s_d + 1)}{v_i v_f} p_f^2. \tag{3.21}$$

It is implied in this equation that we are averaging over all possible spin states of the colliding particles a and b. $|M_{if}|^2$ is summed over all orbital angular momentum states involved; since we shall consider forward and backward reactions at the same value of E_0, these factors will be common in the two cases.

In perturbation theory as stated above, M_{if} has the form $\langle f| H' |i \rangle$ where H' is the Hamiltonian operator acting between initial and final states, i and f. H' is a hermitian operator, which means that

$$\langle f| H' |i \rangle = \langle i| H' |f \rangle^*. \tag{3.22}$$

Hence

$$|M_{if}|^2 = |M_{fi}|^2,$$

so that detailed balance is guaranteed in a *perturbation* calculation, such as can be used in electromagnetic and weak interactions. However, (3.14) is a strong interaction, so that (3.22) does not apply. Detailed balance in this case is guaranteed by assuming invariance of the interaction under time reversal and space inversion. Time reversal interchanges final and initial states, but reverses all momenta and spins; space inversion then changes back the signs of momenta but leaves spins unchanged (see Table 3.6). Thus, writing $M_{if} = \langle f| T |i \rangle$, where T is a suitable transition matrix operator, we have

$$\langle f(p_c, p_d, s_c, s_d)| T |i(p_a, p_b, s_a, s_b) \rangle \xrightarrow[\text{+ space inversion}]{\text{time reversal}}$$

$$\langle i(p_a, p_b, -s_a, -s_b)| T |f(p_c, p_d, -s_c, -s_d) \rangle. \tag{3.23}$$

Summing over all $(2s + 1)$ spin projections, running from $-s$ to $+s$, we have again

$$M_{if}^2 = M_{fi}^2.$$

Applying (3.21) to (3.14), detailed balance gives us

$$\frac{\sigma(pp \rightarrow \pi d)}{\sigma(\pi d \rightarrow pp)} = 2 \frac{(2s_\pi + 1)(2s_d + 1)}{(2s_p + 1)^2} \frac{p_\pi^2}{p_p^2}, \qquad (3.24)$$

where p_π and p_p refer to the known values of final state momenta in the center-of-momentum system (CMS) for forward and backward reactions at fixed CMS energy. Thus, using $s_p = \frac{1}{2}$ and $s_d = 1$, the observed ratio measures s_π. The factor 2 comes in because we must respect the antisymmetry of the total wave function of the two protons in the back reaction. If their relative angular momentum l is even, the space wave function is symmetric, and the spin wave function has to be antisymmetric, i.e. the spins must be antiparallel. If l is odd, the spins must be parallel—see Eq. (3.43). Thus, half the possible spin states are disallowed by the Pauli Principle. Another way to phrase this result is that for the phase-space factor (3.17) we integrated over 4π solid angle, whereas integrating over only one hemisphere gives all the states when a pair of identical particles is involved.

The differential cross section $d\sigma/d\Omega$ for the reaction $p + p \rightarrow \pi^+ + d$ was measured by Cartwright *et al.* (1953), using 340 MeV incident protons, corresponding to a pion CMS kinetic energy $T_\pi = 21.4$ MeV. The total cross section for the inverse reaction $\pi^+ + d \rightarrow p + p$ had previously been measured by Clark *et al.*, with $T_\pi = 23$ MeV, giving $\sigma = (4.5 \pm 0.8) \times 10^{-27}$ cm^2, and by Durbin *et al.*, with $T_\pi = 25$ MeV, giving $\sigma = (3.1 \pm 0.3) \times 10^{-27}$ cm^2. Integration of the Cartwright measurements predicted for this reaction $\sigma = (3.0 \pm 1.0) \times 10^{-27}$ cm^2 for $s_\pi = 0$, and $\sigma = (1.0 \pm 0.3) \times 10^{-27}$ cm^2 for $s_\pi = 1$, using (3.24). These data clearly show that $s_\pi = 0$. Later experiments yielded a more precise value:

$$2s_\pi + 1 = 1.00 \pm 0.01.$$

It may be remarked that the existence of the decay

$$\pi^0 \rightarrow 2\gamma$$

proves that s_π must be integral (since $s_\gamma = 1$) and even. A photon has only two possible spin states, either parallel or antiparallel to the direction of motion. Taking the common line of flight of the photons in the π^0 rest frame as the quantization axis, the z-component of total photon spin can have the value $s_z = 0$ or 2. Suppose $s_\pi = 1$, then only $s_z = 0$ is possible. In this case the two-photon amplitude must behave under spatial rotations like the polynomial $P_1^m(\cos\theta)$ with $m = 0$, where θ is the angle relative to the z-axis. Under a 180° rotation about an axis normal to z, $\theta \rightarrow \pi - \theta$; since $P_1^0 \propto \cos\theta$, it therefore changes sign. For $s_z = 0$, the situation corresponds to two right-circularly polarized photons, or to two left-circularly polarized photons, traveling in opposite directions. The above rotation is therefore equivalent to interchange of two photons with identical spin polarization, for which the wave function must be symmetric. Thus, odd integral spin for the neutral pion is forbidden.

3.5 PARITY OF THE PION

It has been found necessary to assign an extra quantum number, the intrinsic parity, to the pion. The reasons for, and the meaning of, this concept are discussed in the next section. The intrinsic parity of the charged pion has been determined from observations of absorption of slow negative pions in deuterium. The reactions

$$\pi^- + d \rightarrow n + n \qquad (3.25a)$$
$$\rightarrow 2n + \gamma \qquad (3.25b)$$

occur, the ratio of (3.25a) to (3.25b) being approximately $2:1$. In particular, no γ-rays are found consistent with the reaction $\pi^- + d \rightarrow 2n + \pi^0$. This situation may be contrasted with capture in hydrogen, where, as we have seen in Chapter 1, both the reactions

$$\pi^- + p \rightarrow n + \gamma$$
$$\rightarrow n + \pi^0$$

occur. The existence of (3.25a) proves that the pion has odd parity. In this reaction, the capture of the pion by the deuteron takes place from an atomic s-state; this is proved by studies of mesic X-rays as well as direct calculations, which show that capture from a p-state is very much slower. Since $s_d = 1$ and $s_\pi = 0$, the initial state in (3.25a) has $J = 1$. There are thus two possibilities for the two-neutron state; either $l = 1$ with spins antiparallel, or $l = 0$ (or 2) with spins parallel. The second possibility is excluded for two identical fermions since it corresponds to a total wave function which is symmetric. Thus we must have $l = 1$ and a final state parity $(-1)^l = -1$. By conservation of parity, that of the initial state is also odd. As indicated below, neutrons and protons are, by convention, assigned an intrinsic parity $+1$. The deuteron therefore has even parity, the orbital state of the neutron and proton in the deuteron being predominantly s-wave (with a little d-wave). Thus we must assign the negative pion an odd intrinsic parity. This result depends, of course, on assigning the *same* parity to neutron and proton, and this is the conventional choice.

The parity of the neutral pion can be established from observations on the polarization of the γ-rays from π^0-decay. Using arguments precisely the same as in Section 3.14 for the 2γ-decay of the singlet state of positronium, an odd (even) parity assignment would predict the polarization vectors of the γ-rays to be predominantly orthogonal (parallel). It is difficult to measure this directly.

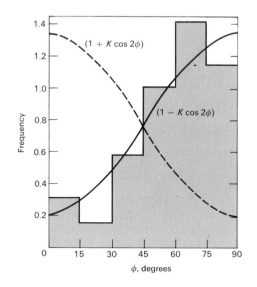

Fig. 3.2 Plot of weighted frequency distribution of angle ϕ between planes of polarization of the pairs in the "double Dalitz decay," $\pi^0 \to (e^+ + e^-) + (e^+ + e^-)$. For a scalar π^0, this should have the form $(1 + K \cos 2\phi)$ and for a pseudoscalar π^0, $(1 - K \cos 2\phi)$. (After Plano *et al.*, 1959.)

However, a γ-ray may "internally convert" to an $e^+ e^-$ pair ("Dalitz decay") with a branching ratio of order $\alpha = 1/137$ (see Fig. 1.7). Thus, the "double Dalitz decay,"

$$\pi^0 \to (e^+ + e^-) + (e^+ + e^-),$$

occurs with a branching ratio of order $\alpha^2 \sim 10^{-4}$. The plane of each electron-positron pair lies predominantly in the plane of the corresponding E-vector, and it has thus been shown that the E-vectors tend to be mutually orthogonal, and therefore that the pion has odd parity (see Fig. 3.2).

A meson which has zero spin and odd intrinsic parity, i.e. $J^P = 0^-$, has to be represented by a wave function which has the space transformation properties under inversion and rotations of a pseudoscalar. Such mesons are therefore called *pseudoscalar* mesons. Similarly, particles of $J^P = 0^+$ are referred to as *scalar*, $J^P = 1^-$ as *vector*, and $J^P = 1^+$ as *pseudovector* or *axial-vector*.

3.6 INTRINSIC PARITY OF PARTICLES AND ANTIPARTICLES

In discussing the process of pion absorption in deuterium, (3.25a), we were forced to introduce the concept of intrinsic parity for the pion. This is -1 if neutron and proton have the *same* intrinsic parity (whether even or odd). The last assumption is only reasonable, since the neutron and proton are to be regarded as different charged states of one particle, the nucleon. One may well ask what "odd intrinsic parity" for the pion really means physically. To answer this, one might invoke some sort of internal structure. For example, Fermi suggested that a pion be regarded as a very strongly bound combination of nucleon and antinucleon. Thus a nucleon and antinucleon in a ^1S-state have $J^P = 0^-$ if one takes into account the opposite intrinsic parity of particle and antiparticle (see Eq. 3.26). This leaves one not much wiser. It is better to regard intrinsic parity simply as a quantum number to assign the pion in order to pre-serve parity conservation in strong processes involving pions. Note particularly that to assign an intrinsic parity quantum number to a pion *in isolation* has no physical meaning whatsoever, any more than to assign a charge quantum number $\pm 1e$ or 0 has any meaning. To say that a pion has charge $+1e$ is really a statement about how the pion will behave in interaction with something else (e.g. an electric field). Similarly, the statement that the pion parity is odd is just a shorthand way of describing the space-reflection properties of interactions in which a pion is produced or absorbed For example, in the reactions $p + p \rightarrow p + n + \pi^+$, or $\pi^0 \rightarrow 2\gamma$, we are compelled to assign the pion this extra quantum number if, as we believe, the space-reflection properties of the initial and final states are to be the same. It is therefore intuitively no more unreasonable to assign a parity number to a pion than a charge number. The difference is only that the concept of intrinsic electric charge is familiar to most people, while intrinsic parity is something new.

The absolute parity of the pion, for example in (3.25a), arises because bosons can be created or destroyed *singly*. On the other hand, baryon number is absolutely conserved, the number of nucleons on each side of a reaction is always the same, and the overall *relative* parity of initial and final states is quite independent of the sign chosen for the nucleon parity.

The *relative* parity of particle and antiparticle is not conventional. For example, one can produce a proton-antiproton pair in the following reaction,

$$p + p \rightarrow p + p + (\bar{p} + p),$$

so that, just as in the case of single pion production or absorption, the intrinsic parity of a nucleon-antinucleon pair is a measurable quantity. If we describe fermions by the Dirac equation, it can be shown that fermion and antifermion should have opposite intrinsic parity (see Appendix B). As indicated in Section 3.14, this prediction has been verified experimentally by observations on the γ-rays from annihilation of positronium. For bosons, particle and antiparticle have the same parity. Therefore, the rule is:

Fermions: Parity of antiparticle $= (-1)$ parity of particle,

Bosons: Parity of antiparticle $= (+1)$ parity of particle.

$$(3.26)$$

The intrinsic parity of the Λ-hyperon is conventionally taken to be even, as for nucleons. Since Λ-hyperons and kaons always occur in association in strong interactions (for example in $p + p \rightarrow K^+ + \Lambda + p$), only the relative parity of kaon plus Λ-hyperon (relative to the nucleon) can be measured; it is found to be odd (see Section 7.1.2), and thus kaons are, by the above convention, assigned odd parity.

The parity of the Σ- and Ξ-hyperons is not conventional. We note that if the Σ and Λ have the same parity, the electromagnetic transition

$$\Sigma^0 \to \Lambda + \gamma$$

must be magnetic dipole ($M1$). Observations on the Dalitz decays, $\Sigma^0 \to \Lambda e^+ e^-$, confirm this. The Ξ-parity (relative to the nucleon) is assumed to be even. In principle it could be measured by observations on the capture process from an S-orbit:

$$\Xi^- + p \to \Lambda + \Lambda,$$

which is analogous to (3.25a), with two identical particles in the final state. Observations on the angular distribution of the pions from the Λ-decays could decide whether the two Λ's have spins antiparallel (^1S-state, even Ξ-parity) or parallel (^3P-state, odd Ξ-parity).

Finally, we note that the intrinsic parity of the photon is odd. A photon will have the transformation properties of the electromagnetic vector potential A, which depends on circulating currents. These clearly change sign under the parity operation.

3.7 CONSERVATION OF PARITY — EXPERIMENTAL TESTS IN STRONG AND ELECTROMAGNETIC INTERACTIONS

The evidence for conservation of parity in strong and electromagnetic interactions is extremely strong. On the other hand, as described in a later chapter, parity conservation is violated in weak interactions. Here we may just mention the *degree* of parity violation in weak interactions. In such processes, the matrix element consists of a superposition of amplitudes of even and odd parities. In the leptonic weak interactions at least, described by the so-called pure V-A theory, the odd and even amplitudes have the same magnitude. This is called the principle of maximal parity violation. A fuller discussion is given in Chapter 4.

At the time (1957) that parity violation was first observed in the weak decays, an intensive search also began for evidence of any violation in the other interactions. If parity is not conserved, the total amplitude will be made up of a parity-conserving amplitude ψ_c and one of the opposite parity ψ_{nc}:

$$\psi = \psi_c + f\psi_{nc}.$$

Upper limits exist on the degree of parity violation, $|f|^2$. In strong interactions, a test was made by Tanner (1957) by means of the reaction

$$p + F^{19} \to O^{16} + He^4. \tag{3.27}$$

Both O^{16} and He^4 are "closed shell" nuclei with spin-parity 0^+. Thus, the combination on the right-hand side of (3.27) can only exist in the states $0^+, 1^-, 2^+, \ldots$, corresponding to $l = 0, 1, 2, \ldots$. The experiment consisted of looking for evidence of the above reaction at a bombarding energy corresponding to the 13.2 MeV excited state of the compound nucleus Ne^{20*}. This state was known to be of $J^P = 1^+$. No evidence was found for the decay of this resonant level to O^{16} plus an α-particle, and gave an estimate $|f|^2 < 10^{-8}$. Many similar tests have yielded the same result. In electromagnetic interactions, the selection rules for atomic transitions provide strong evidence for a high degree of parity conservation, setting limits $|f|^2 < 10^{-12}$ in such interactions.

It is, however, important to emphasize that all particles (except photons) which undergo strong or electromagnetic interactions can also interact weakly. Thus, the Hamiltonian describing the interaction will always have a small parity-violating component, because of the weak interaction. Evidence for such parity violation was first presented in 1964/65 in electromagnetic γ-transitions in nuclei. One method, used by Abov *et al.*, was to observe the angular asymmetry of γ-rays emitted from the excited nucleus Cd^{114*}, formed when a beam of polarized thermal neutrons (Section 3.17) was absorbed in a cadmium target. A small asymmetry was measured, relative to the direction of neutron spin-polarization. As discussed in Section 4.2.3, such a forward-backward asymmetry implies noninvariance under space inversion. Another technique, employed by Boehm and Kankeleit, was to detect the circular polarization, in an analyzing magnet, of γ-rays from a particularly favorable electromagnetic transition in Ta^{181}. A net circular polarization again corresponds to parity violation, since it defines a "screw sense." A similar experimental technique had previously been used in determining the neutrino helicity in β-decay, and is described in Section 4.3.2. Both the experiments detected small asymmetries or circular polarizations, consistent with a parity-violating intensity $|f|^2 \sim 10^{-12}$. More recent work has confirmed that the limit on the degree of parity violation observed is indeed consistent with that calculated from the weak interaction contribution. Unfortunately, at present (1970) the results of different experiments are inconsistent and it is not clear if any have established a significant effect.

To summarize, therefore, all evidence shows that the "pure" strong and electromagnetic interactions are absolutely parity conserving.

3.8 ISOSPIN AND ITS CONSERVATION

3.8.1 Charge Symmetry and Charge Independence in Nuclear Physics

The *charge symmetry* of nuclear forces is illustrated by the existence of mirror nuclei. Li^7 and Be^7 are examples of a mirror pair. The ground states have binding energies differing by only ~ 1.5 MeV, and this can be accounted for by

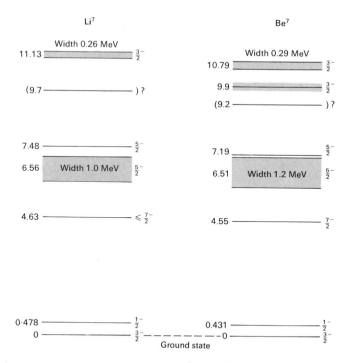

Fig. 3.3 Comparison of the excited states of Be⁷ and Li⁷. The numbers on the left-hand side refer to energies in MeV, and those on the right-hand side to the spin-parity assignments.

the extra Coulomb energy of Be^7. The excited states show a strong similarity (see Fig. 3.3).

This equivalence in binding due to specifically *nuclear* forces (i.e. with the small Coulomb and magnetic effects subtracted) would follow if the strong p–p and n–n interactions were identical. Each nucleus consists of a Li^6 core plus an extra neutron or proton, giving an extra n–n or p–p "bond" respectively. The number of n–p "bonds" is identical, so we learn nothing about n–p forces relative to n–n or p–p.

Charge independence goes further by postulating that all pairs of interactions of neutrons and protons are identical:

$$n\text{–}n \equiv p\text{–}p \equiv n\text{–}p$$

(always for pairs in the same state of angular momentum and spin). Consider a nuclear configuration consisting of a C^{12} core + two nucleons:

$$\text{C}^{12} \begin{cases} +n\text{–}n = \text{C}^{14}, \\ +p\text{–}p = \text{O}^{14}, \\ +n\text{–}p = \text{N}^{14}. \end{cases}$$

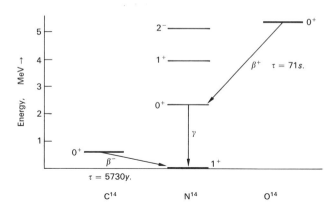

Fig. 3.4 Energy levels of the triad of states O^{14}, N^{14}, and C^{14}. Heavy lines denote the ground states.

C^{14} and O^{14} are mirror nuclei, and their binding energies are consistent with charge symmetry. The ground state of N^{14} does *not* fit; however, the 2.3 MeV level of $N^{14}*$ fits in well with the C^{14} and O^{14} ground states and has the same spin-parity (Fig. 3.4). We note that $N^{14}*$ is produced by β^+-decay of O^{14} and this is a "super-allowed" transition—which means the nuclear configurations are identical; this is to be contrasted with the very slow β^--transition from C^{14} to the N^{14} ground state.

We may also note at this point that $N^{14}*$ is *not* produced in the reaction

$$O^{16} + d \rightarrow N^{14} + He^4, \tag{3.28}$$

for reasons that will become clear shortly.

Other evidence for charge independence comes from the near equality of the scattering length and potential in p–p and p–n scattering in the singlet spin state.

3.8.2 Isospin of the Nucleon

Heisenberg (1932) first suggested that neutron and proton should be treated as different charge states of one particle, the nucleon—precisely as a neutron with spin up and a neutron with spin down are different spin states of one particle. Hence the name isospin or I-spin (replacing the more ancient usage, isotopic spin or isobaric spin).

Indeed, the formal description of isospin operators and wave functions is identical with that for ordinary spin and angular momentum. The basic idea in nonformal language is that one can visualize an "isospin space," the isospin of a state being represented by a vector I in this space. Notice the precise analogy with the angular momentum vector in real space; conservation of angular momentum is expressed by invariance of the "length" of this vector under rotations of the coordinate axes. The Cartesian components of angular momentum are denoted by J_x, J_y, and J_z. The corresponding components of isospin are usually referred to as I_1, I_2, and I_3—an unfortunate change of nomenclature, but one to which we should adhere.

Since the nucleon has two charge states, we assign it an isospin $I = \frac{1}{2}$ with components

$$I_3 = +\tfrac{1}{2} \qquad \text{for the proton,}$$
$$I_3 = -\tfrac{1}{2} \qquad \text{for the neutron,}$$

(3.29)

and the electric charge given by

$$\frac{Q}{e} = \frac{B}{2} + I_3,$$

(3.30)

where $B(= 1)$ is the baryon number. As far as strong interactions are concerned, everything is specified by I and we have no way to differentiate between the substates of neutron and proton—they are degenerate. Only when we "turn on" the electromagnetic interactions do we differentiate between the two states of the nucleon, by providing a quantization axis (for example, the direction of an applied electric field), and, consequently, well-defined and distinguishable eigenvalues of I_3. Since Q and B are absolutely conserved, so also is I_3, in both strong and electromagnetic interactions.

Conservation of isospin in nuclear interactions means that the latter depend on I and not on I_3. It should be emphasized that the hypothesis of conservation of isospin is a much stronger and more general statement than that nuclear forces are charge independent. The latter is a consequence of the former, but is true only for a pair of similar particles of isospin $\frac{1}{2}$. As we shall see, isospin is conserved in the pion-nucleon interaction, but the interaction is *not* charge independent. Charge independence follows when, by some symmetry argument, the total isospin of the system is limited to a unique value, and is then equivalent to the statement that, for fixed I, the interaction is independent of I_3. Returning to (3.29) it is clear that, for a nucleus, $I_3 = (Z - N)/2$, where Z and N are the numbers of protons and neutrons respectively. In our triad above $(C^{14}, O^{14}, N^{14*})$, we have *one* nuclear state of $I = 1$ and components $I_3 = -1 \,(C^{14})$, $I_3 = +1 \,(O^{14})$, and $I_3 = 0 \,(N^{14*})$. O^{16}, d, and He^4 all have $I_3 = 0$ and, it also turns out, $I = 0$, so the state N^{14*} could not be produced in reaction (3.28) if isospin is conserved.

An example of conservation of isospin is provided by observations on the reaction

$$d + d \rightarrow He^4 + \pi^0$$

| I | 0 | 0 | 0 | 1 |
| I_3 | 0 | 0 | 0 | 0. |

As indicated above, both the deuteron and α-particle have $I = 0$, and indeed they have no known isobars, while the π^0 has $I = 1$, $I_3 = 0$ [Eq. (3.31)]. Searches for this reaction have been made, and it is found that the cross section is at least two orders of magnitude smaller than that expected if isospin were conserved. The process cannot take place as a *strong* reaction; it can obviously proceed as an *electromagnetic* transition, with a correspondingly smaller cross section, since I_3 is conserved.

3.9 ISOSPIN ASSIGNMENTS FOR PIONS AND STRANGE PARTICLES

As we have seen, the pion exists in three charge states of closely similar mass. Since I_3 has $(2I + 1)$ projections on the z-axis, we therefore set $I_\pi = 1$. Since $B = 0$, $Q/e = I_3$, and thus we have

$$
\begin{array}{cc}
 & I_3 \\
\pi^+ & +1 \\
\pi^0 & 0 \\
\pi^- & -1.
\end{array}
\tag{3.31}
$$

For the strange particles, note first that the Λ-hyperon has no charged counterpart, implying $I_\Lambda = 0$. Thus, (3.30) has to be modified for strange particles. The values of I and I_3 involved in Λ-decay are as follows:

$$
\begin{array}{cccc}
 & \Lambda & \rightarrow p & + \pi^- \\
I & 0 & \frac{1}{2} & 1 \\
I_3 & 0 & \frac{1}{2} & -1.
\end{array}
\tag{3.32}
$$

This is a *weak* decay process, and neither I nor I_3 are conserved on the two sides of the equation. From the fact that the Λ-hyperon and neutral kaon are observed to be produced in association in strong interactions of pions with protons, in the early diffusion cloud chamber experiments, we can assign the kaon half-integral isospin. $I_K = \frac{1}{2}$ is the simplest choice:

$$
\begin{array}{ccccc}
 & \pi^- & + p & \rightarrow \Lambda & + K^0 \\
I & 1 & \frac{1}{2} & 0 & \frac{1}{2} \\
I_3 & -1 & \frac{1}{2} & 0 & -\frac{1}{2}.
\end{array}
\tag{3.33}
$$

The correct charge for the neutral kaon is obtained if we set

$$
\frac{Q}{e} = I_3 + \tfrac{1}{2},
\tag{3.34}
$$

implying that the K^+-meson, of $I_3 = +\frac{1}{2}$, is the charged member of a kaon doublet. This assignment also forbids the strong decay of a kaon according to the scheme $K^+ \rightarrow \pi^+ + \pi^- + \pi^+$; as already seen, this is a weak decay

TABLE 3.2 Isospin and strangeness assignments for particles decaying by weak or electromagnetic interactions

			I_3				
B	S	I	-1	$-\frac{1}{2}$	0	$+\frac{1}{2}$	$+1$
1	0	$\frac{1}{2}$			n	p	
1	-1	0			Λ		
0	0	1	π^-		π^0		π^+
0	$+1$	$\frac{1}{2}$		K^0		K^+	
0	-1	$\frac{1}{2}$		K^-		$\overline{K^0}$	
1	-1	1	Σ^-		Σ^0		Σ^+
1	-2	$\frac{1}{2}$		Ξ^-		Ξ^0	
1	-3	0			Ω^-		
0	0	0			η		

process. The K^--meson does not fit into this scheme. It was therefore necessary to postulate a second K-doublet, with

$$\frac{Q}{e} = I_3 - \tfrac{1}{2}, \tag{3.35}$$

which predicts a second neutral kaon of $I_3 = +\frac{1}{2}$—called $\overline{K^0}$. K^+ and K^- are considered as particle and antiparticle, as are K^0 and $\overline{K^0}$. The interesting phenomena associated with K^0 and $\overline{K^0}$ are discussed in Section 4.11.

Gell-Mann (1953) and Nishijima (1955) pointed out that the formulae (3.30), (3.34), and (3.35) could be more elegantly expressed by introducing the *strangeness* quantum number, according to the formula

$$\frac{Q}{e} = \frac{B}{2} + \frac{S}{2} + I_3. \tag{3.36}$$

The assignment of strangeness S follows from the isospin assignments. Nucleons and pions clearly must have $S = 0$, the Λ-hyperon $S = -1$, the K^0, K^+ doublet $S = +1$, and the $K^-, \overline{K^0}$-doublet $S = -1$. These assignments are shown in Table 3.2.

An example of conservation of strangeness in the reaction

$$K^- + p \to \Lambda + \pi^0$$

$$S \quad -1 \quad\ 0 \ -1 \quad\ 0$$

$$I_3 \quad -\tfrac{1}{2} \ +\tfrac{1}{2} \quad 0 \quad\ 0$$

is shown in Fig. 3.5.

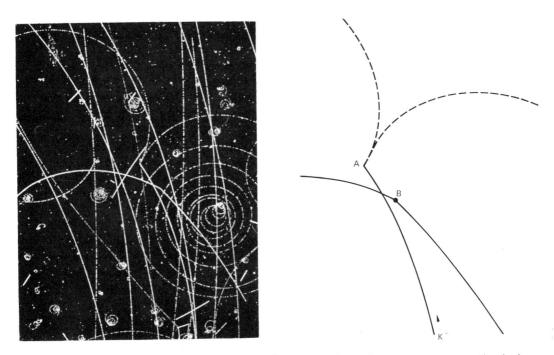

Fig. 3.5 Example of the reaction $K^- + p \to \Lambda + \pi^0$ occurring when a K^--meson comes to rest in a hydrogen bubble chamber, at the point "A." The neutral pion undergoes Dalitz decay, $\pi^0 \to e^+ e^- \gamma$. The Λ-hyperon decays ($\Lambda \to \pi^- + p$) at the point "B." (Courtesy, CERN Information Service.)

The Σ-hyperons fit into a charge triplet, $I = 1$. This assignment fits into the observed strong production reactions

$$\pi^{\pm} + p \to \Sigma^{\pm} + K^+$$

I	1	$\frac{1}{2}$	1	$\frac{1}{2}$
I_3	± 1	$\frac{1}{2}$	± 1	$\frac{1}{2}$
S	0	0	-1	1,

and accounts for the fact that the decay $\Sigma^+ \to n + \pi^+$ is weak. The predicted neutral member, Σ^0, was not finally identified until 1959. It undergoes the electromagnetic decay mode

$$\Sigma^0 \to \Lambda + \gamma$$

I	1	0	0
I_3	0	0	0
S	-1	-1	0.

The cascade or Ξ^--hyperon was first observed in the early cosmic ray work with cloud chambers; the $S = -2$ assignment followed from the production in association with a pair of K^0-mesons, and from the fact that the decay

$$\Xi^- \to \Lambda + \pi^-$$

was weak. The expected neutral counterpart Ξ^0 was detected in 1959. An example of Ξ^--decay is shown in Fig. 3.6.

The last two entries in the table are of comparatively recent vintage. η decays electromagnetically according to the scheme $\eta \to 3\pi^0$, 2γ, $\pi^+\pi^-\pi^0$, etc. The Ω^--particle decays weakly in the modes $\Xi^- + \pi^0$, $\Lambda + K^-$, etc. These particles are important in that they completed unitary symmetry multiplets and provided strong evidence in favor of the SU3 classification scheme, which is discussed in Chapter 6.

The electromagnetic interactions do not conserve I and thus differentiate between the members of isospin multiplets. Thus, electromagnetic splitting of the degeneracy appears as a mass difference between the charged components

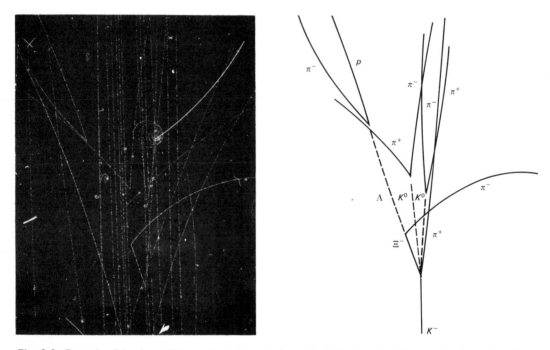

Fig. 3.6 Example of the decay $\Xi^- \to \pi^- + \Lambda$ in a hydrogen bubble chamber. The reaction is produced by an incident 10 GeV/c K^--meson:

$$K^- + p \to \Xi^- + \pi^+ + K^0 + K^0 + \overline{K^0}$$

Strangeness S: -1 0 -2 0 $+1$ $+1$ -1.

The final $\overline{K^0}$ is not observed in this picture. Both K^0-mesons decay in the mode $K^0 \to \pi^+\pi^-$. (Courtesy, CERN Information Service.)

TABLE 3.3 Mass differences in isospin multiplets

	$\Delta m(\mathrm{MeV}/c^2)$	$m_{av}(\mathrm{MeV}/c^2)$	$\Delta m/m \times 10^3$
$n - p$	1.3	939	1.4
$\Sigma^0 - \Sigma^+$	3.1	1190	2.6
$\Sigma^- - \Sigma^0$	4.9	1195	4.1
$\Xi^- - \Xi^0$	6.5	1318	4.9
$K^0 - K^\pm$	3.9	495	7.9
$\pi^\pm - \pi^0$	4.6	140	33

of a multiplet, of order $\Delta m/m \sim \alpha \sim 10^{-2}$. The actual numbers for the long-lived baryons and mesons are given in Table 3.3. Note that particle and antiparticle must have identical masses, by the *CPT* theorem (Section 3.17), and thus $m_{\pi^+} \equiv m_{\pi^-}$, $m_{K^+} \equiv m_{K^-}$. $m_{\Sigma^+} \neq m_{\Sigma^-}$, since Σ^+ and Σ^- are both baryons, rather than baryon and antibaryon.

3.10 ISOSPIN FUNCTIONS FOR A PAIR OF NUCLEONS

Just as for the description of particles of spin $\frac{1}{2}$, we can formally express the isospin wave function of a nucleon as a two-component matrix

$$\chi = \begin{vmatrix} 1 \\ 0 \end{vmatrix} = p,$$

$$\chi = \begin{vmatrix} 0 \\ 1 \end{vmatrix} = n, \tag{3.37}$$

where, for brevity, p and n signify the proton (or isospin up) state and neutron (or isospin down) state, respectively. Just as in the formalism for spin $\frac{1}{2}$, we introduce an isospin operator τ which has Cartesian components given by the 2×2 matrix operators

$$\tau_1 = \frac{1}{2} \begin{vmatrix} 0 & 1 \\ 1 & 0 \end{vmatrix}, \quad \tau_2 = \frac{1}{2} \begin{vmatrix} 0 & -i \\ i & 0 \end{vmatrix}, \quad \tau_3 = \frac{1}{2} \begin{vmatrix} 1 & 0 \\ 0 & -1 \end{vmatrix}, \tag{3.38}$$

which, apart from the factor $\frac{1}{2}$, are identical with the Pauli spin matrices, and obey the same commutation relations. If we apply the τ_3-operator to the wave functions (3.37) we obtain

$$\tau_3 p = \frac{1}{2} \begin{vmatrix} 1 & 0 \\ 0 & -1 \end{vmatrix} \begin{vmatrix} 1 \\ 0 \end{vmatrix} = \frac{1}{2} \begin{vmatrix} 1 \\ 0 \end{vmatrix} = \frac{1}{2}p, \tag{3.39}$$

and

$$\tau_3 n = -\frac{1}{2}n,$$

expressing the fact that the eigenvalues of τ_3 are $+\frac{1}{2}$ and $-\frac{1}{2}$ for the two charge states of the nucleon. When we say that a nucleon has an isospin $\tau = \frac{1}{2}$, we really mean that there are $(2\tau + 1)$ eigenvalues of the *operator* τ_3. The eigenvalue of τ^2 should then be $\tau(\tau + 1) = \frac{3}{4}$. We verify this from (3.38):

$$\tau^2 = \tau_1^2 + \tau_2^2 + \tau_3^2 \qquad \text{where} \qquad \tau_1^2 = \tau_2^2 = \tau_3^2 = \frac{1}{4}\begin{vmatrix} 1 & 0 \\ 0 & 1 \end{vmatrix}, \qquad (3.40)$$

so,

$$\tau^2 \chi = \frac{3}{4}\begin{vmatrix} 1 & 0 \\ 0 & 1 \end{vmatrix} \chi = \frac{3}{4}\chi.$$

Other familiar combinations of the Pauli operators are the "raising" and "lowering" operators

$$\tau_+ = \tau_1 + i\tau_2 = \begin{vmatrix} 0 & 1 \\ 0 & 0 \end{vmatrix},$$

$$(3.41)$$

$$\tau_- = \tau_1 - i\tau_2 = \begin{vmatrix} 0 & 0 \\ 1 & 0 \end{vmatrix}.$$

Applying these to (3.37) one obtains

$$\tau_+ p = 0, \qquad \tau_+ n = p,$$

$$(3.42)$$

$$\tau_- p = n, \qquad \tau_- n = 0.$$

Thus the operator τ_+ transforms a neutron into a proton state, and τ_- transforms a proton into a neutron—in other words, the operators $\tau_\pm$ flip the sign of the third component of isospin of the nucleon.

We now consider two nucleons, labeled (1) and (2). The total isospin operator $I = \tau(1) + \tau(2)$. $I = 0$ or 1 obviously, where the eigenvalues of the operator I^2 are $I(I + 1)$, equal to 0 or 2, and of I_3 are $-1, 0$, and $+1$. We write $\chi(I, I_3)$ as the isospin wave function. Two of the possible combinations are straightforward:

2 protons	$\chi(1, 1) = p(1)p(2)$	S,	(3.43a)
2 neutrons	$\chi(1, -1) = n(1)n(2)$	S.	(3.43b)

These states have $I_3 = \pm 1$ and hence must have $I = 1$. They are obviously symmetric under interchange of particles (1) and (2), hence they are labeled "S." What about p, n combinations? Clearly $p(1)n(2)$ is no use, since

$$n(1)p(2) \neq \pm n(2)p(1).$$

The answer is to find χ-functions with definite exchange symmetry, since the *total* two-nucleon wave function, of which χ, as well as space and spin functions, will be a part, must be antisymmetric, as explained in Section 1.2.1. We can however write $p(1)n(2)$ as the sum or difference of two terms:

$$\chi(1, 0) = \frac{1}{\sqrt{2}} [p(1)n(2) + n(1)p(2)] \qquad S, \qquad (3.43c)$$

$$\chi(0, 0) = \frac{1}{\sqrt{2}} [p(1)n(2) - n(1)p(2)] \qquad A. \qquad (3.43d)$$

The $\sqrt{2}$ factor is for normalization; the three components $\chi(1, 1)$, $\chi(1, -1)$, and $\chi(1, 0)$ must have equal weight; (3.43c) is symmetric and (3.43d) is anti-symmetric under $1 \leftrightarrow 2$ interchange. It is plausible that (3.43c), with the same symmetry as (3.43a) and (3.43b), should be the $I_3 = 0$ member of the $I = 1$ triplet, leaving (3.43d) as the antisymmetric $I = I_3 = 0$ isospin singlet state. This assignment is readily confirmed by application of the Pauli spin operators. The total isospin operator is $I = \tau(1) + \tau(2)$, where $\tau(1)$ acts only on nucleon (1) and $\tau(2)$ only on nucleon (2). For example, the I_3-eigenvalue of the p–p combination is given by

$$I_3[p(1)p(2)] = [\tau_3(1) + \tau_3(2)]p(1)p(2)$$
$$= [\tau_3(1)p(1)]p(2) + [\tau_3(2)p(2)]p(1)$$
$$= (\tfrac{1}{2} + \tfrac{1}{2})p(1)p(2),$$

and is therefore $+1$, as in (3.43a). Similarly, the eigenvalues of I^2 and I_3 may be checked for the other combinations (3.43a) to (3.43d). An equivalent demonstration is to apply the raising or lowering operators $I^\pm$ (see Appendix C) to the states 3.43(a), (b), (c) or (d). Using the formula

$$I^\pm \chi(I, I_3) = \sqrt{I(I + 1) - I_3(I_3 \pm 1)} \cdot \chi(I, I_3 \pm 1),$$

it is readily demonstrated that applying I^+ to the n–p combination (3.43c) gives (3.43a), and that I^- gives the state (3.43b). Thus (3.43a, b, and c) transform one into the other by rotations in isospin space. However, $I^\pm$ applied to (3.43d) gives zero, so that this is an $I = 0$ singlet which cannot transform into any of the others by isospin rotations.

The total wave function for the two-nucleon state may be written

$$\psi_{total} = \Phi \text{ (space)} \times \alpha \text{ (spin)} \times \chi \text{ (isospin)}. \qquad (3.44)$$

This factorization is valid if we neglect possible interactions between spin and isospin or isospin and space coordinates. The justification is that we obtain results consistent with experiment. Applying (3.44) to the deuteron, we note that α is symmetric, since the neutron and proton spins are parallel and thus the spin function must be analogous to (3.43a) or (3.43b). The space wave function Φ has symmetry $(-1)^l$, where the relative orbital angular momentum of the two nucleons is $l = 0$ (with a small $l = 2$ admixture). Thus Φ is also symmetric. Since, by the extended Pauli Principle, ψ_{total} for two nucleons must be antisymmetric, it follows that χ is antisymmetric. From (3.43d), it follows that $I = 0$—the deuteron is an isosinglet. As an example, consider the reactions

$$\text{(i) } p + p \rightarrow d + \pi^+ \qquad \text{(ii) } p + n \rightarrow d + \pi^0$$

| I | 1 | 0 | 1 | | 0 or 1 | 0 | 1. |

In each case the final state is of $I = 1$. On the left-hand side, we have a pure $I = 1$ state in reaction (i), but $50\%\ I = 0$ and $50\%\ I = 1$ in reaction (ii). Conservation of isospin means that either reaction can only proceed through the $I = 1$ channel. Consequently, $\sigma(\text{ii})/\sigma(\text{i}) = \frac{1}{2}$, as is observed.

The isospin functions for a nucleon-antinucleon pair are discussed in Section 6.2.

3.11 ISOSPIN IN THE PION-NUCLEON SYSTEM

An important application of isospin conservation arises in the strong interactions of nonidentical particles, which will generally consist of mixtures of different isospin states. The classical example of this is pion-nucleon scattering. Since $I_\pi = 1$ and $I_N = \frac{1}{2}$, one can have $I_{\text{total}} = \frac{1}{2}$ or $\frac{3}{2}$. If the strong interactions depend only on I and not on I_3, then the $3 \times 2 = 6$ pion-nucleon scattering processes can all be described in terms of two isospin amplitudes.

Of the six elastic scattering processes,

$$\pi^+ p \rightarrow \pi^+ p \tag{a}$$

and

$$\pi^- n \rightarrow \pi^- n \tag{b}$$

have $I_3 = \pm\frac{3}{2}$, and are therefore described by a pure $I = \frac{3}{2}$ amplitude. Clearly, at a given bombarding energy, (a) and (b) will have identical cross sections, since they differ only in the sign of I_3.

The remaining interactions,

$$\pi^- p \rightarrow \pi^- p,$$
$$\pi^- p \rightarrow \pi^0 n,$$
$$\pi^+ n \rightarrow \pi^+ n,$$
$$\pi^+ n \rightarrow \pi^0 p,$$

have $I_3 = \pm\frac{1}{2}$ and therefore $I = \frac{1}{2}$ or $\frac{3}{2}$. There is now no principle of exchange symmetry, which restricts the isospin to one of the two values, and therefore one has a mixture. The weights of the two amplitudes in the mixture are given by Clebsch-Gordan coefficients (alternatively known as vector-coupling or Wigner coefficients). Their derivation is given in Appendix C. An alternative method, which can be applied to the present case, has been given by Feynman. It is not only elegant, but illuminates the physical meaning of charge independence and we reproduce it here.

The force between nucleons is charge independent. This force may be thought of as being due to exchange of virtual particles between the nucleons (just as the electrostatic force between two charges can be thought of in terms of exchange of virtual photons). There is no unique "carrier" of the strong force analogous to the photon in the electromagnetic case. Many processes, such as pion exchange, ρ-meson exchange, $K\bar{K}$ exchange, etc., may contribute. It is very plausible to assume that the *contribution* from single pion exchange alone is charge independent—if it were not, charge independence would be an accident depending on chance cancellation of forces due to all possible exchange mechanisms, which individually were not charge independent. This possibility is discounted as too remote.

The p–p, p–n, and n–n interactions contributed by single pion exchange are represented diagrammatically in Fig. 3.7, where a, b, and c are unknown coupling constants. In the last diagram, the virtual π^- can of course be replaced by a π^+ with arrow reversed—we do not distinguish the two. If the p–p, n–n, and n–p forces are equal, we have

$$a^2 = b^2 = ab + c^2.$$

Thus, $a = \pm b$. Furthermore, c cannot be zero, since it is known that the last diagram exists and leads to charge-exchange p–n scattering. One must then have

$$b = -a,$$

$$c = \pm\sqrt{2}\,a,$$

where the last sign is arbitrary.

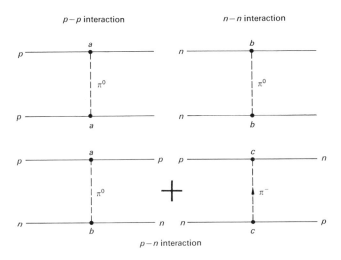

Figure 3.7

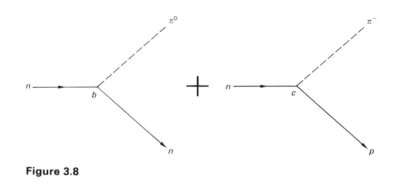

Figure 3.8

Because of its strong interaction, a nucleon can virtually dissociate into a nucleon plus one or more pions, a hyperon-kaon pair, and so on. The possible virtual states of the neutron involving a nucleon plus one pion are shown in Fig. 3.8.

The initial state has $I = \frac{1}{2}, I_3 = -\frac{1}{2}$; the amplitudes in the final state can be written $(n\pi^0)$ and $(p\pi^-)$. Thus,

$$\chi(\tfrac{1}{2}, -\tfrac{1}{2}) = b(n\pi^0) + c(p\pi^-)$$
$$= b[(n\pi^0) \pm \sqrt{2}\,(p\pi^-)].$$

Normalizing the wave function, i.e. with $\chi\chi^* = 1$, we obtain

$$\chi(\tfrac{1}{2}, -\tfrac{1}{2}) = \sqrt{\tfrac{1}{3}}\,(n\pi^0) - \sqrt{\tfrac{2}{3}}\,(p\pi^-). \tag{3.45}$$

Similarly, the proton state will be

$$\chi(\tfrac{1}{2}, \tfrac{1}{2}) = -\sqrt{\tfrac{1}{3}}\,(p\pi^0) + \sqrt{\tfrac{2}{3}}\,(n\pi^+), \tag{3.46}$$

where the sign of c has been chosen to conform to the usual notation.

To find the coefficients for the $I = \frac{3}{2}, I_3 = \pm\frac{1}{2}$ states, we make use of the orthogonality and normalization conditions; for example,

$$\chi(\tfrac{1}{2})\chi(\tfrac{3}{2})^* = 0; \qquad (n\pi^0)(n\pi^0)^* = 1; \qquad (p\pi^-)(n\pi^0)^* = 0.$$

Thus, if

$$\chi(\tfrac{3}{2}, \tfrac{1}{2}) = A(p\pi^0) + B(n\pi^+),$$
$$\chi(\tfrac{3}{2}, -\tfrac{1}{2}) = C(p\pi^-) + D(n\pi^0),$$

one finds

$$A^2 + B^2 = C^2 + D^2 = 1,$$
$$A = B\sqrt{2}, \qquad D = C\sqrt{2},$$

and thus

$$B = C = (\pm)\sqrt{\tfrac{1}{3}}, \qquad D = A = (\pm)\sqrt{\tfrac{2}{3}}.$$

TABLE 3.4 Clebsch-Gordan coefficients in pion-nucleon scattering

Pion	Nucleon	$I = \frac{3}{2}$				$I = \frac{1}{2}$	
		$I_3 = \frac{3}{2}$	$\frac{1}{2}$	$-\frac{1}{2}$	$-\frac{3}{2}$	$\frac{1}{2}$	$-\frac{1}{2}$
π^+	p	1					
π^+	n		$\sqrt{\frac{1}{3}}$			$\sqrt{\frac{2}{3}}$	
π^0	p		$\sqrt{\frac{2}{3}}$			$-\sqrt{\frac{1}{3}}$	
π^0	n			$\sqrt{\frac{2}{3}}$			$\sqrt{\frac{1}{3}}$
π^-	p			$\sqrt{\frac{1}{3}}$			$-\sqrt{\frac{2}{3}}$
π^-	n				1		

We thus obtain Table 3.4 for the Clebsch-Gordan coefficients, the signs following the usual notation (Condon and Shortley 1951).

We can now calculate the relative cross sections for the following three processes, at a fixed energy:

$$\pi^+ + p \rightarrow \pi^+ + p \quad \text{(elastic scattering),} \quad \text{(a)}$$
$$\pi^- + p \rightarrow \pi^- + p \quad \text{(elastic scattering),} \quad \text{(b)} \qquad (3.47)$$
$$\pi^- + p \rightarrow \pi^0 + n \quad \text{(charge exchange).} \quad \text{(c)}$$

The cross section is proportional to the square of the matrix element connecting initial and final states, i.e.

$$\sigma \propto \langle \psi_f | H | \psi_i \rangle^2 = M_{if}^2,$$

where H is an isospin operator, having a value $H = H_1$ if it operates on initial and final states of $I = \frac{1}{2}$, and $H = H_3$ for states of $I = \frac{3}{2}$. By conservation of isospin, there is no operator connecting initial and final states of different isospin. Let

$$M_1 = \langle \psi_f(\tfrac{1}{2}) | H_1 | \psi_i(\tfrac{1}{2}) \rangle,$$
$$M_3 = \langle \psi_f(\tfrac{3}{2}) | H_3 | \psi_i(\tfrac{3}{2}) \rangle.$$

Reaction (a) involves pure states of $I = \frac{3}{2}, I_3 = \pm\frac{3}{2}$. Therefore,

$$\sigma_a = K \times |M_3|^2,$$

where K is some constant.

Referring to our table, in reaction (b) we may write

$$\psi_i = \psi_f = \sqrt{\tfrac{1}{3}}\,\chi(\tfrac{3}{2}) - \sqrt{\tfrac{2}{3}}\,\chi(\tfrac{1}{2}).$$

Therefore

$$\sigma_b = K\langle \psi_f | H_1 + H_3 | \psi_i \rangle^2$$
$$= K\,|\tfrac{1}{3}M_3 + \tfrac{2}{3}M_1|^2.$$

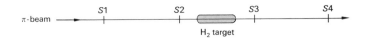

Fig. 3.9 Schematic drawing of measurement of the total pion-proton cross section.

For reaction (c), one has

$$\psi_i = \sqrt{\tfrac{1}{3}}\,\chi(\tfrac{3}{2}) - \sqrt{\tfrac{2}{3}}\,\chi(\tfrac{1}{2}),$$

$$\psi_f = \sqrt{\tfrac{2}{3}}\,\chi(\tfrac{3}{2}) + \sqrt{\tfrac{1}{3}}\,\chi(\tfrac{1}{2}),$$

and thus

$$\sigma_c = K \left| \frac{\sqrt{2}}{3} M_3 - \frac{\sqrt{2}}{3} M_1 \right|^2.$$

The cross section ratios are then

$$\sigma_a : \sigma_b : \sigma_c = |M_3|^2 : \tfrac{1}{9} |M_3 + 2M_1|^2 : \tfrac{2}{9} |M_3 - M_1|^2. \tag{3.48}$$

The limiting situations, if one or other isospin amplitude dominates under the experimental conditions, are

$$\left.\begin{array}{ll} M_3 \gg M_1; & \sigma_a : \sigma_b : \sigma_c = 9:1:2, \\ M_1 \gg M_3; & \sigma_a : \sigma_b : \sigma_c = 0:2:1. \end{array}\right\} \tag{3.49}$$

Numerous experimental measurements have been made of the total and differential pion-nucleon cross sections. The earliest and simplest experiments measured the attenuation of a collimated, monoenergetic $\pi^{\pm}$-beam in traversing a liquid hydrogen target. Thus, in the sketch of Fig. 3.9, one would measure, for a given number of coincidences in the counters S1 and S2, the change in rate of counters S3 and S4 with the target both full and empty. The results of such measurements are shown in Fig. 3.10. For both positive and negative pions, there is a strong peak in σ_{total} at a pion kinetic energy of 200 MeV. The ratio $(\sigma_{\pi^+ p}/\sigma_{\pi^- p})_{\text{total}} = 3$, proving that the $I = \tfrac{3}{2}$ amplitude dominates this region. This bump is referred to as a *resonance*, the N* (1236)—(1236 MeV being the invariant pion-nucleon mass). The width of this state at half-height is 120 MeV. Because the spin-parity turns out to be $J^P = \tfrac{3}{2}^+$, and $I = \tfrac{3}{2}$, it is often referred to as the (3, 3) resonance. A discussion of this pion-nucleon state is given in Section 7.4. As Fig. 3.10 indicates, more pion-nucleon resonances are observed—for example, there is an $I = \tfrac{1}{2}$ resonance at 1525 MeV. In general, in any one region of pion-nucleon invariant mass, several amplitudes will contribute, and one cannot simply interpret a bump in cross section as signifying a unique resonant state. Only for the first (3, 3) resonance is such an interpretation unambiguous.

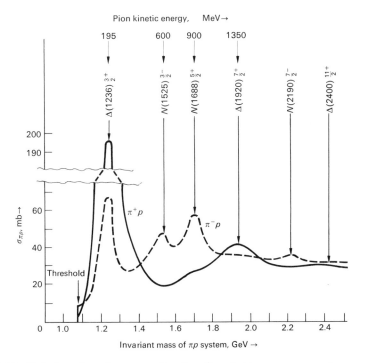

Fig. 3.10 Variation of total cross section for π^+ and π^- mesons on protons, with incident pion energy. The symbol "Δ" refers to resonances of $I = \frac{3}{2}$; "N" refers to $I = \frac{1}{2}$. The positions of only a few of the known states, together with their spin-parity assignments, are given.

3.12 CHARGE CONJUGATION INVARIANCE

As the name implies, the operation of charge conjugation reverses the sign of charge and magnetic moment of a particle (leaving all other coordinates unchanged). Symmetry under charge conjugation in classical physics is evidenced by the invariance of Maxwell's equations as a result of change in sign of the charge and current density and also of E and H. In relativistic quantum mechanics the term "charge conjugation" also implies the interchange of particle and antiparticle. For baryons and leptons, a reversal of charge entails a change in sign of the baryon number or lepton number. For example, a proton becomes an antiproton under the C (charge conjugation)-operation:

	Proton $\xrightarrow{\;c\;}$	Antiproton
Charge	$+1e$	$-1e$
Baryon number	$+1$	-1
Magnetic moment	$+\mu$	$-\mu$
Spin	σ	σ

Note that $\pm\mu$ means that the magnetic moment is parallel (antiparallel) to the spin vector.

Experimental evidence for invariance of the strong and electromagnetic interactions under the C-operation is good but not overwhelming; it is discussed in Section 3.15. It is true to say that, at the present time, no experimental evidence exists which positively requires violation of C-invariance in these interactions, and any possible violation is below the 1 % level.

On the contrary, weak interactions violate charge conjugation invariance, just as they violate invariance under the parity operation. These features are discussed in detail in Chapter 4, but it is perhaps worth mentioning them briefly at this point. The noninvariance of weak interactions under the parity and charge conjugation operations is exemplified by the longitudinal polarization of neutrinos (v) and antineutrinos ($\bar{v}$) emitted in β-decay, in company with positrons and electrons respectively. Neutrinos have spin $\frac{1}{2}$ and zero (or almost zero) mass, so that, as pointed out in Section 1.2.2, the spin vector $\boldsymbol{\sigma}$ must be either parallel or antiparallel to the momentum vector $\boldsymbol{p}$. Experimentally, for neutrinos it is found that $\boldsymbol{\sigma}$ is in the direction $-\boldsymbol{p}$, so that the neutrino is "left-handed." This is shown in Fig. 3.11, where $\boldsymbol{p}$ is along the negative z-axis in (a). Under the parity operation, which in this case corresponds to the inversion $z \rightarrow -z$, the polar vector $\boldsymbol{p}$ changes sign, whereas, as Table 3.6 indicates, the axial vector $\boldsymbol{\sigma}$ remains unchanged. This situation is shown in diagram (b). It corresponds to a right-handed neutrino, which does not exist in nature. This tells us that the weak interaction is not invariant under space inversions. One can contrast this with photons emitted in electromagnetic processes. Both

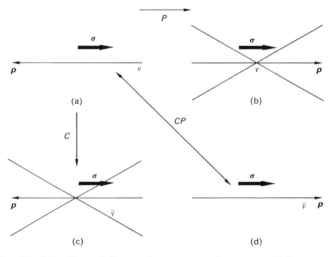

Fig. 3.11 Results of the C- and P-operations on neutrino states. Only states (a) and (d) are observed in nature.

right-handed and left-handed photons exist, but they occur with equal probability, so that the interaction is invariant under a spatial inversion.

Similarly, if we apply the charge conjugation operation to the v-state (a), we obtain a left-handed $\bar{v}$, as in (c). This also does not exist. However, if, in addition, we make the spatial inversion of this state, we end up with a right-handed antineutrino (d), which *is* observed. Thus the weak interactions are not invariant under *C* or *P* separately, but do exhibit *CP*-invariance. Although we come to this conclusion by considering neutrino states, it may be remarked that it is a general property of all weak interactions, whether they involve neutrinos or not.

3.13 EIGENSTATES OF THE *C*-OPERATOR

Consider the operation of charge conjugation performed on a charged pion wave function, which we write as $|\pi^{\pm}\rangle$:

$$C\,|\pi^+\rangle \rightarrow |\pi^-\rangle \neq \pm\,|\pi^+\rangle.$$

In this operation, an arbitrary phase may enter; this is not important for the present discussion. We note that $|\pi^+\rangle$ and $|\pi^-\rangle$ are *not* *C*-eigenstates. However, for a neutral system, the charge conjugation operator may have a definite eigenvalue. Thus, for the neutral pion,

$$C\,|\pi^0\rangle = \eta\,|\pi^0\rangle,$$

since the π^0 transforms into itself; η is a constant. Clearly, repeating the operation gives us $\eta^2 = 1$, so

$$C\,|\pi^0\rangle = \pm 1\,|\pi^0\rangle.$$

To find the sign, we note that electromagnetic fields are produced by moving charges (currents) which change sign under charge conjugation. As a consequence, the photon has $C = -1$. Since the charge conjugation quantum number is multiplicative, this means that a system of n photons has C-eigenvalue $(-1)^n$. The neutral pion undergoes the decay

$$\pi^0 \rightarrow 2\gamma,$$

and thus has even C-parity:

$$C\,|\pi^0\rangle = +\,|\pi^0\rangle. \tag{3.50}$$

The decay

$$\pi^0 \rightarrow 3\gamma$$

will therefore be forbidden if the electromagnetic interactions are invariant under *C*. Experimentally, the branching ratio

$$\frac{\pi^0 \rightarrow 3\gamma}{\pi^0 \rightarrow 2\gamma} < 5 \times 10^{-6}. \tag{3.51}$$

This limit, though impressive, is not as significant as at first appears, since, even if C were violated, the expected branching ratio would be of the same order of magnitude.

3.14 POSITRONIUM DECAY

We now consider the restrictions imposed by C-invariance on the states of *positronium*, which undergoes annihilation in the modes

$$e^+e^- \rightarrow 2\gamma, 3\gamma.$$

The bound state of electron and positron possesses energy levels similar to the hydrogen atom (but with exactly half the spacing, because of the factor 2 in the reduced mass). We write down the total wave function of the positronium state as the product of three wave functions depending on the spin, space, and charge coordinates:

$$\psi(\text{total}) = \Phi(\text{space}) \ \alpha(\text{spin}) \ \chi(\text{charge}). \tag{3.52}$$

For the moment, we assume that we have two identical fermions (rather than fermion and antifermion) labeled (1) and (2), and consider how the functions ψ, α, and Φ behave under particle interchange. If the spins of electron and positron are parallel (total spin $S = 1$), then as Eq. (3.43) indicates, α is symmetric under interchange of the spin coordinate of particle (1) with that of particle (2). Similarly, for antiparallel spins, $S = 0$, α is antisymmetric. The space wave function Φ will be expressed in spherical harmonics and will thus have symmetry $(-1)^l$, where l is the orbital angular momentum quantum number, as in (3.13). Under interchange of charge coordinates, let χ acquire a factor C. Thus the factors entering under the coordinate interchanges are:

$$\text{spin interchange} \qquad (-1)^{S+1};$$
$$\text{space interchange} \qquad (-1)^l;$$
$$\text{charge interchange} \qquad C.$$

These three operations in succession are equivalent to total interchange of particle (1) with particle (2). Because of the Pauli principle, ψ must be antisymmetric under this process. Thus

$$(-1)^{S+1}(-1)^l C = -1,$$

or

$$C = (-1)^{l+S} \tag{3.53}$$
$$= (-1)^n,$$

where n is the number of photons into which the positronium decays. If, as indicated below, positronium decays only from the ground state $l = 0$, the two possibilities are:

  i) Singlet state $^1S_0$ of $J = 0$, n even, decaying to 2 γ-rays;
  ii) Triplet state $^3S_1$ of $J = 1$, n odd, decaying to 3 γ-rays.

Decay into 4 γ-rays and 5 γ-rays is less probable by a factor $\alpha^2 \sim 10^{-4}$. We may note that 2γ-decay of the triplet state is also forbidden by angular momentum conservation. (The proof is left as an exercise.) The annihilation rate to 2 γ-rays has been calculated to be

$$\frac{1}{\tau(2\gamma)} = 4\pi r_e^2 c \, |\psi(0)|^2, \tag{3.54}$$

where $r_e = e^2/mc^2$ is the classical electron radius, and $\psi(0)$ is the amplitude of the electron-positron radial wave function at the origin. From the solution of the Schrödinger equation for the ground state of the hydrogen atom, we know that

$$|\psi(0)|^2 = \frac{1}{\pi a^3}, \tag{3.55}$$

where a is the Bohr radius. States of angular momentum l contain a factor r^l in the radial wave function—hence, for all except the ground state, they vanish at the origin. Remembering the factor 2 for the reduced mass effect, the Bohr radius in positronium will have a value

$$a = \frac{2r_e}{\alpha^2}. \tag{3.56}$$

From (3.54), (3.55), and (3.56), and using the values $r_e = 2.8 \times 10^{-13}$ cm, $\alpha^{-1} = \hbar c/e^2 = 137$, one obtains for the mean lifetime,

$$\tau(2\gamma) = \frac{2r_e}{c\alpha^6} = 1.25 \times 10^{-10} \text{ sec.} \tag{3.57}$$

For the 3γ-decay, the annihilation rate is slower by a factor of order α. The calculated value is

$$\tau(3\gamma) = \frac{9\pi}{4(\pi^2 - 9)} \times \frac{\tau(2\gamma)}{\alpha} = 1.4 \times 10^{-7} \text{ sec.} \tag{3.58}$$

Both the 2γ- and 3γ-periods have been detected in the work of Deutsch, who measured the annihilation rates of positrons stopping in gases. The long period due to the 3γ-mode was found to be $(1.45 \pm 0.15) \times 10^{-7}$ sec, in agreement with (3.58).

The discussion of the quantum numbers of positronium given above does not quite tell the whole story. Electron and positron are in fact not two fermions, but fermion and antifermion, and therefore must have opposite intrinsic parity. This fact will not affect the foregoing analysis, but it does mean that, for annihilation from the $^1S_0$-state, the 2γ-wave function must have odd parity.

Let k, $-k$ be the momentum vectors of the two photons, e_1 and e_2 their polarization vectors (E-vectors). The initial state has $J = 0$, and the simplest independent scalars one can form to include both E-vectors and satisfy requirements of exchange symmetry for identical bosons, are

$$\psi(2\gamma) = a(e_1 \cdot e_2),$$

or

$$\psi(2\gamma) = b(e_1 \wedge e_2) \cdot k, \qquad (3.59)$$

where a and b are constants. The first product is a scalar and is therefore even under space inversions. The second is a pseudoscalar (the product of a polar vector with an axial vector) and therefore has the odd parity required. If $\psi(2\gamma)$ is to be finite, e_1 and e_2 cannot be parallel. Thus the planes of polarization of the two photons should be preferentially at right angles. Experimentally, this can be checked by observing the angular distribution of the Compton scattering of the γ-rays, which depends strongly on polarization, being more probable in a plane normal to the electric vector. This is clear if one considers that the incident photon sets up oscillations of the target electron in the direction of the E-vector. In the subsequent dipole radiation of the scattered photon, the radiated intensity is greatest normal to E.

The argument used above represents the properties of each photon by a momentum vector k and a plane polarization vector e normal to k, giving the direction of the associated electric (or magnetic) field. An alternative and equivalent discussion is possible in terms of the spin vectors of the photons. As mentioned in Chapter 1, a photon may be either right-handed (R) or left-handed (L) according to whether the spin points in the direction of motion or against it. This corresponds to the classical description of circularly polarized light with rotating e-vectors. Let us denote by $(R, +k)$ right-handed photons, amplitude R, traveling in the direction $+k$; and $(L, -k)$ left-handed photons, amplitude L, in the direction $-k$. Since we are discussing single photons, $RR^* = LL^* = 1$. The combinations $(R, +k; L, -k)$ or $(L, +k; R, -k)$ of two photons obviously correspond to states of $J = +2$ or -2, which do not interest us here. The combinations $(R, +k; R, -k)$ or $(L, +k; L, -k)$ have $J = 0$ as we require in the positronium problem, but neither of these is an eigenstate of the parity operator. However, linear superpositions of these states have definite parity:

$$\phi = [(R, +k; R, -k) + (L, +k; L, -k)],$$
$$P\phi = [(L, -k; L, +k) + (R, -k; R, +k)]$$
$$= +\phi,$$
$$\chi = [(R, +k; R, -k) - (L, +k; L, -k)],$$
$$P\chi = -[-(L, -k; L, +k) + (R, -k; R, +k)]$$
$$= -\chi,$$

The state χ has odd parity, and consists of the combination $a_+ = R - L$ in direction $+\boldsymbol{k}$ and $a_- = R + L$ in direction $-\boldsymbol{k}$. Combining these circular polarizations, it is seen that we obtain two waves with plane polarizations. These planes are orthogonal, with the product $a_+ a_- = R^2 - L^2 = 0$.

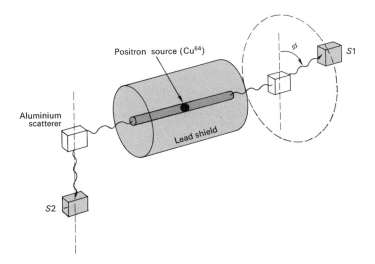

Fig. 3.12 Sketch of the method used by Wu and Shaknov (1950) to measure the relative orientations of the polarization vectors of the two photons emitted in decay of $^1S_0$-positronium. S1 and S2 are anthracene counters, recording the γ-rays after Compton scattering by aluminum cubes. Their results proved that fermion and antifermion have opposite intrinsic parity, as predicted by the Dirac theory.

The experimental set-up used by Wu and Shaknov (1950) is shown in Fig. 3.12. The coincidence rate of γ-rays scattered by aluminum blocks was measured in the anthracene counters S1 and S2 as a function of their relative azimuthal angle, ϕ. The expected correlation has the form

$$(1 + \alpha \sin^2 \phi).$$

The observed azimuthal asymmetry ratio equals

$$\text{rate}(\phi = 90°)/\text{rate}(\phi = 0°) = 2.04 \pm 0.08,$$

compared with a theoretically expected value of 2.00. Such experiments therefore bear out the prediction of preferentially orthogonal polarizations of the γ-rays, and thus demonstrate the correctness of the assumption that fermions and antifermions have opposite intrinsic parities. Similar arguments have been made to demonstrate that the neutral pion has odd intrinsic parity (Section 3.5).

Opposite parity for baryons and antibaryons implies that for proton-antiproton annihilation at rest (from an S-state) into two neutral kaons,

$p\bar{p} \rightarrow K_1^0 + K_2^0$ is allowed, while $p\bar{p} \rightarrow 2K_1^0$, or $2K_2^0$ is forbidden. This is verified experimentally (see Problem 3.13).

3.15 EXPERIMENTAL TESTS OF *C*-INVARIANCE

In principle, an obvious test of *C*-invariance in the *strong* interactions would be obtained by comparing the cross section for a reaction with that in which all the particles are replaced by their antiparticles. Thus, the $\pi^+ p$ elastic scattering cross section should be identical to the $\pi^- \bar{p}$ cross section. Since this is experimentally inaccessible, other tests have been made. For example, bubble chamber experiments have compared the rates and spectra of positive and negative pions, in the reactions

$$p + \bar{p} \rightarrow \pi^+ + \cdots ,$$

and

$$p + \bar{p} \rightarrow \pi^- + \cdots .$$

No discernible difference was obtained between the shapes and intensities of the positive and negative pion spectra, allowing the experiments to set an upper limit to the *C*-violating amplitude f in strong interactions:

$$f < 0.01. \tag{3.60}$$

Observations on $p\bar{p}$-annihilations into K^-, K^+ plus other particles, gave limits on *C*-violation, as determined from positive and negative kaon spectra, of $<2\%$.

The search for evidence of *C*-violation in *electromagnetic* interactions has been intensive in recent years. The reason for this is that such violation had been suggested as a possible source of the violation of *CP*-invariance in weak K_L-decay (Section 4.12). A suitable place to look for effects is in the following three decay modes of the η-meson, which decays by an electromagnetic transition (Section 3.16):

$$\eta \rightarrow \pi^+ \pi^- \pi^0 \tag{3.61}$$
$$\rightarrow \pi^+ \pi^- \gamma \tag{3.62}$$
$$\rightarrow \pi^0 e^+ e^-. \tag{3.63}$$

The η-meson is an eigenstate of *C*-parity of value $+1$, since $\eta \rightarrow 2\gamma$ with a large (35%) branching fraction. If we interpret (3.63) as the decay $\eta \rightarrow \pi^0 \gamma$ with internal conversion of the γ-ray, it is forbidden, since $C_\eta = +1, C_\gamma = -1$, and $C_{\pi^0} = +1$. The decay (3.63) could, however, take place by a higher order process, conserving C, but only with a very low branching ratio (estimated to be $\sim 10^{-8}$). Thus, if (3.63) occurred with a higher branching ratio than this, one would have definite evidence for *C*-violation. Experimentally, (3.63) has never

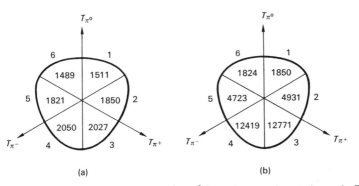

Fig. 3.13 (a) Dalitz plot of the decay $\eta \rightarrow \pi^+ \pi^- \pi^0$ from the experiment shown in Figs. 3.14 and 3.15. Kinetic energies of the pions are plotted along three axes, at 120°. The curve is the boundary imposed by energy/momentum conservation. Charge conjugation invariance implies equal numbers of events to left and right of the center line. Numbers of events in each sextant are given (total 10,000 events). (b) Similar plot from an experiment by Gormley *et al.* (1968) (total 37,000 events).

been observed, the upper limit being

$$\frac{R(\eta \rightarrow \pi^0 e^+ e^-)}{R(\eta \rightarrow \text{all})} < 10^{-4}. \tag{3.64}$$

The interpretation of this limit in terms of the level of a possible *C*-violating amplitude is unfortunately not straightforward. Even if *C* is violated, one expects the decay mode (3.63) to be strongly suppressed for other reasons, and it is hard to evaluate these factors theoretically. Due to this circumstance, the

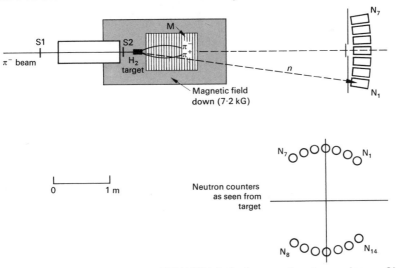

Fig. 3.14 Experimental layout of the CERN-Zürich-Saclay experiment on η-decay. *S1, S2,* beam defining scintillators. *M*, magnet spark chamber, recording the π^+ and π^- mesons from η-decay. N_1 to N_7, ring of neutron time-of-flight counters. (After Cnops *et al.*, 1966.)

main experimental effort has gone into an examination of decay modes (3.61) and (3.62). In these modes, violation of C-invariance would manifest itself in an asymmetry of the energy spectra of the π^+- and π^--particles. For the three pion decay process, a Dalitz plot (Section 7.1) can be made. The coordinates of each event are the kinetic energies of the three pions, in the η rest frame, measured along axes inclined at 120° (Fig. 3.13). A right-left asymmetry in the Dalitz plot would be evidence for C-violation. The numbers of events in each sextant obtained in two recent experiments are shown. The latest experiment (Gormley *et al.*, 1968) indicates a right-left asymmetry of $1.5 \pm 0.5\%$.

The safest conclusion to be drawn from these results is that C-invariance in electromagnetic interactions is confirmed to at least this level of accuracy.

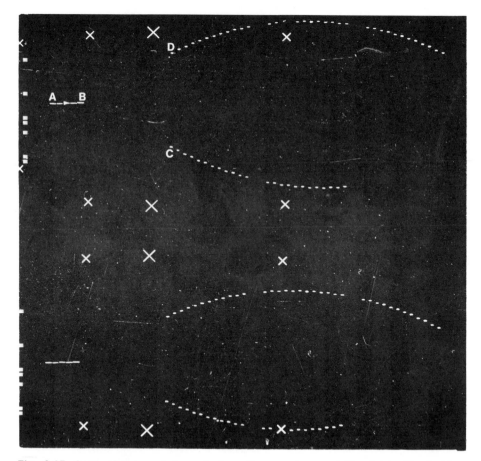

Fig. 3.15 Spark chamber photograph, from the experiment of Fig. 3.14. The incident negative pion track A appears on the left, entering the hydrogen target at B. C and D are tracks of π^+ and π^- in the magnetized spark chamber. From the two stereo views, the tracks may be reconstructed in space, and the momenta evaluated. (Courtesy, CERN Information Service.)

Attention is drawn, in the two sets of data in Fig. 3.13, to the widely different trends in Dalitz plot population as a function of π^0-energy. This is presumably due to different types of experimental selection bias. With such large systematic effects as a function of the π^0-energy, it seems hard to exclude the possibility of very small ones, at the 1% level, producing small apparent asymmetries in the distribution of the π^+ and π^- decay products.

Figure 3.14 indicates the experimental set-up used in a CERN experiment to measure η-decay asymmetry. The method employed an incident negative pion beam of momentum 713 MeV/c incident on a liquid hydrogen target:

$$\pi^- + p \rightarrow n + \eta$$
$$\hookrightarrow \pi^+ \pi^- \pi^0.$$

The events leading to an η-meson in the final state may be selected by measuring the angle of emission and time of flight of the neutron. If the time of flight is correct, the magnet spark chamber is triggered, and the π^+ and π^- from the η-decay are recorded, as shown in the example of Fig. 3.15.

3.16 _G_-PARITY AND NUCLEON-ANTINUCLEON ANNIHILATION

The charge conjugation operator C can have eigenvalues only for neutral systems, such as the π^0, γ, η, and e^+e^-. It is useful, however, to be able to formulate selection rules for some charged systems; this can be done for strong interactions by combining the operation of charge conjugation with an isospin rotation. For this purpose, consider the operation

$$G = CR = C \exp{(i\pi I_2)}. \tag{3.65}$$

The operation G consists of a rotation R of $180°$ about the y-axis in isospin space [compare (3.10)], followed by charge conjugation. Applied to a state with a z-component of isospin I_3, this amounts to first flipping $I_3 \rightarrow -I_3$ and then reversing the process, $-I_3 \rightarrow I_3$. It is therefore plausible that charged states may be eigenfunctions of the G-operator. In order to get the eigenvalues, consider an isospin state $\chi(I, I_3 = 0)$. Under isospin rotations, this state behaves precisely like the angular momentum wave function $Y_l^{m=0}(\theta, \phi)$ of (3.12) under rotations in ordinary space (see Fig. 3.1). The R-operation $\exp{(i\pi L_y)}$ implies $\theta \rightarrow \pi - \theta$, $\phi \rightarrow \pi - \phi$, and hence

$$Y_l^0 \xrightarrow[R]{} (-1)^l Y_l^0.$$

Therefore,

$$\chi(I, 0) \xrightarrow[R]{} (-1)^I \chi(I, 0). \tag{3.66}$$

For a state of nucleon and antinucleon, of total spin S and orbital angular momentum l, the effect of the C-operation is to give a factor $(-1)^{l+S}$, just as in the case of positronium. Thus, the effect of the operation $G = CR$ on a neutral

$(I_3 = 0)$ nucleon-antinucleon system will be

$$G |\psi\rangle = (-1)^{l+S+I} |\psi\rangle. \tag{3.67}$$

Since the strong interactions are invariant under isospin rotations, (3.67) must, in fact, be a general formula, and not limited to the case $I_3 = 0$ for which it was derived. Now suppose the G-operator acts on a pion wave function, $|\pi^+\rangle$. R reverses I_3, thus converting $\pi^+ \to \pi^-$, and C flips the charge back, $\pi^- \to \pi^+$. We may then write

$$G |\pi^+\rangle = \pm |\pi^+\rangle,$$
$$G |\pi^-\rangle = \pm |\pi^-\rangle,$$
$$G |\pi^0\rangle = \pm |\pi^0\rangle.$$

The neutral pion must be an eigenstate of C, with eigenvalue $+1$, since it decays in the mode $\pi^0 \to 2\gamma$. From (3.66), the R eigenvalue is $(-1)^I = -1$. Thus

$$G |\pi^0\rangle = - |\pi^0\rangle.$$

The eigenvalue of the G-operator is called the G-parity. While the G-parity of the neutral pion is unambiguous, that of the charged pions is not. They are not eigenstates of C, and in the process of charge conjugation, an arbitrary phase appears, which can be chosen at will. For convenience, however, it is the practice to define the phases so that all members of an isospin triplet have the same G-parity as the neutral member. In the present case we can then write

$$G |\pi\rangle = - |\pi\rangle \tag{3.68}$$

provided we define

$$C |\pi^\pm\rangle = - |\pi^\mp\rangle.$$

For details, the reader is referred to Appendix E. In some books, the operation R is defined as a rotation through π about the x-axis, instead of the y-axis, in isospin space. In that case the result (3.68) obtains if $C |\pi^\pm\rangle = + |\pi^\mp\rangle$. The two ways of defining G are entirely equivalent, since the aim is simply to reverse the value of I_3. The underlying physics is incorporated in the G-parity selection rules discussed below. These result from invariance of the strong interactions under charge conjugation and isospin rotations, and are independent of arbitrary choices of axes or phases.

It may be pointed out in passing that similar arbitrary phases occur in the definitions of the spherical harmonics, Clebsch-Gordan coefficients, etc., in the treatment of angular momentum. In this case, fortunately, the conventions of Condon and Shortley (1951) have been almost universally adopted.

Since the C-operation reverses the sign of the baryon number, it will be apparent that eigenstates of G-parity must have baryon number zero. The charge conjugation quantum number is multiplicative and isospin additive, so that *G*-parity is multiplicative. Thus, for a state of n pions,

$$G |\psi(n\pi)\rangle = (-1)^n |\psi(n\pi)\rangle. \qquad (3.69)$$

Let us now return to the problem of nucleon-antinucleon annihilation. As an example, consider the annihilation of an antiproton with a neutron into two pions:

$$\bar{p} + n \to \pi^0 + \pi^-. \qquad (3.70)$$

Application of the formula

$$Q/e = I_3 + B/2$$

shows that the antiproton has $I_3 = -\frac{1}{2}$ (while the antineutron has $I_3 = +\frac{1}{2}$). The total isospin of a nucleon-antinucleon pair can have the values $I_{total} = 0, 1$. Since, on the left-hand side of (3.70), $I_3 = -1$, the reaction can only proceed through the $I_{total} = 1$ isospin channel. For the two-pion annihilation,

$$G = (-1)^{l+S+I} = +1,$$

so that $(l + S)$ must be odd. There are two possibilities:—

$S = 0$ (singlet state)

For the choice $l = J = 1$, the parity on the left-hand side of (3.70) is $(-1)^{l+1}$, i.e. even [the extra (-1) factor coming from the opposite intrinsic parity of fermion and antifermion]. The parity of the two-pion $l = 1$ state is $(-1)^l$, or odd. Thus the $^1P_1$-state is forbidden by parity conservation. Equally, $l = 0$ is forbidden, since *G*-parity conservation requires $(l + S)$ to be odd. Thus both the singlet states $^1P_1$ and $^1S_0$ are forbidden.

$S = 1$ (triplet state)

The states $^3P_{0,1,2}$ are forbidden since $(l + S)$ must be odd. An S-state is allowed, since $l = 0$, $J = 1$ gives negative parity in both initial and final states. Thus, for values of $l \leqslant 1$, (3.70) can proceed only by annihilation in the $I = 1$, $^3S_1$ state.

We may summarize our discussion of selection rules for the positronium and nucleon-antinucleon systems as follows:

$$e^+e^- \to n\gamma$$

C-parity $(-1)^{l+S} \quad (-1)^n, \qquad (3.71)$

$$N\bar{N} \to n\pi$$

G-parity $(-1)^{l+S+I} \quad (-1)^n. \qquad (3.72)$

In both cases, the parity of the initial state is $P = (-1)^{l+1}$.

The concept of *G*-parity introduces nothing that is not already known from the twin postulates of charge conjugation and isospin invariance; it simply

TABLE 3.5

Particle $\left(\dfrac{\text{Mass}}{\text{MeV}}\right)$	$\pi(140)$	$\rho(765)$	$\omega(783)$	$\phi(1019)$	$f(1260)$	$\eta(549)$	$\eta'(958)$
Spin parity J^P	0^-	1^-	1^-	1^-	2^+	0^-	0^-
Isospin I	1	1	0	0	0	0	0
G-parity	-1	$+1$	-1	-1	$+1$	$+1$	$+1$
Dominant pion decay mode	—	2π	3π	3π	2π	3π	5π

allows some short cuts when we consider selection rules for the decay of meson resonances. The relevant quantum numbers are given in Table 3.5.

Note that the vector mesons ρ, ω, ϕ, and f decay by strong interactions, being resonant states with large widths (3 to 100 MeV), and that the multiplicity of the pion decay modes follows the rule $G = (-1)^n$. On the other hand, the states η and η' are of unmeasurably small widths, and the existence of $\gamma\gamma$-decay modes proves that they decay by electromagnetic transitions. Since $I = 0$, and $C(2\gamma) = +1$, they must have $G = +1$. The strong decay into two pions is forbidden by parity conservation, leaving the three-pion, G-violating electromagnetic decay as the only possibility.

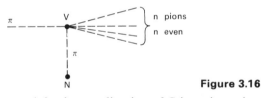

Figure 3.16

A further application of G-invariance is to pion multiplicities in high energy reactions. For example, it is sometimes possible to describe pion-nucleon collisions, with low momentum transfer to the nucleon, in terms of a collision of the incident pion with a single pion from the virtual meson cloud surrounding the nucleon. This is the so-called one-pion exchange model (Fig. 3.16). The initial state of the two pions at vertex V has $G = +1$, so that the final state must contain an even number of pions. (Note that there is no restriction on the number of pions at the bottom vertex, which has baryon number 1 and, therefore, no defined G-parity.) Such a process apparently dominates production of ρ-mesons ($n = 2$) in pion-nucleon collisions (see Section 7.5).

3.17 TIME REVERSAL INVARIANCE, CP-VIOLATION, AND THE CPT THEOREM

Both invariance and noninvariance under time reversal are familiar in classical physics. Thus, Newton's law $F(x) = m \, d^2x/dt^2$ is invariant under reversal of the time coordinate: a film of the trajectory of a projectile in the earth's

TABLE 3.6

Quantity	T	P	
r	r	$-r$	
p	$-p$	$-p$	Polar vector
σ (Spin)	$-\sigma$	σ	Axial vector ($r \wedge p$)
E (Electric field)	E	$-E$	($E = -\partial V/\partial r$)
B (Magnetic field)	$-B$	B	(As for σ, e.g. consider a ring current)
$\sigma \cdot B$	$\sigma \cdot B$	$\sigma \cdot B$	Magnetic dipole moment
$\sigma \cdot E$	$-\sigma \cdot E$	$-\sigma \cdot E$	Electric dipole moment
$\sigma \cdot p$	$\sigma \cdot p$	$-\sigma \cdot p$	Longitudinal polarization
$\sigma \cdot (p_1 \wedge p_2)$	$-\sigma \cdot (p_1 \wedge p_2)$	$\sigma \cdot (p_1 \wedge p_2)$	Transverse polarization

gravitational field looks equally realistic whether we run it forward or backward, provided we neglect air resistance. The situation is quite different for the laws of heat conduction or diffusion, which depend on the first derivative of the time coordinate. These laws were originally formulated on the principle of microscopic reversibility, where individual collisions between atoms or molecules are indistinguishable under the operation of time reversal. Yet a system consisting of many particles evolves on a macroscopic scale in accordance with statistical laws, or the principle of increase of entropy. A film of an order-disorder transition, such as of gas expanding through a nozzle, does not correspond to the real world if run in the reverse direction.

The transformations of common quantities in classical physics under space inversion, P, and time reversal, T, are given in Table 3.6. We note that an "elementary" particle of spin σ would not be expected to possess a static electric dipole moment, if the interaction of the particle with the electromagnetic field is to be invariant under the T- or P-operations. This conclusion depends, of course, on what one means by a "particle." We know, for example, that many molecules possess large electric dipole moments. What we imply by a particle in the present context is an object which is completely specified by its mass, charge, and spin—in other words, it is not supposed to have a complicated structure, described by further quantum numbers, as in a molecule. Then, the spin vector $\boldsymbol{\sigma}$ prescribes the only possible direction in space of any electric (or magnetic) dipole moment (i.e. $\boldsymbol{\mu}_e$ or $\boldsymbol{\mu}_m$ must be either parallel or antiparallel to $\boldsymbol{\sigma}$). If it turned out that a particle, previously described by the spin vector $\boldsymbol{\sigma}$, *did* possess an electric dipole moment, one would be faced with the dilemma of whether to ascribe this to a genuine violation of T- and P-invariance or the need to invoke further quantum numbers in order to specify the new property. As we shall see, this dilemma has not yet arisen.

Experimentally, time reversal invariance may be verified in strong interactions by application of the principle of detailed balance, as explained in Section 3.4. Figure 3.17 shows the results of a test of detailed balance in the reaction

$$p + Al^{27} \rightleftarrows \alpha + Mg^{24},$$

from which it was concluded that the T-violating amplitude was $<0.3\%$ of the T-conserving amplitude.

Another consequence of T-invariance is the so-called polarization-asymmetry equality in p–p elastic scattering. In scattering through an angle

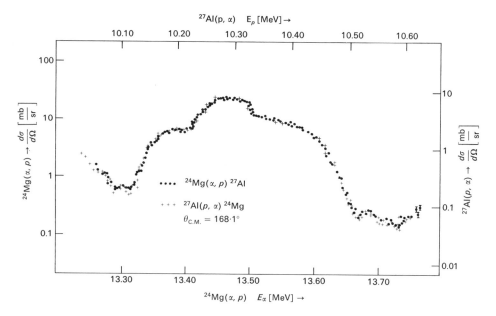

Fig. 3.17 The differential cross sections for the reaction $^{24}Mg(\alpha, p)^{27}Al$ and its inverse, as measured by Von Witsch *et al.* (1968).

θ, from an unpolarized target, an initially unpolarized proton beam will acquire a polarization

$$P(\theta) = \frac{N_+ - N_-}{N_+ + N_-},$$

where N_+ and N_- represent the number of protons with spin up and spin down relative to the scattering plane. On the other hand, if one starts off with a fully (transversely) polarized proton beam, one obtains a right-left asymmetry in the scattering at angle θ in the plane normal to the polarization direction:

$$A(\theta) = \frac{N_R - N_L}{N_R + N_L},$$

where R and L refer to right and left scattering angles. If the strong interactions are invariant under time reversal, it may be shown that

$$A(\theta) = P(\theta).$$

This equality has been verified to an accuracy of $\approx 1\%$.

Experimental tests of T-invariance have acquired considerable importance in recent years, because of the observed breakdown of CP-invariance in K^0-decay in weak interactions. The reason for this is that the T- and CP-transformations are connected by the famous *CPT theorem*, which is one of the most important principles of quantum field theory. This theorem states that *all* interactions are invariant under the succession of the three operations C, P, and T taken in any order. The proof of this theorem is based on very general assumptions and is cherished by theorists in the sense that it is difficult to formulate field theories which are not automatically *CPT*-invariant. However, the *CPT* theorem is not on quite such experimentally solid ground as conservation of energy, for example, and bearing in mind the history of long-respected conservation laws which have fallen by the wayside, we must make experimental checks.

Some consequences of the *CPT* theorem which may be verified experimentally relate to the properties of particles and antiparticles, which should have the same mass and lifetime, and magnetic moments equal in magnitude but opposite in sign. These results would follow from charge conjugation invariance alone, if it held universally. However, weak interactions are not C-invariant, so that the prediction rests on the more general theorem.

The experimental consequences of the *CPT* theorem seem to be well verified. Current results are as shown in Table 3.7. The best experimental limit comes from the $K^0 - \overline{K^0}$ comparison; this is discussed further in Section 4.11. Another consequence of the *CPT* theorem is that particles would be expected to obey the "normal" spin-statistics relationship, i.e. integral spin and half-integral spin particles should follow Bose and Fermi statistics respectively.

TABLE 3.7 Tests of CPT theorem

		Limit on fractional difference
Lifetime	$\tau_{\pi^+} - \tau_{\pi^-}$	$<10^{-3}$
	$\tau_{\mu^+} - \tau_{\mu^-}$	$<2 \times 10^{-3}$
	$\tau_{K^+} - \tau_{K^-}$	$<10^{-3}$
Magnetic moment	$|\mu_{\mu^+}| - |\mu_{\mu^-}|$	$<10^{-6}$
Mass	$M_{\pi^+} - M_{\pi^-}$	$<10^{-3}$
	$M_{\bar{p}} - M_p$	$<8 \times 10^{-3}$
	$M_{K^+} - M_{K^-}$	$<10^{-3}$
	$M_{K^0} - M_{\bar{K}^0}$	$<10^{-14}$

Until 1964, it was believed that all types of interaction were invariant under the combined operation CP. Weak interactions were known to violate C- and P-invariance separately, but to respect the CP-symmetry. In that year, however, it was discovered by Christenson *et al.* (1964) that the long-lived neutral K-particle, which normally decays by a weak interaction into three pions of CP eigenvalue -1, could occasionally decay into two pions, of $CP = +1$.

The origin of this CP-violation is not at present established, but, whatever the process is, it implies, through the CPT theorem, a T-violation also. (Actually, the analysis of K^0-decay establishes T-violation independent of the validity of CPT.) It is likely that the effect arises from a specific type of weak (or superweak) interaction, but because of the uncertainty in origin, searches have been made for evidence of T-violation in both weak and electromagnetic interactions.

As indicated in Table 3.6, T-violation would imply a transverse polarization of the muon in the weak decay $K^+ \to \pi^0 + \mu^+ + \nu$ (relative to the π^0, ν plane). Several recent experiments have failed to detect any transverse polarization. This would have been revealed as an up-down asymmetry of the decay electron relative to the above plane, in the subsequent decay $\mu^+ \to e^+ + \nu + \bar{\nu}$. Similar results are obtained from the (strangeness conserving) decays of polarized neutrons (Section 4.4).

3.18 ELECTRIC DIPOLE MOMENT OF THE NEUTRON

We previously noted that the existence of an electric dipole moment implies violation of both T- and P-invariance. Since a dipole moment can be measured with great precision, this provides a sensitive test of T-invariance. The best limits have been obtained by Ramsey and co-workers, who sought to measure the electric dipole moment (EDM) of the neutron. Before describing the experiment, let us try to make a guess at the magnitude of the effect, just from

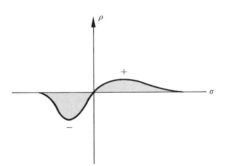

Fig. 3.18 An asymmetric distribution of positive and negative charge density, ρ, in the neutron would give rise to an electric dipole moment.

dimensional arguments. One can write

$$\text{EDM} = \text{charge } (e) \times \text{a length } (l) \times T\text{-violation parameter } (f).$$

The neutron is uncharged, so that the dipole moment could result from an asymmetry between positive and negative charge clouds, of net value zero, relative to the spin direction $\boldsymbol{\sigma}$ (Fig. 3.18). Since P-invariance is also violated, we must somehow bring in the weak interactions; the natural length involving the weak interaction is $l = GM$, where M is some chosen mass—the obvious one being the nucleon mass M—and $G = 10^{-5}/M^2$ is the weak interaction coupling constant. We have used units $\hbar = c = 1$ here. Then

$$\text{EDM} = e \times f \times 10^{-5}/M \sim 10^{-19}f \, (e \text{ cm}), \tag{3.73}$$

where, for $1/M$, we put the proton Compton wavelength $\hbar/Mc = 2 \times 10^{-14}$ cm. Detailed models do not give substantially different results from this simple estimate. If the source of T-violation is in the electromagnetic interactions, for example, then from the η-decay results one might assume $f < 10^{-2}$. Thus, one could expect the maximum value of the electric dipole moment to be of order $10^{-21} \, (e \text{ cm})$. Note that the effective length of the dipole is very small compared with the "size" of an elementary particle, of order 10^{-13} cm (see Section 5.6).

In the experiment, a reactor is used as a source of (predominantly) thermal neutrons. In order to make the experiment more sensitive, these neutrons are "cooled" by passage through a narrow, curved tube of highly polished nickel of 1 m radius of curvature (see Fig. 3.19). The critical angle for total internal reflection of neutrons by the tube is inversely proportional to velocity, so that, for a beam of finite divergence, only low velocity neutrons are transmitted with high intensity. The beam emerging from the tube then falls on a polarizing magnet, consisting of a polished, magnetized mirror of cobalt–iron alloy. Reflection at grazing incidence (2°) from this mirror gives a transversely spin-polarized neutron beam, the degree of polarization being 70%. The

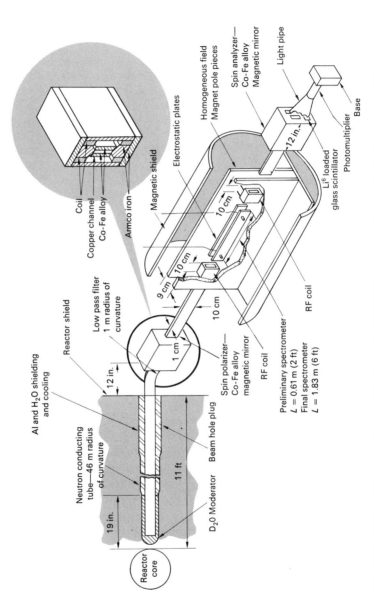

Fig. 3.19 Sketch of the apparatus used by Dress *et al.* (1968) to measure the electric dipole moment of the neutron.

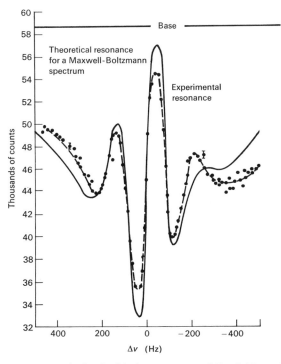

Fig. 3.20 Resonance curve obtained with the apparatus of Fig. 3.19, as the RF frequency is varied about the resonance value. At the steepest part of the curve, the counting rate varies by 1% for a change of frequency $\Delta v = 1$ cps. The full-line curve is that calculated assuming a simple Maxwellian distribution in velocity of the neutron beam.

Maxwellian velocity distribution of these neutrons corresponded to a temperature of 1°K (i.e. a mean velocity v of order 100 m/sec). After traversing a spectrometer, the neutrons are reflected from the analyzing magnet (similar to the polarizer), and recorded in the detector, consisting of a Li^6-loaded glass scintillator, sensitive to neutrons. The transmitted intensity I is a maximum for neutrons which do not suffer depolarization in the spectrometer.

The spectrometer consists firstly of a uniform magnetic field H (≈ 10 G) which causes the neutrons to precess with the Larmor frequency $v_L = \mu H/h$, where μ is the neutron magnetic moment ($v_L \approx 25$ kHz). Secondly, a RF field of frequency v is applied by means of two coils so that, at resonance, when $v = v_L$, spin-flip transitions occur, the neutron beam is partly depolarized, and the transmitted intensity I changes. Two coils are used to give an interferometer effect, producing several maxima and minima in the resonance curve (Fig. 3.20), and providing a rapid change in counting rate with RF frequency. Finally, a reversible electric field E of 100 kV/cm is applied in the direction of the steady magnetic field H.

The experiment consists essentially of sitting in a region of the resonance curve where dI/dv is large, and observing the change in I when the electrostatic field E is reversed. If the neutron possesses an electric dipole moment in the direction of the spin, the field E will produce an additional small precession and consequently a change in I when the frequency v is held constant. dI/dv is proportional to the time spent by the neutron between the RF coils and is thus greatest for large coil separations and low velocities—hence the advantage of using "cold" neutrons. No effect was observed, and the final result of this very precise experiment was to set a limit on the electric dipole moment of

$$\text{EDM of neutron} < 3 \times 10^{-22} \, (e \text{ cm}). \tag{3.74}$$

It may be remarked that this limit is determined by the difficulty of obtaining E and H exactly parallel. If there is a small component $E_\perp$ perpendicular to H, this will induce an extra magnetic field $\Delta H = (v/c)E_\perp$ in the direction of H and hence a spurious effect. Comparing (3.73) with (3.74) we see that the experiment sets an upper limit to the C-violating parameter $f < 10^{-2}$ in electromagnetic processes.

3.19 CONCLUSIONS

Let us briefly summarize the results obtained to date on the subject of C-, P-, and T-violation. There is ample evidence to suggest that strong interactions conserve C, P, and T separately. Electromagnetic interactions conserve P, and any possible violations of C and T are below the 1% level. Weak interactions are found to be invariant under time reversal T, within an accuracy of a few per cent. They violate C- and P-invariance, but except for the K^0-phenomena, conserve the product CP. A definite (10^{-3}) CP-violating amplitude is observed in K^0-decay, but the score of this violation is not finally established, although recent data suggest that it may be due to a new type of superweak interaction. All attempts to detect a CP-violation, or equivalently, a T-violation, in other processes have so far proved unsuccessful.

BIBLIOGRAPHY

Fraser, W. R., *Elementary Particles*, Prentice-Hall, Englewood Cliffs, New Jersey, 1966.

Hamilton, W. D., "Parity violation in electromagnetic and strong interaction processes," *Prog. Nucl. Phys.* **10**, 1 (1969).

Henley, E. M., "Parity and time-reversal invariance in nuclear physics," *Ann. Rev. Nucl. Science* **19**, 367 (1969).

Jackson, J. D., *The Physics of Elementary Particles*, Princeton University Press, Princeton, New Jersey, 1958.

Kemmer, N., J. C. Polkinghorne, and D. Pursey, "Invariance in elementary particle physics," *Rep. Prog. Phys.* **22**, 368 (1959).

Muirhead, A., *The Physics of Elementary Particles*, Pergamon, London, 1965, Ch. 5.

Rowe, E. G. and E. J. Squires, "Present status of C-, P-, and T-invariance," *Rep. Prog. Phys.* **32**, 273 (1969).

Sakurai, J. J., *Invariance Principles and Elementary Particles*, Princeton University Press, Princeton, New Jersey, 1964.

Wick, G. C., "Invariance principles of nuclear physics," *Ann. Rev. Nucl. Science* **8**, 1 (1958).

Williams, W. S. C., *An Introduction to Elementary Particles*, Academic Press, New York and London, 1961.

PROBLEMS

3.1 Using the formulae in Appendix C, deduce the Clebsch-Gordan coefficients for combining two particles each of isospin $\frac{1}{2}$, and thus verify the isospin wave functions for two nucleons given in Eq. (3.43).

3.2 Find a relation between the total cross sections (at a given energy) for the reactions

$$\pi^- p \to K^0 \Sigma^0,$$
$$\pi^- p \to K^+ \Sigma^-,$$
$$\pi^+ p \to K^+ \Sigma^+.$$

3.3 At a given center-of-mass energy, what is the ratio of the cross sections for $p + d \to \text{He}^3 + \pi^0$ and $p + d \to \text{H}^3 + \pi^+$?

3.4 A hypernucleus is one in which a neutron is replaced by a bound Λ-hyperon. $_\Lambda\text{He}^4$ and $_\Lambda\text{H}^4$ are a doublet of mirror hypernuclei. Deduce the ratio of the reaction rates

$$K^- + \text{He}^4 \to {}_\Lambda\text{He}^4 + \pi^-$$
$$\to {}_\Lambda\text{H}^4 + \pi^0.$$

3.5 State which of the following combinations can or cannot exist in a state of $I = 1$, and give the reasons:

a) $\pi^0\pi^0$,
b) $\pi^+\pi^-$,
c) $\pi^+\pi^+$,
d) $\Sigma^0\pi^0$,
e) $\Lambda\pi^0$.

3.6 In which isospin states can (a) $\pi^+\pi^-\pi^0$, (b) $\pi^0\pi^0\pi^0$ exist? [*Hint:* first write down the isospin functions for a pair, e.g. $\pi^0\pi^0$, and then combine with the third pion. Refer to p. 330 for any Clebsch-Gordan coefficients required.]

3.7 Deduce through which isospin channels the following reactions may proceed:

i) $K^+ + p \to \Sigma^0 + \pi^0$,
ii) $K^- + p \to \Sigma^+ + \pi^-$.

Find the ratio of cross sections for (i) and (ii), assuming that one or other channel dominates.

3.8 Write down the quantum numbers (G, I, J^P) of the S- and P-states of the $p\bar{p}$-system able to decay to (a) $\pi^+\pi^-$, (b) $\pi^0\pi^0$, (c) $\pi^0\pi^0\pi^0$. In annihilations at rest, the process $p\bar{p} \to 2\pi^0$ does not appear to occur; what do you conclude from this fact?

3.9 The A1 meson, of $I = 1$, is considered to be a resonant state of a ρ-meson ($I = 1$) and pion ($I = 1$). Thus decay A1 $\rightarrow \rho + \pi$ is dominant. Find the expected branching ratio

$$\frac{\text{A1} \rightarrow \pi^0 \pi^0 \pi^+}{\text{A1} \rightarrow \pi^+ \pi^- \pi^+}.$$

3.10 The ω-meson has G-parity -1, and the ρ-meson has G-parity $+1$. The ω and ρ have the same spin and parity (1^-). The ρ-meson has central mass 765 MeV and is a broad state with a width $\Gamma \sim 120$ MeV, overlapping the ω-state (783 MeV, $\Gamma \sim 12$ MeV). Would you expect the ω- and ρ-states to interfere, and what qualitative effects would any interference have on the $\pi^+ \pi^-$ and $\pi^+ \pi^0$ mass spectra in reactions where both ω and ρ can be produced?

3.11 Show that, for pions with zero relative orbital angular momentum, the combination $\pi^+ \pi^-$ is an eigenstate of $CP = +1$, and $\pi^+ \pi^- \pi^0$ is an eigenstate of $CP = -1$.

3.12 What restrictions does the decay mode $K_1^0 \rightarrow 2\pi^0$ place on (a) the kaon spin, (b) the kaon parity?

3.13 As shown in Chapter 4, the neutral kaons decay from the states K_1^0 and K_2^0, of CP eigenvalues $+1$ and -1 respectively. If, as we believe, $p\bar{p}$ annihilation at rest takes place from an atomic S-state only, show that $p\bar{p} \rightarrow K_1^0 + K_2^0$ occurs, but that $p\bar{p} \rightarrow 2K_1^0$ or $p\bar{p} \rightarrow 2K_2^0$ does not.

Weak Interactions

4.1 LEPTONIC, SEMILEPTONIC, AND NONLEPTONIC INTERACTIONS

The observed weak interactions can be subdivided into three types; leptonic, semileptonic, and nonleptonic. As the name implies, the purely leptonic weak interactions involve only electrons, muons, and neutrinos. Only one example, that of muon decay, has been observed so far:

$$\mu^+ \rightarrow e^+ + v_e + \bar{v}_\mu,$$

and

$$\mu^- \rightarrow e^- + \bar{v}_e + v_\mu.$$

The necessity of postulating two types of neutrinos, v_e and v_μ, and their antiparticles was mentioned in Section 1.3.2 and is discussed in detail in Section 4.9. The leptonic weak interactions, as we shall see, are distinguished by the fact that they obey the pure V-A theory. The semileptonic weak interactions involve both leptons and hadrons, strange or nonstrange. Examples are nuclear β-decay,

$$n \rightarrow p + e^- + \bar{v}_e,$$

the inverse process of neutrino absorption by a nucleon,

$$v_e + n \rightarrow p + e^-,$$

pion decay,

$$\pi^+ \rightarrow \mu^+ + v_\mu,$$

and the β-decay of the Λ-hyperon,

$$\Lambda \rightarrow p + e^- + \bar{v}_e.$$

Finally, the nonleptonic weak interactions occur between hadrons only.

They always involve a change of strangeness of the hadron, $|\Delta S| = 1$. For example,

$$K^+ \to \pi^+ + \pi^+ + \pi^-,$$

and

$$\Lambda \to p + \pi^-.$$

Nonleptonic weak interactions between nonstrange hadrons must also take place. Thus, two nucleons undergo a specifically weak interaction, but it is feeble in comparison with the strong interaction. Only in those cases where a strong interaction is forbidden (when the strangeness changes) can the nonleptonic weak interactions be observed experimentally.

All the weak interactions involve decay lifetimes or collision cross sections typical of the weak coupling (Fermi) constant, $G = 10^{-5}/M_p^2$. We have already mentioned the smallness of this constant, and remarked that, in the weak process of muon capture by a nucleus,

$$\mu^- + p \to n + \nu_\mu,$$

the mean free path of the muon in nuclear matter is of order 10^{14} times the characteristic nuclear scale of length (10^{-13} cm). Quite out of context, we may note however that, on a cosmic scale, the Fermi constant plays an important role in determining the evolutionary rate of young dwarf stars like the sun, and its smallness has, in a very real sense, guaranteed the long life of the solar system. Thermonuclear energy is released in the sun by the conversion of hydrogen to helium via the p–p cycle:

$$
\begin{aligned}
H^1 + H^1 &\to H^2 + e^+ + \nu &&(10^9 \text{ yr}),\\
H^2 + H^1 &\to He^3 + \gamma &&(1 \text{ sec}),\\
He^3 + He^3 &\to He^4 + H^1 + H^1 &&(10^6 \text{ yr}).
\end{aligned}
$$

The overall rate of these reactions is determined by that of the first and slowest step (the estimated mean reaction time in the solar core being given in brackets), and this depends very largely on the magnitude of the β-decay constant, G. It may also be remarked that recent experiments to detect solar neutrinos underground have confirmed the correctness of the above reactions as the solar energy source, on the basis of a null result. (See Problem 4.10.)

4.2 NUCLEAR β-DECAY; A BRIEF REVIEW

The first weak interaction process to be observed was nuclear β-decay. The salient observations were:

a) e^+ and e^- are emitted by radioactive nuclei. The lifetimes are very long (second-years) compared with the nuclear time scale (10^{-22} sec).

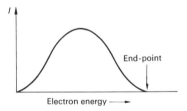

Fig. 4.1 Typical shape of electron energy spectrum in nuclear β-decay ("allowed" transition).

b)   The electron (or positron) energy spectrum is continuous (Fig. 4.1). The total disintegration energy (as measured by the masses of parent and daughter nuclei) is discrete, and corresponds to the maximum possible energy of the electrons (the end-point of the spectrum).

c)   Conservation of energy, momentum, and angular momentum require the existence of a light, uncharged particle of spin $\frac{1}{2}$, called a *neutrino* (Pauli, 1933), and its antiparticle, the antineutrino. Careful measurement of the recoil momentum of the daughter nucleus, as well as the electron energy, established that the neutrino mass was nearly zero (i.e. $E_\nu \simeq cp_\nu$).

d)   The independent existence of neutrinos was demonstrated by Reines and Cowan in 1953–59, in experiments using the intense $\bar{\nu}$-fluxes from a reactor (see Section 4.10).

e)   The prototype β-interactions are described by the equations

$$n \rightarrow p + e^- + \bar{\nu}, \quad \text{negatron emission (i)},$$

$$p \rightarrow n + e^+ + \nu, \quad \text{positron emission (ii)},$$

$$e^- + p \rightarrow n + \nu, \quad K\text{-capture (iii)},$$

$$\bar{\nu} + p \rightarrow n + e^+, \quad \text{antineutrino absorption (iv)}.$$

Reaction (i) occurs for free neutrons (mean lifetime $\tau \sim 10$ min), (ii) occurs only in nuclei where the difference in binding between parent and daughter nuclei exceeds the negative Q-value of the reaction. K-capture (iii) occurs chiefly in heavy nuclei as an alternative to positron emission, or where positron emission is forbidden energetically.

4.2.1 Fermi and Gamow-Teller Transitions

The so-called "allowed" β-transitions are those in which the electron and neutrino carry off zero orbital angular momentum ($l = 0$). Their total angular momentum is then 1 (spins parallel) or 0 (spins antiparallel). The two cases are referred to as Gamow-Teller and Fermi transitions respectively. If we call

J the nuclear angular momentum, we have, indicating spin directions symbolically by vertical arrows:

Fermi	Gamow-Teller
$n \to p + e^- + \bar{\nu}$	$n \to p + e^- + \bar{\nu}$
↑ ↑ ↓ ↑	↑ ↓ ↑ ↑
$\Delta J = 0$	$\Delta J = 1(\vert\Delta J\vert = 0, \pm 1)$
$J_i = 0 \to J_f = 0$ allowed.	$J_i = 0 \to J_f = 0$ forbidden.

There are then three types of allowed β-transitions; pure Fermi, pure Gamow-Teller, and mixed transitions. Examples are:

$$\mathrm{He}^6 \to \mathrm{Li}^6 + e^- + \bar{\nu}; \qquad \Delta J = 1; \qquad\qquad \text{pure Gamow-Teller,}$$
$$J^P \quad 0^+ \qquad 1^+$$

$$\mathrm{O}^{14} \to \mathrm{N}^{14*} + e^+ + \nu; \qquad J = 0 \to J = 0; \qquad \text{pure Fermi,}$$
$$J^P \quad 0^+ \qquad 0^+$$

$$n \to p + e^- + \bar{\nu}; \qquad J \to J(J \neq 0); \qquad \text{mixed.}$$
$$J^P \quad \tfrac{1}{2}^+ \quad \tfrac{1}{2}^+$$

Note that in all cases, the parity of the initial and final nuclear states is the same. The effect of the decay is to flip the isospin of one nucleon ($n \to p$ or vice versa), and, in the Gamow-Teller case, also to flip the nucleon spin, but otherwise to leave the nuclear configuration unaltered.

4.2.2 Fermi Theory of β-decay—the Phase-Space Factor

Since the interaction is known to be weak, the transition probabilities for β-decays can be calculated from first order perturbation theory. The transition probability per unit time is then (Section 3.4)

$$W = \frac{2\pi}{\hbar} G^2 \, |M|^2 \, \frac{dN}{dE_0}, \qquad (4.1)$$

where E_0 is the available energy of the final state, dN/dE_0 is the density of final states per unit energy, and G^2 is a universal constant typical of the β-decay coupling. $|M|^2$ is the square of the matrix element for the transition, which involves an integral over the nuclear (interaction) volume of the four fermion wave functions. For the present we shall regard $|M|^2$ as some constant and calculate the phase-space factor dN/dE_0. This is determined by the number of ways it is possible to share out the available energy $E_0 \to E_0 + dE_0$ between, for example, p, e^-, and $\bar{\nu}$ in the neutron decay (i) above. The quantity dE_0 arises because of the spread in energy of the final state, corresponding to the

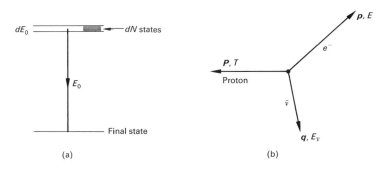

Fig. 4.2 (a) An initial state with a spread in energy dE_0 decays to a final (stable) state of unique energy, with energy release E_0. (b) The momentum vectors in neutron β-decay.

finite lifetime of the initial state. In Fig. 4.2, p, q, and P are the momentum of the electron, neutrino, and proton; E, E_v, and T are their kinetic energies.

Then, in the rest frame of the initial state (neutron):

$$P + q + p = 0,$$
$$T + E_v + E = E_0.$$

Assume $m_v = 0$, so $E_v = qc$. In order of magnitude, $E_0 \sim 1$ MeV, so $Pc \sim 1$ MeV. Thus, if the recoiling nucleon mass is M, its kinetic energy $T = P^2/2M \simeq 10^{-3}$ MeV only, and can be neglected. The nucleon serves to conserve momentum, but we can regard E_0 as shared entirely between electron and neutrino. Thus $qc = E_0 - E$. The number of states in phase space available to an electron, confined to a volume V with momentum $p \rightarrow p + dp$ defined inside the element of solid angle $d\Omega$, is

$$\frac{V\, d\Omega}{h^3} p^2\, dp.$$

If the fermion wave functions are normalized to unit volume, $V = 1$, and we integrate over all space angles, the electron phase-space factor is

$$\frac{4\pi p^2\, dp}{h^3}.$$

Similarly, for the neutrino it is

$$\frac{4\pi q^2\, dq}{h^3}.$$

We disregard any possible correlation in angle between p and q, and treat these two factors as independent, since the proton will take up the resultant momentum. [There is no phase-space factor for the proton since its momentum is now fixed: $P = -(p + q)$.] So the number of final states is

$$d^2N = \frac{16\pi^2}{h^6} p^2 q^2 \, dp \, dq.$$

Also, for given values of p and E the neutrino momentum is fixed,

$$q = (E_0 - E)/c,$$

within the range $dq = dE_0/c$. Hence

$$\frac{dN}{dE_0} = \frac{16\pi^2}{h^6 c^3} p^2 (E_0 - E)^2 \, dp. \tag{4.2}$$

If $|M|^2$ is regarded as a constant, this last expression gives the electron spectrum

$$N(p) \, dp \propto p^2 (E_0 - E)^2 \, dp,$$

and thus if we plot $[N(p)/p^2]^{1/2}$ against E, a straight line cutting the x-axis at $E = E_0$ should result. This is called a *Kurie plot*. For many β-transitions, the Kurie plot is linear, as shown for the decay $H^3 \rightarrow He^3 + e^- + \bar{\nu}$ (mixed). In Fig. 4.3, $F(z, p)$ is a correction factor (~ 1 in this case) to account for the Coulomb effects on the electron wave function. For a nonzero neutrino mass, the plot bends over near the endpoint, and cuts the axis at $E' = E_0 - m_\nu c^2$. This particular decay, with small E_0, has been used to set an upper limit, $m_\nu < 60$ eV.

The total decay rate is obtained by integrating (4.2) over the electron spectrum. As a very crude approximation in some decays we can consider the

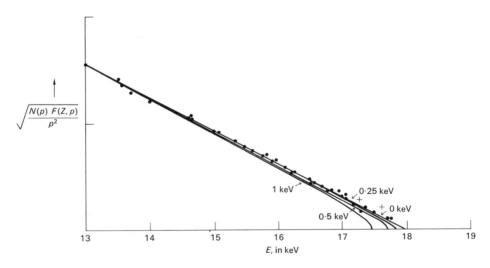

Fig. 4.3 Kurie plot for tritium β-decay. Deviations near the end point for various neutrino mass values are indicated. (After Langer and Moffat, 1952.)

electrons as extreme relativistic, $E \approx pc$ (not for tritium!), whence we obtain

$$N \sim \int_0^{E_0} E^2 (E_0 - E)^2 \, dE = E_0^5/30. \tag{4.3}$$

Under these conditions, the disintegration constant (i.e. the decay rate) varies as the fifth power of the disintegration energy (the Sargent rule).

More generally, and including the Coulomb term, the transition rate will be

$$W = |M|^2 \, G^2 \, \frac{4\pi^2}{h} \frac{16\pi^2}{h^6 c^3} \int_0^{p_{max}} F(z, p) N(p) \, dp.$$

If p, E are expressed in units of the electron mass, $p = p^1 \times mc$ and $E = E^1 \times mc^2$, the integral becomes

$$m^5 c^7 \int_0^{p_{max}^1} F(z, p^1) N(p^1) \, dp^1 = m^5 c^7 f, \tag{4.4}$$

where f is a dimensionless number depending on p_{max}^1 and the daughter nuclear charge z. Then

$$W = |M|^2 \, G^2 \, \frac{64\pi^4 m^5 c^4}{h^7} f \quad \text{transitions/sec.}$$

If t denotes the half-life, $W = (\log_e 2)/t$ so that

$$ft = \frac{K}{|M|^2} \quad \text{where} \quad K = \frac{h^7 \log_e 2}{64 G^2 \pi^4 m^5 c^4}. \tag{4.5}$$

The quantity ft therefore gives information about the matrix element M, which depends on the overlap of the initial and final nuclear wave functions. For allowed transitions, where the overlap is essentially complete, $|M|^2 \approx 1$. One finds that the observed β-decays subdivide into two more or less distinct classes: those of $\log_{10} ft$ of 3 to 4, corresponding to the allowed transitions discussed so far—and those of much larger ft values, with $\log_{10} ft$ of order 10 to 20. The latter are called forbidden transitions, and clearly involve much smaller values of $|M|^2$. They correspond crudely speaking to processes where the "electron-neutrino radiation" carries off orbital angular momentum, $l > 0$. Since the neutrino and electron are relativistic, it is not strictly possible to speak separately of their spin and orbital angular momentum. Only the total angular momentum, J_{lepton}, is a good quantum number. However, a nonrelativistic picture will serve to orient us on magnitude. We note first that the matrix element M will contain the product of four wave functions, ψ_i and ψ_f of the initial and final nuclear states, and u_1 and u_2 of the leptons. If we neglect Coulomb interactions we can represent the u's by the nonrelativistic plane wave form

$$u = e^{i\mathbf{k} \cdot \mathbf{r}} = 1 + i\mathbf{k} \cdot \mathbf{r} + \frac{(i\mathbf{k} \cdot \mathbf{r})^2}{2!} + \cdots, \tag{4.6}$$

where $\hbar k$ is the lepton momentum, and r the space coordinate relative to the nuclear center. One recognizes (4.6) as the multipole expansion of a plane wave. In the interaction region $r \approx R_0 \simeq \hbar/m_\pi c$, while $\hbar k \approx m_e c$, since Q-values in β-decay are of order of MeV. Thus $kr \simeq m_e/m_\pi \simeq 10^{-2}$. So, for an allowed transition (S-wave) the first term of (4.6) dominates and the integral of the ψ's and u's over the interaction volume gives unity. However, for a forbidden transition, for example $l = 1$, the first term cannot contribute, and the lepton wave functions give a factor of order $(kr)^2 = 10^{-4}$. The integral over the nuclear wave functions $\psi_f^* \psi_i$ is also small, since a forbidden transition involves a change in the configuration of the nucleus, including a change in parity. Another consequence is that, for a forbidden transition, the electron spectrum has extra terms [like $(kr)^2$, and therefore p^2 in the above example], which means that in general the Kurie plot (4.2) will no longer be a straight line.

From (4.5), we may calculate G for a pure Fermi transition by taking the observed ft value of O^{14}-decay, 3100 ± 20 sec, and assuming $|M|^2 = 1$. (This value of ft is a mean of recent values and incorporates some corrections.) Actually, we need to double the value of ft since O^{14} is a C^{12}-core plus two protons, either of which can decay. Inserting the numerical constants in (4.5), one gets

$$G = 1.4 \times 10^{-49} \text{ erg cm}^3$$

$$= \frac{10^{-5}}{M_p^2} \quad \text{in units } \hbar = c = 1. \tag{4.7}$$

We have purposely quoted G here to only one decimal place, since a small correction is required in a more refined theory (Section 4.8.2).

We may also compare the O^{14} pure Fermi transition rate with that of the neutron, which is mixed. Call C_F and C_{GT} the coefficients corresponding to the Fermi and Gamow-Teller couplings. The neutron decay rate is then proportional to $C_F^2 + 3C_{GT}^2$. The factor 3 enters because there are $(2J + 1)$ or three possible orientations for the total angular momentum $J = 1$ of the leptons in a Gamow-Teller transition. If we consider only the total event rate, the two contributions add in quadrature. For the O^{14} rate, for the reason stated above, we write $2C_F^2$. Then

$$\frac{(ft)_{O^{14}}}{(ft)_n} = \frac{3100 \pm 20 \text{ sec}}{1080 \pm 16 \text{ sec}} = \frac{C_F^2 + 3C_{GT}^2}{2C_F^2},$$

which yields

$$\lambda = \left| \frac{C_{GT}}{C_F} \right| = 1.25 \pm 0.02. \tag{4.8}$$

Thus if we define the constant G for Fermi transitions, then for the Gamow-Teller transitions it has the slightly larger value λG. Note that the sign of λ is not determined here (actually it is negative). It may also be pointed out that (4.8) involves radiative and other corrections for O^{14}-decay. These can be avoided in experiments with polarized neutrons, which give an independent value of λ (see Section 4.4).

4.2.3 Parity Nonconservation in β-decay

In 1956 Lee and Yang suggested that the weak interactions did not conserve parity (invariance under spatial inversions). They were forced to this conclusion owing to the absence of any direct experimental evidence to the contrary, and because of the so-called "$\tau - \theta$ paradox." The K^+-meson decays in several modes, two of which are:

$$K_{\pi 3}: \quad K^+ \to \pi^+ + \pi^+ + \pi^- \qquad \text{(historically the } \tau\text{-decay),}$$

$$K_{\pi 2}: \quad K^+ \to \pi^+ + \pi^0 \qquad \text{(historically the } \theta\text{-decay).}$$

Analysis of the $K_{\pi 3}$-decay showed that the three pions were in an S-state (the mean kinetic energy per pion is only 25 MeV). Thus the spin-parity of the 3π-state is 0^- (see Chapter 7). In the $K_{\pi 2}$-decay, since $J = 0$, the 2π-parity must be even, i.e. a 0^+-state. If parity were conserved in both processes, then the τ and θ must be different particles. Eventually it became clear that the same particle was responsible for the two decay modes, and thus parity was violated. Could this be the case for weak interactions generally? A large number of experiments were done, showing that Lee and Yang were right. It is instructive to trace some of the history of why it took 22 years (from 1934) to get this basic information. Immediately after Dirac produced his celebrated equation describing relativistic spin $\frac{1}{2}$ particles with mass, Weyl (1929) showed that *zero* mass spin $\frac{1}{2}$ particles would be described by two simpler, decoupled equations, before ever the neutrino was invented. The trouble was that Weyl's equations described left-handed (longitudinally polarized) particles and right-handed antiparticles, or vice versa, and thus violated the parity principle.† They were abandoned because of Pauli's alleged remark, "God could not be only weakly left-handed." The techniques of low temperature necessary to align radioactive nuclei and investigate the possibility of parity violation existed long before 1956, but no one troubled to look since the theorists assured them there was nothing to find; parity must be conserved (as indeed it was, in electromagnetic and strong interactions).

The first experiment, by Wu et al. (1957), employed a sample of Co^{60} at $0.01°K$ inside a solenoid. At this temperature a high proportion of Co^{60} nuclei are aligned. $Co^{60}(J = 5)$ decays to $Ni^{60*}(J = 4)$—a pure Gamow-Teller transition. The relative electron intensities along and against the field direction were measured. The degree of Co^{60} alignment could be determined from observations of the angular distribution of γ-rays from Ni^{60*}. The results were

---

† The Dirac and Weyl equations are discussed in Appendix B.

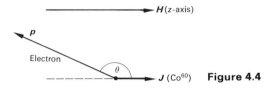

Figure 4.4

consistent with an intensity distribution of the form:

$$I(\theta) = 1 + \alpha \frac{(\boldsymbol{\sigma} \cdot \boldsymbol{p})}{E}$$

$$= 1 + \alpha \frac{v}{c} \cos \theta, \tag{4.9}$$

where $\alpha = -1$; $\boldsymbol{\sigma}$ = unit spin vector in direction $\boldsymbol{J}$; $\boldsymbol{p}$, E are the electron momentum and total energy; and θ is the angle of emission of the electron with respect to $\boldsymbol{J}$. The variation with electron velocity was checked over the range $0.4 < v/c < 0.8$.

The fore-aft asymmetry of $I(\theta)$ implies that the interaction violates parity conservation; for imagine the whole system reflected in a mirror normal to the z-axis. The first term (1) does not change sign under reflection—it is *scalar* (even parity). $\boldsymbol{\sigma}$, being an axial vector, does not change sign, while the polar vector $\boldsymbol{p} \to -\boldsymbol{p}$. Thus the product $\boldsymbol{\sigma} \cdot \boldsymbol{p} \to -\boldsymbol{\sigma} \cdot \boldsymbol{p}$ under reflection—it is called a *pseudoscalar* (odd parity). The fact that one has two such terms, of opposite parity, means that parity is not a well-defined quantum number in weak interactions. Note also that:

i) For a pure Fermi transition, $J_i = J_f = 0$, the nucleus could not be aligned and therefore parity violation could not be demonstrated by this method.

ii) Since $l = 0$ and $\Delta J = 1$ the spins of both electron and neutrino must be aligned along the positive z-axis. For $\theta = \pi$, when $I(\theta)$ is a maximum, the electron must then be left-handedly polarized.

This longitudinal polarization of electrons and positrons is found to be of the same form for Fermi, Gamow-Teller, and mixed transitions. If $\boldsymbol{\sigma}$ is the electron spin vector, the intensity is

$$I = 1 + \alpha \frac{\boldsymbol{\sigma} \cdot \boldsymbol{p}}{E}. \tag{4.10}$$

The polarization or *helicity* H is defined as

$$H = \frac{I_+ - I_-}{I_+ + I_-} = \alpha \frac{v}{c},$$

where I_+ and I_- represent intensities for $\boldsymbol{\sigma}$ parallel and antiparallel to $\boldsymbol{p}$.

Experimentally:

$$\alpha = +1 \text{ for } e^+, \qquad H = +\frac{v}{c},$$

$$= -1 \text{ for } e^-, \qquad H = -\frac{v}{c}. \qquad (4.11)$$

The property of helicity is not limited to leptons. We have already seen that, owing to their zero mass, photons must be longitudinally spin polarized. There are right-handed (or right-circularly polarized) photons and left-handed photons, with helicities +1 and −1 respectively. Parity is conserved in electromagnetic processes involving photons, because the two types of photon are always emitted with equal probability, and one does not therefore observe a *net* circular polarization. On the contrary, in weak interactions, β-processes consist of the emission of *either* electrons with a net left-handed spin polarization, *or* positrons which are predominantly right-handed.

4.3 LEPTON POLARIZATION IN β-DECAY

4.3.1 Measurement of Electron Polarization

Experimentally, the longitudinal polarization of electrons has been observed by a variety of methods. Three of these are sketched diagrammatically in Fig. 4.5. In (a), the longitudinal polarization is turned into a transverse polariza-

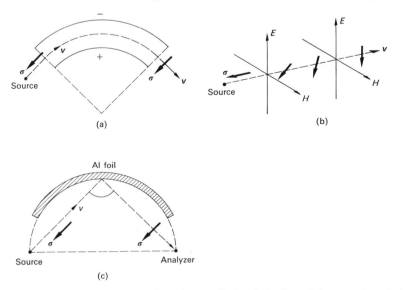

Fig. 4.5 Methods employed to transform longitudinal polarization of electrons from β-decay into transverse polarization. (a) Frauenfelder *et al.* (1957), (b) Cavanagh *et al.* (1957), (c) De Shalit *et al.* (1957).

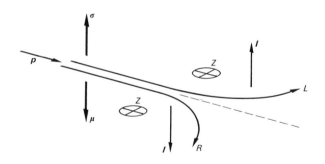

Fig. 4.6 Schematic diagram of scattering of a transversely polarized electron by a nucleus Z.

tion by electrostatic bending through 90°. Provided the electrons are non-relativistic ($v^2 \ll c^2$), the spin direction $\boldsymbol{\sigma}$ is virtually unaltered by the field (as $v \rightarrow c$, $\boldsymbol{\sigma}$ tends to follow the trajectory). A second technique, (b), employs crossed electric and magnetic fields. When $E/H = v/c$, the beam is undeflected, but $\boldsymbol{\sigma}$ precesses about $\boldsymbol{H}$. The value and extent of the fields is adjusted to give a 90° precession. A third method, (c), makes use of the fact that Coulomb scattering in a light element (in this case a 0.5-mm thick aluminum foil) does not affect the spin direction. The arrangement employs a semicircular strip of foil, with source and analyzer at each end of a diameter, resulting in a 90° scattering angle.

In each case the sense of polarization may be analyzed by observing the right-left asymmetry in the scattering by a foil of heavy element. The effect can be seen by a classical argument (Fig. 4.6). Let l be the orbital angular momentum of the electron relative to the scattering nucleus, Z. In the coordinate frame of the electron, the approaching nucleus appears as a positive current, clockwise or anticlockwise according to whether the electron passes to left or right. The magnetic field $\boldsymbol{H}$ associated with this current is therefore in the direction of l. Let $\boldsymbol{\sigma}$ represent the spin vector of the electron; the spin magnetic moment $\boldsymbol{\mu}$ then points in the direction $-\boldsymbol{\sigma}$. It follows that the magnetic energy $\boldsymbol{\mu} \cdot \boldsymbol{H}$ is positive (i.e. one gets a repulsive force) when l and $\boldsymbol{\sigma}$ are parallel, and negative (attractive force) when l and $\boldsymbol{\sigma}$ are antiparallel. Thus the magnetic interaction arising from the spin-orbit coupling adds to the electrical force when $l \cdot \boldsymbol{\sigma}$ is negative, and more electrons are scattered to right than to left. From the right-left asymmetry and a suitable calibration, the degree of transverse polarization can be determined.

The longitudinal polarization of electrons can also be found directly, without spin-twisting, by using a magnetized iron sheet and observing the electron-electron (Møller) scattering. Both electrons have to be observed in coincidence to distinguish the effect from the more intense nuclear scattering. The electron beam from the β-decay source falls obliquely on an iron foil magnetized in its plane, so that p and H are parallel. The scattering is greatest when spins of incident and target electrons are parallel (i.e. $\sigma \cdot H/H = +1$). On reversing the field H, the scattering ratio determines the longitudinal polarization. Another method is to observe the circular polarization of γ-rays from high energy bremsstrahlung when the electrons are stopped in an absorber. The sense of polarization of the forward γ-rays is the same as that initially possessed by the electrons, and is in turn determined by scattering in magnetized iron, as described below. Observations on the polarization of β-rays were carried out in numerous laboratories during 1957–58, and, within the experimental errors, clearly verified the helicity assignments (4.11).

4.3.2 Helicity of the Neutrino

The result (4.11), if applied to a neutrino ($m = 0$), implies that such a particle must be fully polarized, $H = +1$ or -1. In order to determine the type of operators occurring in the matrix element M in (4.1), the sign of the neutrino helicity turns out to be crucial. The neutrino is here defined as the neutral particle emitted together with the positron in β^+-decay, or following K-capture. The antineutrino then accompanies negative electrons in β^--decay. The neutrino helicity was determined in a classic and beautiful experiment by Goldhaber *et al.* in 1958. The steps in this experiment are indicated in Fig. 4.7.

i) Eu^{152} undergoes K-capture to an excited state of Sm^{152}, of $J = 1$, Fig. 4.7(a). In order to ensure conservation of angular momentum, J must be opposite to the spin of electron and neutrino. The recoiling Sm^{152}* therefore has the same sense of longitudinal polarization as the neutrino, Fig. 4.7(b).

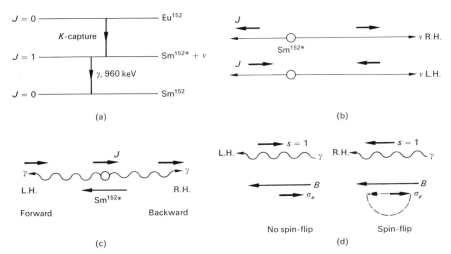

Fig. 4.7 Principal steps in the experiment of Goldhaber *et al.* to determine the neutrino helicity.

ii)   Now, in the transition $Sm^{152*} \to Sm^{152} + \gamma$, γ-rays emitted in the forward (backward) direction with respect to the line of flight of Sm^{152*} will be polarized in the same (opposite) sense to the neutrino, as in Fig. 4.7(c). Thus, the polarization of the "forward" γ-rays is the same as that of the neutrino.

iii)   The next step is to observe resonance scattering of the γ-rays in a Sm^{152} target. Resonance scattering is possible with γ-rays of just the right frequency to "hit" the excited state:

$$\gamma + Sm^{152} \to Sm^{152*} \to \gamma + Sm^{152}.$$

To produce resonance scattering, the γ-ray energy must slightly exceed the 960 keV to allow for the nuclear recoil. It is precisely the "forward" γ-rays, carrying with them a part of the neutrino-recoil momentum, which are able to do this, and which are therefore automatically selected by the resonance scattering.

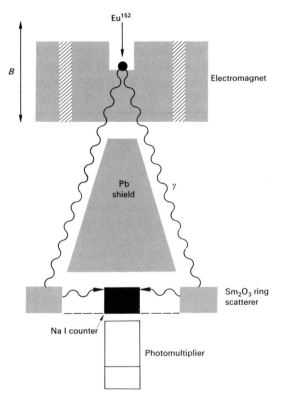

Fig. 4.8 Schematic diagram of apparatus used by Goldhaber *et al.* γ-rays from decay of Sm^{152*}, produced following K-capture in Eu^{152}, may undergo resonance scattering in Sm_2O_3, and are recorded by a sodium iodide scintillator and photomultiplier. The transmission of photons through the iron surrounding the Eu^{152} source depends on their helicity and the direction of the magnetic field B.

iv) The last step is to determine the polarization sense of the γ-rays. To do this, they were made to pass through magnetized iron before impinging on the Sm^{152} absorber. An electron in the iron with spin σ_e opposite to that of the photon can absorb the unit of angular momentum by spin-flip; if the spin is parallel it cannot. This is indicated in Fig. 4.7(d). If the γ-ray beam is in the same direction as the field B, the transmission of the iron is greater for left-handed γ-rays than for right-handed.

A schematic diagram of the apparatus is shown in Fig. 4.8. By reversing B the sense of polarization could be determined from the change in counting rate. When allowance was made for various depolarizing effects, it was concluded that neutrinos were left-handedly spin polarized.

In conclusion, the helicity assignments for the leptons emitted in nuclear β-decay are therefore as follows:

$$\begin{array}{ccccc} \text{Particle} & e^+ & e^- & \nu & \bar{\nu}, \\ \text{Helicity} & +v/c & -v/c & -1 & +1. \end{array} \tag{4.12}$$

4.4 THE V-A INTERACTION

The formulas (4.9), (4.10), and (4.11) have been presented from a purely empirical viewpoint. In order to make theoretical predictions about β-decay, where we are concerned with relativistic particles of spin $\frac{1}{2}$, we should use the Dirac theory. This is described briefly in Appendix B; here we only outline the results which finally come out. In his original theory of β-decay, Fermi (1934) assumed that the matrix element of the interaction would involve a linear combination of the four fermion wave functions, and an appropriate operator O, say. (These wave functions are called spinors, each with four components.) Thus, for neutron decay, one could write, in obvious notation,

$$M = G(\psi_p^* O \psi_n)(\psi_e^* O \psi_\nu), \tag{4.13}$$

where G is the coupling constant typical of β-decay.

The grouping of the nucleon and the lepton wave functions in this way is made more plausible if we write the β-decay process

$$n \rightarrow p + e^- + \bar{\nu}$$

in the equivalent form

$$\nu + n \rightarrow p + e^-.$$

Fermi assumed an analogy between this process and the electromagnetic scattering of two charged particles, for example an electron and a proton:

$$e^- + p \rightarrow p + e^-.$$

The last process can be described as the interaction between two *currents*; an electron current $e(\psi_e^* O \psi_e)$ with a proton current $e(\psi_p^* O \psi_p)$. In the electromagnetic interaction, the operator O is a vector operator (actually $O = \gamma_4 \gamma_\mu$, where the γ's are 4×4 matrices, with index μ running from 1 to 4, which operate on the 4-component spinor wave functions). Fermi assumed also that the operator O in (4.13) would again be a vector operator. The only differences from the electromagnetic case are that, for β-decay, one has a constant G instead of e^2; and that the "weak currents" in (4.13) are assumed to interact at a point, i.e. the weak interaction is of extremely short range. On the contrary, in the electromagnetic case, the interactions of the currents must be integrated over all space, since the Coulomb force is long-range. The current-current interaction is discussed further in Section 4.7.

The vector interaction was satisfactory (prior to the discovery of parity violation in 1956) in describing Fermi transitions. It did not account for Gamow-Teller transitions, since it could not produce a flip-over of the nucleon spin. Within certain requirements of relativistic invariance, one can in fact have five possible independent forms for the operator O in (4.13). They are called scalar (S), vector (V), tensor (T), axial vector (A), and pseudoscalar (P). These names are associated with the transformation properties of the weak currents under space inversions. The S- and V-interactions produce Fermi transitions, while T and A produce Gamow-Teller transitions. The pseudoscalar interaction P is unimportant in β-decay, since it couples spinor components proportional to particle velocity and thus introduces a factor in M^2 of v^2/c^2, where v is the nucleon velocity ($v^2/c^2 \sim 10^{-6}$ in β-decay).

The various operators O_i lead to different predictions on the angular correlations of the leptons in β-decay, and these are shown in Fig. 4.9. In these diagrams, v refers to the velocity of the positron.

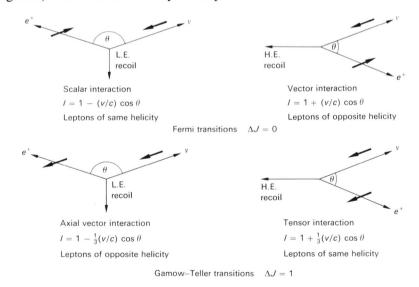

Fig. 4.9 Angular distributions in β-decay predicted by various types of interaction. The thick arrows denote lepton spin directions which result from angular momentum conservation assuming the neutrino is always left-handed. L.E. and H.E. represent low and high energy recoils respectively.

We note that, since in a Fermi transition the total angular momentum of the leptons is zero, the S-interaction must produce leptons of the same helicity, and the V-interaction, leptons of opposite helicity. In Gamow-Teller transitions, the total lepton angular momentum is $J = 1$, so that the T- and A-interactions produce leptons of the same and opposite helicities respectively. The foregoing experiments on lepton polarization, summarized in (4.12), show therefore that only the V- and A-interactions can produce the helicities actually observed. A similar conclusion is arrived at by studying the shape of the momentum spectra of the nuclear recoils in Fermi and Gamow-Teller transitions. We may also remark that the factor $\frac{1}{3}$ in the $\cos\theta$ term for A- and T-couplings arises because the total angular momentum ($J = 1$) of the leptons possesses three possible orientations in space, and this reduces the angular correlation. Thus we can write in place of (4.13), for a general β-interaction,

$$M = G \sum_{i=V,A} C_i(\psi_p^* O_i \psi_n)(\psi_e^* O_i \psi_\nu), \qquad (4.14)$$

where C_V and C_A are appropriate coefficients. For example, for a pure Fermi transition, $C_A = 0$. The matrix element (4.14) is, however, a scalar quantity. It implies that parity is conserved, and that, for each diagram in Fig. 4.9, we should add a second diagram with lepton spins reversed, so that there is no net lepton polarization. This is clearly wrong. We need to add another term to (4.14) which will give us a pseudoscalar quantity, so that the matrix element contains both scalars and pseudoscalars, and thus has no well-defined parity. We can do this simply by adding a similar expression, but with the operator in one bracket (the lepton bracket) multiplied by a matrix γ_5 ($\gamma_5 = \gamma_1\gamma_2\gamma_3\gamma_4$). Thus one obtains

$$M = G \sum_{i=V,A} (\psi_p^* O_i \psi_n)[\psi_e^* O_i(C_i + C_i'\gamma_5)\psi_\nu]. \qquad (4.15)$$

If, as all the evidence suggests, the interaction is invariant under time reversal, it can be shown that C_i and C_i' must be real coefficients; furthermore, if a neutrino (or antineutrino) is to be completely spin polarized, $H = \pm 1$, one must have $C_i' = \pm C_i$. Since scalar and pseudoscalar terms then occur with equal magnitude, this is called the principle of *maximum parity violation*. For this case, the operator $(1 \pm \gamma_5)$ acting on the neutrino wave function in (4.15) projects out one sign of helicity:

$$(1 + \gamma_5) \rightarrow \text{L.H. neutrino state (R.H. antineutrino)},$$
$$(1 - \gamma_5) \rightarrow \text{R.H. neutrino state (L.H. antineutrino)}.$$

Since the experiment in Section 4.3.2 shows the neutrino to be left-handed, we need to take the first expression, corresponding to $C_i' = C_i$. This type of weak interaction is called the *two-component neutrino theory*, and was proposed independently by Lee and Yang, by Landau, and by Salam. It also correctly

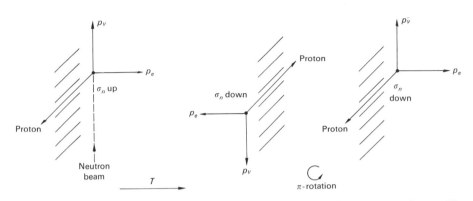

Fig. 4.10 Experiment to test time reversal invariance in polarized neutron decay. The electron and proton from neutron decay are detected in coincidence by counters to either side of the neutron beam. A slit system is used near the proton counter to ensure that the anti-neutrino travels preferentially in the neutron beam direction. The neutron spin σ_n is arranged normal to the beam and to the line joining the counters. Time reversal (see Table 3.6), followed by a 180° rotation of the figure, amounts to inversion of σ_n and therefore a change in sign of the D term in (4.17). If the decay is invariant under time reversal, D must be zero. This is investigated experimentally by reversing σ_n and observing the change in counting rate. (After Konopinski, 1959.)

predicts the helicities $\pm v/c$ for the charged leptons in (4.12). We may therefore rewrite (4.15) as

$$M = G \sum_{i=V,A} C_i(\psi_p^* O_i \psi_n)[\psi_e^* O_i(1 + \gamma_5)\psi_\nu]. \qquad (4.16)$$

It remains to determine the ratio C_A/C_V and its sign. The sign can be determined by observing the interference terms between the V- and A-interactions in mixed transitions, and this has been accomplished by determination of angular correlations in the decay of *polarized* neutrons (in the decay of unpolarized neutrons, the V-A interference term is washed out in the sum over initial neutron spin). A classic series of experiments on polarized neutron decay have been carried out at Chalk River (1958), at the Argonne National Laboratory (1960, 1969), and in Moscow (1968). The principle of these experiments is to employ a polarized neutron beam (Section 3.18), to detect the proton and electron from neutron decay in coincidence, to either side of the neutron beam, and to vary the direction of neutron polarization relative to the counters in order to measure the various correlation coefficients in (4.17). The reader is referred to the review article by Konopinski (1959) for details. The measurement of the T-violating term D is illustrated in Fig. 4.10. Essentially the directions of three vectors may be determined experimentally; the neutron spin vector σ_n and the momentum vectors p and q of electron and antineutrino, the latter being deduced from observation of the recoil proton in the decay

$n \to p + e^- + \bar{\nu}$. The intensity or counting rate must then have the general form

$$I = 1 + a\left[\frac{q \cdot p}{qE}\right] + \left[A\frac{p}{E} + B\frac{q}{q}\right] \cdot \sigma_n + D\left[\frac{p \wedge q}{qE}\right] \cdot \sigma_n, \quad (4.17)$$

where E is the total energy of the electron. The various coefficients a, A, B, and D are given by

$$
\begin{aligned}
a &= (C_V^2 - C_A^2)/(C_V^2 + 3C_A^2), &\text{(a)} \\
A &= [-2C_A^2 - (C_V C_A^* + C_V^* C_A)]/(C_V^2 + 3C_A^2), &\text{(b)} \\
B &= [+2C_A^2 - (C_V C_A^* + C_V^* C_A)]/(C_V^2 + 3C_A^2), &\text{(c)} \\
D &= i(C_V C_A^* - C_V^* C_A)/(C_V^2 + 3C_A^2), &\text{(d)}
\end{aligned}
\qquad (4.18)
$$

where for generality we assume the coefficients C are complex numbers. As indicated from Table 3.6 and Fig. 4.10, the D term is the only one in (4.17) which changes sign under time reversal, and we see from (4.18d) that the D term must be zero (i.e. the interaction is T-invariant) if C_A and C_V are relatively real. Experimentally, it is found that D is consistent with zero, i.e.

$$C_A = |C_A/C_V| \times C_V e^{i\phi}$$

where $\phi = 0°$ or $180°$. Thus, $C_A \approx \pm C_V$. For these two possibilities, the values of A and B expected from (4.18), together with the observed values, are given in Table 4.1.

TABLE 4.1 Correlation coefficients in polarized neutron decay

	$C_A = +C_V$	$C_A = -C_V$	Experimental value
a	0	0	-0.1 ± 0.1[a]
A	-1	0	-0.11 ± 0.02[b] -0.115 ± 0.008[c]
B	0	$+1$	$+0.88 \pm 0.15$[b] $+1.01 \pm 0.04$[c]
D	0 (for T-invariance)		-0.04 ± 0.07[b] $+0.01 \pm 0.01$[d]

[a] Robson, *Phys. Rev.* **100**, 933 (1955).

[b] Burgy, Krohn, Novey, Ringo, and Telegdi, *Phys. Rev.* **120**, 1829 (1960).

[c] Christensen, Krohn, and Ringo, *Phys. Lett.* **28B**, 411 (1969).

[d] Erozolimsky, Bonlarenko, Mostovoy, Obinyakov, Zacharova, and Titov, *Phys. Lett.* **27B**, 557 (1968).

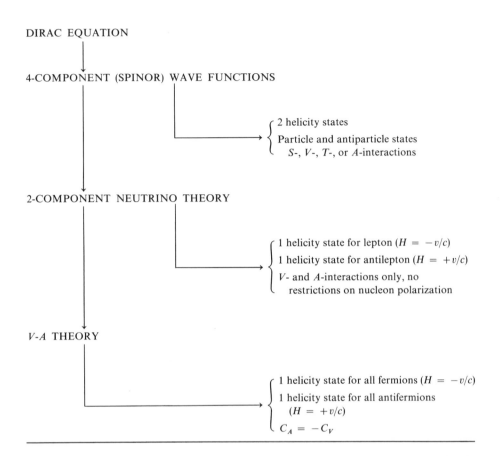

DIRAC EQUATION

4-COMPONENT (SPINOR) WAVE FUNCTIONS

2 helicity states
Particle and antiparticle states
S-, V-, T-, or A-interactions

2-COMPONENT NEUTRINO THEORY

1 helicity state for lepton ($H = -v/c$)
1 helicity state for antilepton ($H = +v/c$)
V- and A-interactions only, no restrictions on nucleon polarization

V-A THEORY

1 helicity state for all fermions ($H = -v/c$)
1 helicity state for all antifermions ($H = +v/c$)
$C_A = -C_V$

These results show clearly that C_A and C_V are of opposite sign. This is expressed briefly by saying that the interaction is $V - A$. The departures of A and B from 0 and 1 are accounted for by the fact that C_A and C_V, as we already know, are not quite equal. The values of C_A/C_V from (4.8) and from the experiment on polarized neutron decay are:

$$C_A/C_V$$

ft value of O^{14} and neutron $\pm(1.25 \pm 0.02)$,

Polarized neutron decay -1.26 ± 0.02.   (4.19)

The latter value is inherently more reliable since it does not depend on radiative and other corrections required to arrive at the ft value for O^{14}-decay.

If we set $C_A = -\lambda C_V$, the β-decay interaction (4.16) may be shown to assume the form (see Appendix B):

$$M = GC_V[\psi_p^* O_V(1 + \lambda\gamma_5)\psi_n][\psi_e^* O_V(1 + \gamma_5)\psi_\nu],   (4.20)$$

where $O_V = \gamma_4\gamma_\mu$. We then note that, in the fictitious situation $\lambda = 1$ (pure V-A theory), the operators in the lepton and nucleon brackets are identical. This means that *all* fermions in the β-decay would have helicity $H = -v/c$ and all antifermions would have helicity $H = +v/c$, where v is the fermion velocity. This situation then differs from that of the pure vector interaction originally proposed by Fermi, (4.13), only in the inclusion of the projection operator $(1 + \gamma_5)$ in all terms. It is surmised that the departure of actual β-decay from the predictions of the pure V-A theory of Feynman and Gell-Mann (1958) and Marshak and Sudarshan (1958), is brought about by the effects of redistribution of the "weak charge" of the nucleon arising from the strong interactions; we explore this point in more detail later. In summary, the steps in the argument leading from the Dirac to the V-A theory are outlined in the flow diagram on p. 146.

4.5 PARITY VIOLATION IN Λ-DECAY

Parity nonconservation is a general property of weak interactions not solely associated with leptonic processes. The "τ-θ paradox," associated with the nonleptonic decay of kaons into pions, has already been mentioned. Another nonleptonic decay is that of the Λ-hyperon, for which the dominant decay modes are

$$\Lambda \to \pi^- + p, \qquad \Lambda \to \pi^\circ + n. \qquad (4.21)$$

The Λ-hyperon does in fact also undergo leptonic β-decay, $\Lambda \to p + e^- + \bar{\nu}$, with a small branching ratio.

Parity violation can be demonstrated by considering the decay of Λ-hyperons produced in the associated production process

$$\pi^- + p \to \Lambda + K^\circ. \qquad (4.22)$$

In this process, the Λ can be (and in general is) spin polarized. Parity conservation in such a strong interaction implies that the Λ must be polarized with spin **σ** *transverse* to the production plane, i.e. of the form $\boldsymbol{\sigma} \propto (\boldsymbol{p}_\Lambda \wedge \boldsymbol{p}_K)$, which does not change sign under inversion (Fig. 4.11). Note that spin polarization *in* the production plane in general does change sign, and is not allowed.

In practice the mean transverse polarization

$$P_\Lambda = (N\uparrow - N\downarrow)/(N\uparrow + N\downarrow) \approx 0.7$$

for an incident pion momentum in (4.22) just above 1 GeV/c. In the decay process (4.21), let us define the direction of **σ** as the z-axis of the Λ rest frame (Fig. 4.12). In this frame, the distribution in angle of emission (θ, ϕ) of the pion or proton will depend on their orbital angular momentum, l.

Since $J_\Lambda = \frac{1}{2}$, $J_z = \pm\frac{1}{2}$, we can have either $l = 0$ (proton and Λ-spins parallel) or $l = 1$ (spins antiparallel). Thus, we generally expect a combination

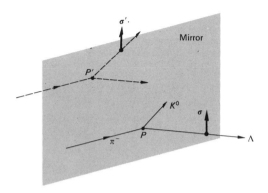

Fig. 4.11 Transverse polarization of the Λ-hyperon in the reaction $\pi^- + p \to \Lambda + K^0$ is invariant under spatial inversion, and therefore allowed for a strong reaction.

of s- and p-waves. Call m_1 the z-component of the proton spin vector, m_2 the z-component of l. In the s-wave case, $m_2 = 0$ and the angular momentum wave function is $Y_m^l = Y_0^0$. Thus, for $J_z = +\frac{1}{2}$, the total wave function is the product

$$\psi_s = a_s Y_0^0 \chi^+, \tag{4.23}$$

where a_s denotes the s-wave amplitude and χ^+ the proton spin-up state of $m_1 = +\frac{1}{2}$. For the p-wave, $m_1 + m_2 = J_z = \frac{1}{2}$, with either $m_1 = +\frac{1}{2}$ and $m_2 = 0$, or $m_1 = -\frac{1}{2}$ and $m_2 = +1$.

Referring to Table 3.4, p. 101, for the Clebsch-Gordan coefficients for adding $J = 1$ and $J = \frac{1}{2}$, the entry for $J_{\text{total}} = m = +\frac{1}{2}$ gives

$$\psi_p = a_p[\sqrt{\tfrac{2}{3}}Y_1^1 \chi^- - \sqrt{\tfrac{1}{3}}Y_1^0 \chi^+]. \tag{4.24}$$

Here, a_s and a_p are, in general, complex amplitudes. Thus if *both* s- and p-waves

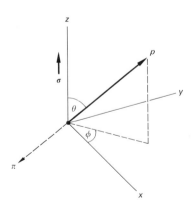

Fig. 4.12 Definition of axes and directions in Λ-decay.

are present, the total amplitude will be

$$\psi = \psi_s + \psi_p = \left[a_s Y_0^0 - \frac{a_p}{\sqrt{3}} Y_1^0\right]\chi^+ + \left[a_p\sqrt{\tfrac{2}{3}}Y_1^1\right]\chi^-.$$

Recalling the orthogonality of the spin states χ^+ and χ^-, the angular distribution becomes

$$\psi\psi^* = \left(a_s Y_0^0 - \frac{a_p}{\sqrt{3}} Y_1^0\right)\left(a_s Y_0^0 - \frac{a_p^*}{\sqrt{3}} Y_1^0\right) + a_p^2(\sqrt{\tfrac{2}{3}}Y_1^1)^2,$$

where one of the phases must be arbitrary and we take a_s to be real. Also, $Y_0^0 = 1$, $Y_1^0/\sqrt{3} = \cos\theta$, $\sqrt{\tfrac{2}{3}}Y_1^1 = -\sin\theta$, so that

$$\psi\psi^* = |a_s|^2 + |a_p|^2\cos^2\theta + |a_p|^2\sin^2\theta - a_s\cos\theta\,[a_p + a_p^*]$$
$$= |a_s|^2 + |a_p|^2 - 2a_s\,\mathrm{Re}\,a_p^*\cos\theta. \tag{4.25}$$

Setting

$$\alpha = \frac{2a_s\,\mathrm{Re}\,a_p^*}{|a_s|^2 + |a_p|^2},$$

the angular distribution has the form

$$I(\theta) = 1 - \alpha\cos\theta. \tag{4.26}$$

The angle θ is defined relative to $\boldsymbol{\sigma}$; physically, one can only measure relative to the normal to the production plane, so that if we redefine θ in this way, the above result becomes

$$I(\theta) = 1 - \alpha P\cos\theta,$$

where P is the average polarization; experiment shows that $\alpha P \approx -0.7$. Thus, the parity violation in the Λ-decay is manifested as an up-down asymmetry of the decay pion (or proton) relative to the production plane.

Note that (4.26) has the same form as (4.9); that the parity violation parameter α is finite only if *both* a_s and a_p are finite; and thus that the parity violation arises from interference of the s (even) and p (odd) waves.

4.6 PION AND MUON DECAY

The lepton helicities first observed in 1957 in nuclear β-decay were detected simultaneously in the decay of pions and muons. We recall that the pion and muon decay schemes are

$$\pi^+ \rightarrow \mu^+ + \nu, \tag{4.27}$$

$$\mu^+ \rightarrow e^+ + \nu + \bar{\nu}. \tag{4.28}$$

Since the pion has spin zero, the neutrino and muon must have antiparallel spin vectors, as shown in Fig. 4.13. If the neutrino has helicity $H = -1$, as in

Fig. 4.13 Sketches indicating sense of spin polarization in pion and muon decay.

β-decay, the μ^+ must have negative helicity. In the subsequent muon decay, the positron spectrum is peaked in the region of the maximum energy, so the most likely configuration is that shown—the positron having positive helicity. In fact, the positron spectrum has the shape indicated in Fig. 4.14. This was originally of interest from the point of view of the identity of the ν and $\bar{\nu}$. The two neutrinos cannot be identical, otherwise the Pauli Principle would forbid them to be in the same quantum state, and the positron spectrum would go to zero at E_{max}. It turns out that although they are ν and $\bar{\nu}$ they are different kinds of neutrinos anyhow; one is ν_e and the other $\bar{\nu}_\mu$ (see below). In the experiments, positive pions decayed in flight and those decay muons projected in the forward direction—thus with negative helicity—were selected. These μ^+ were stopped in a carbon absorber and the e^+ angular distribution relative to the original muon line-of-flight was observed. This latter will be antiparallel to the muon spin-vector, if there is no depolarization of the muons in coming to rest (true in carbon). The angular distribution observed was of the form

$$\frac{dN}{d\Omega} = 1 - \frac{\alpha}{3} \cos \theta, \tag{4.29}$$

where θ is the angle between $\boldsymbol{p}_\mu$, the initial muon momentum vector, and $\boldsymbol{p}_e$, the electron momentum vector, and $\alpha = 1$ within the errors of measurement. The same value of α was found for μ^+ and μ^-. Equation (4.29) is exactly the form predicted by the V-A theory. The helicity of the electrons (positrons) was also measured and shown to be $\mp v/c$.

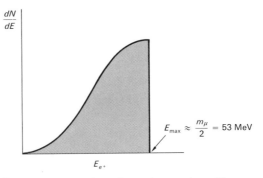

Fig. 4.14 Shape of positron energy spectrum from decay of positive muons (see also Fig. 4.16).

4.6.1 The $\pi \to \mu$ and $\pi \to e$ Branching Ratios

The decay (4.27) does not look like a 4-fermion interaction. However, if we imagine a nucleon-antinucleon intermediate state, we can draw the diagram of Fig. 4.15. It is difficult to calculate the matrix element directly—one obtains divergences when integrating over momentum in the intermediate $p\bar{n}$-state—but what we are interested in now is the ratio

$$\frac{\pi \to e + v}{\pi \to \mu + v}, \tag{4.30}$$

where everything to the left of the dotted line in Fig. 4.15 is common to both decays.

The π^+ has zero spin and odd parity, so that the virtual transition

$$\pi^+ \to p + \bar{n} \to e^+ + v; \qquad J^P = 0^-,$$

where nucleon and antinucleon have odd relative parity, may be thought of as the decay

$$p \to n + e^+ + v, \qquad \Delta J = 0, \text{ parity change.}$$

In the terminology of β-decay, the last reaction is a forbidden transition, in which the overall change in nuclear angular momentum is zero, resulting from spin-flip and the emission of the final nucleon in an orbital state $l = 1$—hence odd parity. Such a transition can only be produced by the axial vector (A) and pseudoscalar (P) couplings. There is no evidence for P-coupling in β-decay, but this might be ascribed to the very difficulty of generating a nucleon of $l = 1$ at the low energies available in such processes. Therefore, we must consider it as a possibility in the higher energy process of pion decay. As we have seen (Fig. 4.9), the A-coupling tends to produce leptons of opposite helicity. The P-coupling (not shown in Fig. 4.9) produces leptons of the same helicity. On the other hand, conservation of angular momentum compels the leptons to have the same helicity (Fig. 4.13). Thus, for A-coupling, we expect for the matrix element $M^2 \propto 1 + (-v/c)$—there being no factor $\frac{1}{3}$ here (as in Fig. 4.9) because we are considering a $\Delta J = 0$ transition instead of $\Delta J = 1$, with three projections of the lepton angular momentum in an arbitrary direction. For P-coupling, one finds instead $M^2 \propto (1 + v/c)$.

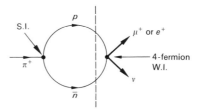

Fig. 4.15 The diagram for the decay $\pi^+ \to \mu^+ + v$, or $\pi^+ \to e^+ + v$, via a nucleon-antinucleon intermediate state.

The branching ratio (4.30) then follows if one includes the phase-space factor of (4.1). This is

$$\frac{dN}{dE_0} = \text{const.} \, p^2 \frac{dp}{dE_0} \, .$$

Here, p is the momentum of the charged lepton in the pion rest frame, v is its velocity and m its rest mass. The neutrino momentum is then $-p$.

$-p$ $\xleftarrow{\hspace{1.5cm}}$ neutrino, $\underset{m_\pi}{\bullet}$ charged lepton $\xrightarrow{\hspace{1cm}}$ p, m, v
neutrino,
$m = 0$

In units $c = 1$, the total energy is

$$E_0 = m_\pi = p + \sqrt{p^2 + m^2}.$$

Hence

$$\frac{dp}{dE_0} = \frac{(m_\pi^2 + m^2)}{2m_\pi^2},$$

$$1 + \frac{v}{c} = 2m_\pi^2/(m_\pi^2 + m^2),$$

$$1 - \frac{v}{c} = 2m^2/(m_\pi^2 + m^2).$$

Thus, for A-coupling the decay rate is given by

$$p^2 \frac{dp}{dE_0}\left(1 - \frac{v}{c}\right) = \frac{m^2}{4}\left(1 - \frac{m^2}{m_\pi^2}\right)^2,$$

and for P-coupling the decay rate is given by

$$p^2 \frac{dp}{dE_0}\left(1 + \frac{v}{c}\right) = \frac{m_\pi^2}{4}\left(1 - \frac{m^2}{m_\pi^2}\right)^2.$$

The predicted branching ratios become, with the approximation $m_e^2/m_\pi^2 \ll 1$,

A-coupling: $\quad R = \dfrac{\pi \rightarrow e + v}{\pi \rightarrow \mu + v} = \dfrac{m_e^2}{m_\mu^2}\dfrac{1}{(1 - m_\mu^2/m_\pi^2)^2}$

$$= 1.28 \times 10^{-4}, \tag{4.31a}$$

P-coupling: $\quad R = \dfrac{\pi \rightarrow e + v}{\pi \rightarrow \mu + v} = \dfrac{1}{(1 - m_\mu^2/m_\pi^2)^2} = 5.5. \tag{4.31b}$

The dramatic difference in the branching ratio for the two types of coupling just stems from the fact that angular momentum conservation compels the electron or muon to have the "wrong" helicity for A-coupling. The phase-space factor for the electron decay is greater than for muon decay, but the $(1 - v/c)$ term chops down hard on the more relativistic (i.e. lighter) lepton.

The first experiments to measure the ratio

$$R = \frac{\pi \rightarrow e + \nu}{\pi \rightarrow \mu + \nu}$$

in fact suggested a limit $R < 10^{-4}$. When Feynman and Gell-Mann put forward the V-A theory in 1958, they noted the discrepancy, and suggested that the early experiments were wrong. They were! The experiments were then repeated by several groups, and the results were consistent with the ratio 4.31(a).

The presently accepted experimental ratio is

$$R_{\text{exp}} = (1.25 \pm 0.03) \times 10^{-4}$$

compared with the V-A prediction, including radiative corrections, of

$$R_{\text{theor.}} = 1.23 \times 10^{-4}.$$

This result was a major triumph for the V-A theory, and proves that the pseudoscalar coupling is zero or extremely small. Figure 4.16 shows a typical electron spectrum observed from positive pions stopping in an absorber in one of the experiments. The rare $\pi \rightarrow e + \nu$ process yields electrons of unique energy, about 70 MeV. They are accompanied by the much more numerous electrons from the decay sequences $\pi \rightarrow \mu + \nu, \mu \rightarrow e + \nu + \bar{\nu}$. The spectrum

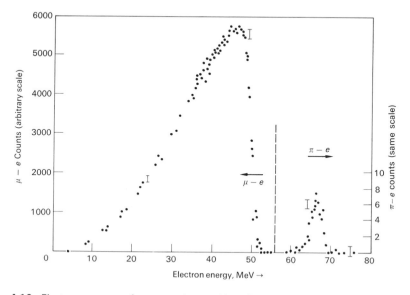

Fig. 4.16 Electron spectrum from stopping positive pions, as measured in the experiment of Anderson *et al.* (1960). The broad distribution extending up to 53 MeV is from $\mu^+ \rightarrow e^+ + \nu + \bar{\nu}$. The narrow peak around 70 MeV is from $\pi^+ \rightarrow e^+ + \nu$ decay. Note the change in vertical scale for these very rare events.

from the muon decay extends up to 53 MeV. The rejection of electrons from $\pi \to \mu \to e$ events is based on momentum, timing (the mean life of the pion being 2.5×10^{-8} sec, while that of the muon is 220×10^{-8} sec), and the absence of a muon pulse in the counters.

The formulas for R in (4.31) apply equally to the branching ratio

$$K \to e + v/K \to \mu + v,$$

if we use m_K in place of m_π. Since $m_K \sim 3m_\pi$, the electron is even more relativistic, and the ratio predicted for A-coupling is therefore smaller, $R = 2 \times 10^{-5}$. The average ratio from two recent experiments is $(2 \pm 0.5) \times 10^{-5}$, confirming the conclusions from pion decay.

4.7 THE CURRENT-CURRENT INTERACTION

4.7.1 Electron-Muon Universality, and Conservation of Vector Current

In deriving the branching ratios (4.31) in pion decay, we tacitly assumed equal weak couplings for the electron and muon decay modes. This hypothesis is known as *muon-electron universality*, and is clearly demonstrated in this case by the agreement between experiment and the prediction of the *V-A* theory.

Before enquiring further into what this means, let us look into electron-proton scattering, an electromagnetic interaction. We have already seen that it is convenient to consider this as an interaction between two electric currents via an intermediate virtual photon—see Fig. 4.17. Such a description is made possible essentially because fermions are conserved—two fermions in, two fermions out. In the scattering of two electrons *or* an electron and proton, we note that each of the currents carries the *same* unit of electric charge (e), and that this charge is conserved. This is not a trivial observation, as we shall see in discussing the case of weak currents. There is *prima facie* no reason why the proton and electron should have identical charges. Let us imagine that we can "turn off" the strong interactions. Then it is plausible to regard the proton as a "heavy electron" (like the muon). We would have a "bare" proton of charge e and magnetic moment equal to the Dirac value $eh/4\pi mc$. Now "turn on" the strong interactions. Then the proton immediately shows differences in behavior from the electron—for example, it develops a large anomalous magnetic moment. One can think of the strong interaction "structure" as being

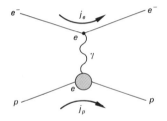

Fig. 4.17 The scattering of an electron by a proton, via an intermediary γ-ray, can be interpreted as an interaction between two currents, j_e and j_p.

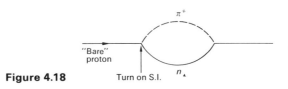

Figure 4.18   Turn on S.I.

due to virtual emission and reabsorption of pions, or other particles, as in Fig. 4.18. Thus, although the *distribution* of electric charge is affected by the strong interactions, the *total* charge remains the same. In other words, the electromagnetic current j is conserved by the strong interactions.

Now consider the weak interaction. As indicated previously, we can again think of the β-decay process as an interaction of two currents. Figure 4.19 shows this for μ-decay. The current J carries a "weak charge," $\sqrt{G}$, instead of an electric charge, and there are three other important differences:

i) The weak currents are electric charge-changing ($n \rightarrow p$, $e^- \rightarrow \nu$) when they interact, so the intermediary particle W (if there is one) besides having unit spin like the photon, also carries an electric charge.

ii) Two weak currents interact effectively only at a *point* in space-time—as distinct from the electromagnetic case, where the interaction energy is obtained by integrating over all space, and as we know, many partial waves are involved.

iii) The electromagnetic interactions are pure vector, whereas the weak interactions are A and V, i.e. $J = J_A + J_V$.

The infinite range of the electromagnetic interaction is associated with the fact that the rest mass of the intermediary particle coupling the currents—the photon—is zero. The point-like interaction of the weak currents can be associated with the large mass of the intermediary boson W. At present (1971) this particle has not been observed experimentally, and its mass, if it exists, exceeds 2 GeV/c^2.

As we have drawn it, the weak charge, $\sqrt{G}$, of the muon and the electron are the same, and we have seen that this is verified experimentally by the $\pi \rightarrow e / \pi \rightarrow \mu$ branching ratio. At one time, it was thought that all elementary

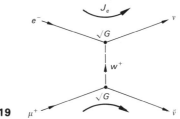

Figure 4.19 μ^+

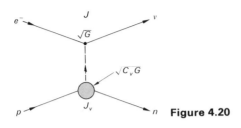

Figure 4.20

particles might have had the same unit of weak charge, just as they have the same electric charges—the so-called Universal Fermi Interaction (now dead).

Consider now the comparison between muon decay and a pure Fermi (V) nuclear transition, $O^{14} \rightarrow N^{14*} + e^+ + \nu$ (Fig. 4.20).

From the lifetime of these decays and the phase-space factors, one can compute the coupling constants involved. In muon decay it can be shown that the decay rate is $1/\tau_\mu = G^2 m_\mu^5/(192\pi^3)$, and inserting the observed lifetime gives the value of G (appropriate to muon decay):

$$G = \frac{1.02 \times 10^{-5}}{M_p^2}. \tag{4.32}$$

In O^{14}-decay, the ft value yields $G^2 C_V^2$, where C_V is an appropriate coefficient for vector β-decay. The experimental values yield

$$C_V = 0.98. \tag{4.33}$$

Thus the couplings in muon decay and in vector nucleon decay are very nearly (but not quite) the same. Let us for the moment neglect the 2% discrepancy. So the "vector weak charge" of the nucleon is unaffected by the strong interactions. Thus, the strong reaction properties of the nucleon, resulting in pion emission and reabsorption, $p \rightleftarrows n + \pi$, do not affect the "lepton-emitting power" of the nucleon; the vector weak current J_V is conserved by the strong interactions, just as is the electromagnetic current. This conclusion is called the *conserved vector current (CVC) hypothesis*.

One application of CVC is to the β-decay of the pion, Eq. (4.34). Consider the O^{14}-transition in the light of one of the possible virtual transitions of the nucleon, such as single pion emission. Part of the time, the proton exists in the state $(n + \pi^+)$; the neutron "core" cannot emit positron and neutrino, because of charge conservation, so that emission of leptons is entirely the responsibility of the π^+ and will be determined by the "weak charge" (vector part) of the pion. If this were not the same as that of the "bare" proton before the strong interactions were turned on, the result $C_V = 1.0$ would not hold. Thus the hypothesis also predicts that the coupling constants in the decays

$$\pi^+ \rightarrow \pi^0 + e^+ + \nu; \qquad J^P = 0^- \rightarrow J^P = 0^-; \qquad \text{pure Fermi} \quad (4.34a)$$

and

$$O^{14} \rightarrow N^{14*} + e^+ + \nu; \qquad J^P = 0^+ \rightarrow J^P = 0^+; \qquad \text{pure Fermi} \quad (4.34b)$$

should be the same, and hence they should have identical ft values (matrix elements). A very rough estimate of the decay rate (4.34) can be obtained by application of the Sargent rule (Eq. 4.3)—see Problem 4.2 at the end of the chapter. The observed branching ratio $(\pi^+ \to \pi^0 + e^+ + v)/(\pi^+ \to \mu^+ + v)$ is 1.02×10^{-8}, in perfect agreement with that expected from CVC.

4.7.2 Nonconservation of Axial Weak Current

As we know from (4.8) and (4.19), the axial vector current J_A is *not* conserved by the strong interactions, and the axial vector part of the weak charge of a nucleon is different from that of a lepton; $C_A = 1.25$. For a long time it has been speculated that the difference of C_A from unity—the divergence of

Figure 4.21

the axial current of the nucleon—is dominated by the process of single pion exchange. In other words, if the process of Fig. 4.21 did not exist, J_A would be a conserved current. Attempts have been made to calculate the magnitude of this term, and one gets good agreement with experiment (within $\sim 2\%$).

4.8 THE DECAYS OF STRANGE PARTICLES

4.8.1 Nonleptonic Decays; the $\Delta I = \frac{1}{2}$ Rule

Although weak interactions violate isospin conservation, there appears to be a selection rule

$$\Delta I = \tfrac{1}{2}$$

which is generally obeyed in the decays of strange particles (applying only to the nonleptonic decays, which have well-defined I-spin in the final state). The evidence for this rule comes, for example, from the branching ratios for Λ-decay:

$$\Lambda \to p + \pi^-$$
$$\to n + \pi^0.$$

Since $I_\Lambda = 0$, the $\Delta I = \frac{1}{2}$ rule states that the nucleon and pion must be in an $I = \frac{1}{2}$ state. Referring to the table of Clebsch-Gordan coefficients (p. 330) the rule predicts that

$$\frac{\text{Rate } \Lambda \to n\pi^0}{\text{Rate } \Lambda \to n\pi^0 + \text{Rate } \Lambda \to \pi^- p} = \frac{1}{3}. \tag{4.35}$$

The observed ratio is 0.36 ± 0.02, in excellent agreement with (4.35) when phase-space factors are included, which bring the expected value up to 0.345.

The reason for proposing the $\Delta I = \frac{1}{2}$ rule arose from a comparison of the decay modes of neutral and charged kaons to two pions:

$$\text{Decay rate}$$

$$K^0 \to \pi^+ + \pi^- \qquad 7.7 \times 10^9 \text{ sec}^{-1}, \qquad (4.36a)$$

$$K^+ \to \pi^+ + \pi^0 \qquad 1.74 \times 10^7 \text{ sec}^{-1}. \qquad (4.36b)$$

The rate of reaction (4.36a) is about 400 times that of (4.36b). Since in each case the parent kaon has $J = 0$, then the boson pair has a symmetric space wave function, and hence a symmetric isospin wave function. The kaon has $I = \frac{1}{2}$, and the pair of pions has I integral and even (0 or 2). For the $\pi^+\pi^-$ state, we have $I_3 = 0$ and therefore $I = 0$ is allowed, whereas for the $\pi^+\pi^0$ state, $I_3 = 1$ and thus only $I = 2$ is possible. The $\Delta I = \frac{1}{2}$ rule thus allows (4.36a) and forbids (4.36b)—where one needs $\Delta I = \frac{3}{2}$ or $\frac{5}{2}$. The reason why $K^+ \to \pi^+ + \pi^0$ is allowed at all (if $\Delta I = \frac{1}{2}$ is valid), even if it is largely suppressed, is not clear at present.

4.8.2 Leptonic Decays: $\Delta Q = \Delta S$ Rule; Cabibbo Angle

For leptonic decays of strange particles, the isospin of the final state (containing leptons) cannot be specified, but empirically it appears that the rule $\Delta Q = \Delta S$ (i.e. the change in charge of hadron equal to the change in strangeness) is valid. From the relation (3.36),

$$Q = I_3 + \tfrac{1}{2}(B + S),$$

it follows that $\Delta I_3 = \frac{1}{2}$ if $\Delta Q = \Delta S = 1$.

As examples of the $\Delta Q = \Delta S$ rule, the decay

$$\Sigma^- \to n + e^- + \bar{\nu} \qquad (\Delta Q = \Delta S = +1)$$

is observed, but not

$$\Sigma^+ \to n + e^+ + \nu \qquad (\Delta S = +1, \Delta Q = -1).$$

Note that the process $\Delta Q = -\Delta S$ would allow $\Delta I_3 = \frac{3}{2}$ in the nonleptonic decays.

The Universal Fermi Interaction we mentioned earlier fails conspicuously when one considers the hyperon leptonic decay modes

$$\Lambda \to pe^-\bar{\nu}, \qquad \Sigma^- \to ne^-\bar{\nu}, \qquad \Xi^- \to \Lambda e^-\bar{\nu}.$$

The rates for these decays are roughly 20 times smaller than those expected if the couplings are the same as for the strangeness-conserving decays. Gell-Mann and Levy (1960) and Cabibbo (1963) proposed a way out of this difficulty. In the framework of unitary symmetry, dealt with in Chapter 6, the baryon states of spin-parity $\frac{1}{2}^+(n, p, \Sigma^+, \Sigma^-, \Sigma^0, \Lambda, \Xi^0, \Xi^-)$ form an octet. In the limit of exact symmetry, all these states would be degenerate, with identical masses and weak coupling. In nature the symmetry is broken, the eight states being split according to the charge, or I_3, and the strangeness S. Thus the physical states n and p, with $S = 0$; Σ^+, Σ^-, Σ^0, and Λ, with $S = -1$; and Ξ^- and Ξ^0, with $S = -2$, have different masses. In this splitting, there is no *a priori* way to determine how the weak coupling is divided. Cabibbo postulated that for

$$\Delta S = 0 \text{ decays, weak coupling} = G \cos \theta,$$

$$\Delta S = 1 \text{ decays, weak coupling} = G \sin \theta.$$

Thus $J_{\text{hadronic}} = J_{(\Delta S = 0)} \cos \theta + J_{(\Delta S = 1)} \sin \theta$, where θ is an arbitrary angle, to be determined by experiment. Note that universality now appears by assigning the same coupling, G, to the leptonic current and to the hadronic current for the octet as a whole.

Thus, the coupling of the strangeness-conserving $n - p$ current (as in nuclear β-decay) to the lepton current would give a matrix element proportional to $G \cos \theta$; while in the decay $\Sigma^- \to ne^-\bar{\nu}$, the strangeness-changing $\Sigma^- - n$ weak current gives a matrix element proportional to $G \sin \theta$. The Cabibbo angle θ can be found by comparing observed $\Delta S = 0$ and $\Delta S = 1$ decay rates in unitary multiplets. For the decays of kaons and pions in the 0^- meson octet, one obtains for the amplitude ratios, when phase-space factors are divided out:

$$\left(\frac{K^+ \to \mu^+ + \nu}{\pi^+ \to \mu^+ + \nu} \right)_A \to \tan \theta_A = 0.275 \pm 0.003, \tag{4.37}$$

$$\left(\frac{K^+ \to \pi^0 + e^+ + \nu}{\pi^+ \to \pi^0 + e^+ + \nu} \right)_V \to \tan \theta_V = 0.251 \pm 0.008. \tag{4.38}$$

As we saw above, the decays (4.37) are axial-vector transitions, and hence are labeled A as involving axial currents, whereas those in (4.38) involve only vector weak currents and are labeled V. The fact that θ_A and θ_V are approximately equal was also predicted by the Cabibbo theory.

There are two important consequences of these results. First, the $\Delta S = 1$ baryonic decays are suppressed relative to the $\Delta S = 0$ decays. For example, the rate for $\Sigma^- \to n + e^- + \bar{\nu}$ is reduced relative to $n \to p + e^- + \bar{\nu}$ by a factor $\tan^2 \theta$, and this clears up the discrepancy noted above. Secondly, the coupling constant for Fermi transitions in β-decay becomes $G \cos \theta$ rather than G. If we define the vector weak coupling in β-decay as GC_V, as before, one obtains from (4.33) a value for the Cabibbo angle

$$\tan \theta = 0.22 \pm 0.02$$

in reasonable agreement with (4.38). Thus, the Cabibbo theory appears to account for most of the discrepancy between the Fermi constant value obtained from μ-decay and from β-decay. It is obvious that the Cabibbo angle is a fundamental quantity, but why it has the particular value (4.38) is not at present clear.

4.9 CONSERVATION OF LEPTON NUMBER AND MUON NUMBER

According to the two-component neutrino hypothesis, neutrino and anti-neutrino are distinguished by having helicities of -1 and $+1$ respectively. If the mass of the neutrino is identically zero, this assignment is unique. (If the mass were finite, the neutrino velocity $v < c$, and it would be possible to transform into a frame with velocity exceeding v, in which the helicity would be reversed in sign.) Apart from the difference in helicity, the nonidentity of neutrino and antineutrino is also established from studies of the process of double β-decay. In favorable cases, it is possible for one nucleus to transform into another by the emission of two electrons and two antineutrinos, as a second order weak process:

$$Z \rightarrow (Z + 2) + 2e^- + 2\bar{\nu}. \tag{4.39}$$

However, if we regard neutrino and antineutrino as identical, we can imagine a two-step process:

$$Z \rightarrow (Z + 1) + e^- + \bar{\nu} \, (\equiv \nu),$$
$$\nu + (Z + 1) \rightarrow (Z + 2) + e^-,$$

or,

$$Z \rightarrow (Z + 2) + 2e^-, \tag{4.40}$$

in which the neutrino from the first step is absorbed in the second step. The result is neutrinoless double β-decay. Both (4.39) and (4.40) are second order processes. In (4.40), however, the disintegration energy is shared between two electrons only, and this process turns out to have a much shorter lifetime. Experiments indicate that (4.39) is the correct process, and thus that neutrino and antineutrino are not identical. For example, the observed lower limit to the lifetime for Ca^{48} double β-decay is 10^{19} yr, compared with theoretical estimates of $3 \times 10^{15 \pm 2}$ yr for the neutrinoless process (4.40), and $4 \times 10^{20 \pm 2}$ yr for (4.39). Another way of confirming this result was to show that the process

$$\bar{\nu} + Cl^{37} \rightarrow Ar^{37} + e^-$$

is not observed with antineutrinos from a reactor. If there had been no distinction between ν and $\bar{\nu}$, a measurable event rate would have been expected. As described in Section 1.3.2, these considerations are formalized by assigning the leptons e^-, μ^-, and ν a lepton number $L = +1$, and the antileptons e^+, μ^+, and $\bar{\nu}$ a lepton number $L = -1$, and postulating that the lepton number is absolutely conserved. Note that L is an additive quantum number.

The suspicion that there might be *two* types of neutrino (and antineutrino) associated with electrons and muons respectively, was aroused because certain decays like

$$\mu^+ \rightarrow e^+ + \gamma,$$
$$\mu^+ \rightarrow e^+ + e^- + e^+$$

had never been observed, although no known selection rule forbade them. They could be forbidden, however, by postulating a conserved *muon number* as well as a lepton number:

$$
\begin{array}{ccccc}
 & \mu^- & \nu_\mu & \mu^+ & \bar{\nu}_\mu \\
\text{Muon number} = & +1 & +1 & -1 & -1 \\
\end{array}
$$
$$= 0 \text{ for all other particles.} \qquad (4.41)$$

It is then necessary to distinguish between a neutrino ν_e accompanying an electron and one accompanying a muon, ν_μ. Examples are

$$\pi^+ \rightarrow \mu^+ + \nu_\mu,$$
$$n \rightarrow p + e^- + \bar{\nu}_e,$$
$$\mu^+ \rightarrow e^+ + \nu_e + \bar{\nu}_\mu.$$

In 1962/63 experiments at Brookhaven and CERN confirmed this hypothesis. A high energy (25 to 30 GeV) proton beam was used to produce pions from a target, which were focused by a magnetic "horn" and traversed a tunnel where a small fraction decayed. Neutrons and muons were removed by a thick (20-m) steel shield, leaving a beam of neutrinos (Fig. 4.22). The interactions of

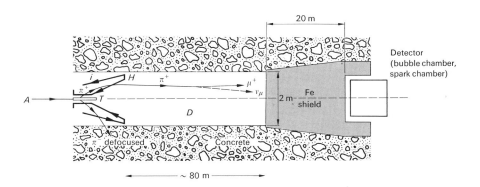

Fig. 4.22 Schematic diagram indicating principle of CERN high energy neutrino experiments (1963). A pencil proton beam *A* falls on a rod target *T*. Pion and kaon secondaries from the target are focused by means of a magnetic horn *H*. This consists of a conical sheet conductor through which flows a pulsed current of about 0.5 million A, thus producing an azimuthal magnetic field. Some 5% of the pions and kaons undergo decay in the tunnel *D*. Strongly interacting particles and muons are filtered out by a steel shield some 20-m thick. In later experiments, additional pulsed reflectors were included to enhance the neutrino flux. The decay tunnel length is about 80-m.

the type

$$v_\mu + n \rightarrow \mu^- + p \qquad (4.42)$$

were observed, but not

$$v_\mu + n \rightarrow e^- + p. \qquad (4.43)$$

A limit on the mass of v_μ can be obtained from the kinematics of $\pi \rightarrow \mu + v_\mu$ and is currently $m_{v_\mu} < 1$ MeV (much poorer than the limit $m_{v_e} < 60$ eV, from tritium β-decay). This duality of electron and muon, and of electron and muon neutrino, is quite one of the most baffling of nature's extravaganzas.

Examples of high energy neutrino reactions in spark chambers and bubble chambers are given in Figs. 4.24, 4.25, and 4.26.

4.10   NEUTRINO INTERACTIONS

4.10.1   Reactor Experiments

Reactions like (4.42) are phenomenally difficult to observe on account of the very small cross section. The first observation of the interactions of free neutrinos was made by Reines and Cowan (1959). They employed a nuclear reactor as a source. Because uranium fission fragments are neutron-rich, they undergo radioactive β-decay, emitting negative electrons and antineutrinos. The flux of antineutrinos attainable is up to 10^{13} cm^{-2} sec^{-1}. Per fission, one obtains of order six antineutrinos, with a broad spectrum centered around 1 MeV. The reaction

$$\bar{v}_e + p \rightarrow n + e^+ \qquad (4.44)$$

was observed, using a target of cadmium chloride ($CdCl_2$) and water. The positron produced in this reaction rapidly comes to rest by ionization loss, and forms positronium which annihilates to γ-rays, in turn producing fast electrons by Compton effect. The electrons are recorded in a liquid scintillation counter. The timescale for this process is of order 10^{-9} sec, so that the positron gives a so-called "prompt" pulse. The function of the cadmium is to capture the neutron after it has been moderated (i.e. reduced to thermal energy by successive elastic collisions with protons) in the water—a process which delays the γ-rays coming from eventual radiative capture of the neutron in cadmium, by several microseconds. Thus the signature of an event consists of two pulses microseconds apart. Figure 4.23 shows schematically the experimental arrangement. Two hundred litres of target are sandwiched in two layers between three tanks of liquid scintillator, which are viewed by banks of photomultipliers. Reines and Cowan observed a significant increase in comparing the coincidence rate with reactor on and off (the background in the experiment is attributed to cosmic rays). They also carried out various tests, such as diluting the water with heavy water, and proving that the rate varied as the normal water concentration (on account of the binding energy, the protons in the deuterium have a much smaller cross section).

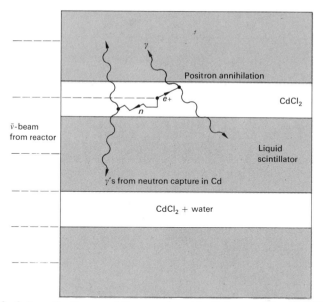

Fig. 4.23 Schematic diagram of the experiment by Reines and Cowan (1959), detecting the interactions of free antineutrinos from a reactor.

The expected magnitude of the cross section (4.44) is readily calculated. Using (3.16) for the transition rate, and integrating over solid angle, we have

$$W = \frac{2\pi}{\hbar} |M|^2 \frac{4\pi p^2 \, dp}{h^3 \, dE_0}.$$

From the relations

$$\sigma = \frac{W}{v_i}, \qquad \frac{dp}{dE_0} = \frac{1}{v_f},$$

it follows that

$$\sigma = \frac{1}{\pi \hbar^4} |M|^2 \frac{p^2}{v_i v_f},$$

where v_i and v_f are the relative velocities of the particles in the initial and final states, and $p_i = p_f = p$ is the numerical value of their momenta, all measured in the center-of-mass frame. In (4.44), the leptons are relativistic, while the nucleons have very low velocity. Thus $v_i \approx v_f \approx c$, and $pc \approx E$, the laboratory neutrino energy (the laboratory and center-of-mass reference frames are essentially one and the same). Setting $|M|^2 = (3C_A^2 + C_V^2)G^2 \approx 4G^2$, since we are dealing with a mixed transition, as in (4.8), and using natural units $\hbar = c = 1$, one obtains

$$\sigma = \frac{4}{\pi}(GM_p)^2\left(\frac{p}{M_p}\right)^2; \qquad G = 10^{-5}/M_p^2$$

$$= \frac{10^{-10}}{M_p^2}\left(\frac{E}{M_p}\right)^2. \tag{4.45}$$

Here, $1/M_p$ in the first term represents the proton Compton wavelength, $\lambda_p = 2 \times 10^{-14}$ cm. For $E = 1$ MeV, $(E/M_p) \approx 10^{-3}$, so that

$$\sigma \approx 10^{-44} \text{ cm}^2. \tag{4.46}$$

The cross section observed by Reines and Cowan was consistent with this estimate. This cross section corresponds to a mean free path for antineutrinos in water of about 10^{21} cm or 1000 light years. We note that (4.45) predicts a cross section rising with phase-space, as the square of the center-of-mass momentum. This is a consequence of the s-wave nature of an allowed β-transition.

4.10.2 High Energy Experiments: Intermediate Boson

The reaction (4.42) typically involves neutrinos of energy about 1 GeV, thus $pc \approx 1$ GeV and $\sigma \sim 10^{-38}$ cm^2. Above 1 GeV neutrino energy, the cross section for (4.42) flattens off because of form-factor effects, associated with the "size" of the nucleon. Weak and electromagnetic form-factors are discussed in Chapter 5.

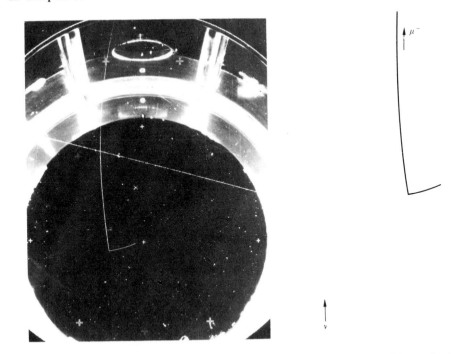

Fig. 4.24 Event attributed to an "elastic" neutrino interaction $v_\mu + n \rightarrow p + \mu^-$ in the CERN heavy liquid bubble chamber, filled with freon (CF$_3$Br). The negatively charged particle passes out of the chamber without interaction; it is attributed to a muon because, in many events of this type, the observed interaction length of the negative particles is very much larger than for strongly interacting particles. The original interaction takes place in a heavy nucleus, and one observes a proton plus a short nuclear fragment.

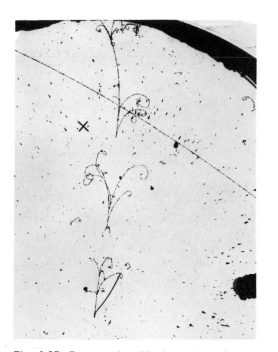

Fig. 4.25 Event produced by interaction of an electron-neutrino, ν_e: $\nu_e + n \rightarrow p + e^-$. The incident beam consists mostly of muon-neutrinos ν_μ, with a very small admixture of ν_e ($\sim\frac{1}{2}\%$) from the 3-body decays in flight, $K^+ \rightarrow \pi^0 + e^+ + \nu_e$. The high energy electron secondary is recognized by the characteristic shower it produces by the processes of bremsstrahlung and pair-production. The chamber diameter is 1.1 m, and the radiation length in CF_3Br is 0.11 m.

The relative numbers of events giving electron and muon secondaries is consistent with the calculated fluxes of ν_e and ν_μ in the beam, and thus confirms conservation of muon number. (Courtesy, CERN Information Service.)

A difficulty with the Fermi theory of the 4-fermion point interaction arises if we consider elastic scattering of neutrinos by electrons (not yet observed experimentally):

$$\nu_e + e^- \rightarrow e^- + \nu_e.$$

Here there are no form-factors, both particles being "point-like." Thus, even to the highest energies, the cross section has the form

$$\sigma_{\text{Fermi}} = \frac{4G^2}{\pi} p^2. \tag{4.47}$$

As before, p is the CMS momentum. A simple calculation gives $p^2 \simeq \frac{1}{2}m_e E$, so that the cross section rises linearly with the laboratory neutrino energy, E. On the other hand, the maximum value of the elastic scattering cross section,

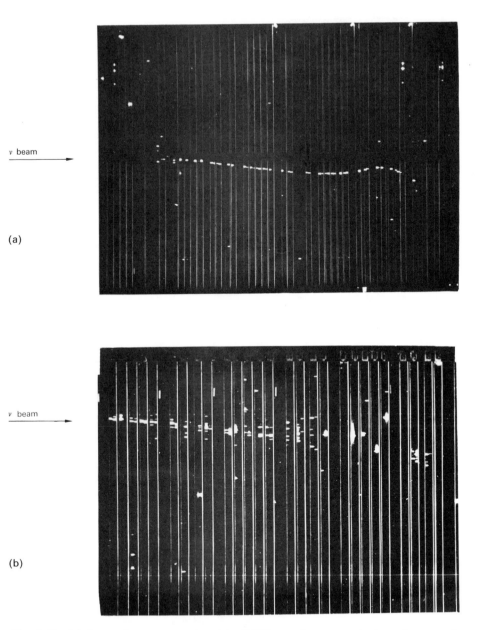

ν beam →

(a)

ν beam →

(b)

Fig. 4.26 (a) A neutrino event produced in CERN experiments employing a large spark chamber array, consisting of many plates of iron, brass, and aluminum. The muon track is characteristic, traversing many plates without interaction before coming to rest inside the chamber volume. (b) An event attributed to an electron neutrino, ν_e. The appearance of the scattered sparks due to the electron shower may be compared with the rectilinear muon track in event (a). (Courtesy, CERN Information Service.)

from ordinary wave theory, is (Section 7.2)

$$\sigma_{max} = \pi \lambda^2 \sum_{l} (2l + 1), \tag{4.48}$$

where λ is the de Broglie wavelength of the particles in the center-of-mass frame, and we sum over all partial waves of angular momentum l which contribute. For a point-like or s-wave interaction therefore, we obtain, in units $\hbar = c = 1$,

$$\sigma_{max} = \frac{\pi}{p^2} .$$

Thus

$$\sigma_{Fermi} > \sigma_{max}$$

when

$$p > G^{-1/2} \approx 300 M_p. \tag{4.49}$$

This means that, when the center-of-mass momentum exceeds about 300 GeV/c, the predicted cross section violates the wave theory limit, i.e. more particles are being scattered out into the final state than enter in the initial state. It is clear therefore that the Fermi theory is wrong in the limit of very high energy. One possible remedy is to "spread" the weak interaction from the point-like structure by postulating an intermediate spin 1 particle, the W-boson, coupling the (v_e, e) and (e, v_e) currents, as in Fig. 4.27(c). The effect of this would be to give $\sigma_{max} \sim G^2 M_W^2$ and would solve this particular problem, provided $M_W < 300$ GeV. It may be remarked that the problem described above is not by any means the only thing wrong with the theory currently employed to describe the weak interactions, when one tries to extrapolate it to the high energy domain, or to convert it into a field theory like quantum electrodynamics. Such matters are, however, beyond the scope of this book.

In conclusion, we may summarize as follows: The original Fermi theory of the point interaction of four fermions has been suitably modified to account for parity violation (by including both axial and vector currents of opposite

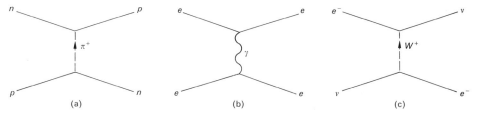

Fig. 4.27 Comparison of exchange mechanisms mediating the different types of interaction: (a) neutron-proton scattering mediated by pion exchange; (b) electron-electron (Møller) scattering via photon exchange; (c) neutrino-electron scattering via exchange of the proposed intermediate boson W of the weak interaction.

phase), and, by the Cabibbo prescription, to accommodate both strangeness-conserving and strangeness-violating processes. In this modified form it accounts very well indeed for a wide range of weak interaction processes.

4.11 K^0-DECAY

The strangeness scheme of Gell-Mann and Nishijima necessitates the introduction of two different neutral K-mesons, the K^0 with strangeness $S = +1$, and the $\overline{K^0}$ with strangeness $S = -1$. This follows from the assignment of isospin $\frac{1}{2}$ for the kaons (3.34) and the relation

$$\frac{Q}{e} = I_3 + \frac{S}{2} + \frac{B}{2}. \tag{4.50}$$

Thus the two kaon isospin doublets are

I_3	$+\frac{1}{2}$	$-\frac{1}{2}$
S		
$+1$	K^+	K^0
-1	$\overline{K^0}$	K^-

The K^0-particle can be produced in association with a hyperon or by charge exchange:

$$\pi^- + p \to \Lambda + K^0 \tag{4.51}$$
$$S \quad 0 \qquad 0 \quad -1 \quad +1,$$

$$K^+ + n \to K^0 + p \tag{4.52}$$
$$S \quad +1 \quad 0 \quad +1 \quad 0,$$

while, since no $S = +1$ baryons appear to exist, $\overline{K^0}$ can be produced only by charge exchange, or in pairs with K^0 or K^+:

$$K^- + p \to \overline{K^0} + n \tag{4.53}$$
$$S \quad -1 \quad 0 \quad -1 \quad 0,$$

$$\pi^+ + p \to K^+ + \overline{K^0} + p \tag{4.54}$$
$$S \quad 0 \qquad 0 \quad +1 \quad -1 \quad 0.$$

We note that, using a pion beam, the threshold energy for (4.51) is less than for (4.54), so that it is experimentally simple to produce a pure K^0-source. K^0 and $\overline{K^0}$ are particle and antiparticle, and can be transformed one into the other by the operation of charge conjugation, which reverses the value of I_3 and the strangeness, S. As far as production in strong interactions is concerned, K^0 and $\overline{K^0}$ are eigenstates of the strangeness operator, with eigenvalues $+1$ and -1.

Neutral kaons are observed experimentally by their decay into other particles. Being the lightest strange particles, they can only decay weakly, into nonstrange particles, violating conservation of strangeness. For example, both K^0 and $\overline{K^0}$ can decay into two (or three) pions. The observation of such decays tells us nothing directly about whether the initial state was K^0 or $\overline{K^0}$—one has to look at the production mechanism (4.51) to (4.54) to decide that. The measured lifetime of the $\pi^+\pi^-$ decay mode is of order 10^{-10} sec and precisely the same for K^0 and $\overline{K^0}$. Fermi allegedly complained therefore that K^0 and $\overline{K^0}$ were really indistinguishable—simply figments of the imagination to fit the magic formula (4.50). The solution to this problem was provided by Gell-Mann and Pais (1955), with spectacular predictions. We first note that, since K^0 and $\overline{K^0}$ can *both* decay into two pions (or three pions), they can transform into one another via virtual intermediate pion states:

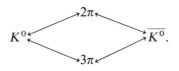

These virtual transitions involve $\Delta S = 2$, and thus are two-stage (or second order) weak interactions. Therefore, they are extremely weak, but nevertheless imply that, if one has a pure K^0-state at time $t = 0$, at any later time one will have a superposition of *both* K^0 and $\overline{K^0}$. This is a situation peculiar to neutral kaons. It does not happen for any other strange neutral particles, either because they decay by strong interactions ($\Delta S = 0$), if they are heavy enough, or because an absolute conservation rule forbids it. For example, the lambda and antilambda hyperon cannot "mix" because they have not only opposite strangeness, but opposite baryon number, and the latter is always conserved:

$$\Lambda \rightarrow \pi^0 + n \longleftrightarrow\!\!\!\!\!| \longrightarrow \pi^0 + \bar{n} \leftarrow \overline{\Lambda}$$

S	$+1$	0	0	-1
B	$+1$	$+1$	-1	$-1.$

The actual physical neutral kaon state observed at any finite distance from the source must therefore be written as

$$|K(t)\rangle = \alpha(t)\,|K^0\rangle + \beta(t)\,|\overline{K^0}\rangle.$$

The coefficients $\alpha(t)$ and $\beta(t)$ can be determined if we ask what are the eigenstates of the weak interaction responsible for the K^0-decay. In our discussion of weak interactions up to now, we have concluded that they are invariant under the *CP*-transformation, turning left-handed neutrinos into right-handed antineutrinos, and so on. Gell-Mann and Pais therefore argued that one should

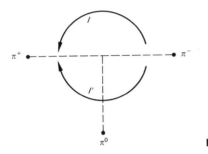

π^+ •━━━━━┳━━━━━• π^-

π^0

Figure 4.28

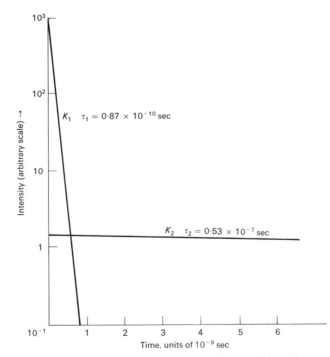

Fig. 4.29 Decay curves showing the components of a neutral kaon beam. The short-lived component K_1 decays to two pions, of CP-eigenvalue $+1$. The long-lived component K_2 has both nonleptonic and leptonic decay modes; the dominant pionic decay is to three pions, of CP-eigenvalue -1.

look for CP-eigenstates. If, in the K^0 rest frame, we consider CP acting on a K^0-state, this has the same effect as C alone, since the kaon has spin zero. Thus

$$CP|K^0\rangle \rightarrow |\overline{K^0}\rangle,$$
$$CP|\overline{K^0}\rangle \rightarrow |K^0\rangle. \qquad (4.55)$$

Clearly, $|K^0\rangle$ and $|\overline{K^0}\rangle$ are not CP-eigenstates. However, if we form the linear combinations

$$|K_1\rangle = \frac{1}{\sqrt{2}}(|K^0\rangle + |\overline{K^0}\rangle)$$

and

$$|K_2\rangle = \frac{1}{\sqrt{2}}(|K^0\rangle - |\overline{K^0}\rangle),$$

then

$$CP|K_1\rangle \rightarrow \frac{1}{\sqrt{2}}(|\overline{K^0}\rangle + |K^0\rangle) = |K_1\rangle,$$

$$\qquad (4.56)$$

$$CP|K_2\rangle \rightarrow \frac{1}{\sqrt{2}}(|\overline{K^0}\rangle - |K^0\rangle) = -|K_2\rangle.$$

Thus $|K_1\rangle$ and $|K_2\rangle$ so defined are CP-eigenstates, with eigenvalues $+1$ and -1 respectively. We note that the relative phases of $|K^0\rangle$ and $|\overline{K^0}\rangle$ in (4.55) are arbitrary. $|K^0\rangle$ and $|\overline{K^0}\rangle$ are related by the C-operation, but we may always introduce an arbitrary phase factor, i.e. $C|K^0\rangle = e^{i\lambda}|\overline{K^0}\rangle$. This phase is, in principle, unobservable in the strong interactions, which cannot connect two states of different strangeness. The factor $1/\sqrt{2}$ in (4.56) is for normalization of the wave functions.

It is now apparent that, although $|K^0\rangle$ and $|\overline{K^0}\rangle$ could not be differentiated by their decay modes, the combinations $|K_1\rangle$ and $|K_2\rangle$ can. First consider a $2\pi^0$-state. This can only exist for $J = l$ even, since the neutral pions are spinless and indistinguishable, and the angular momentum wave function is therefore invariant under a 180° rotation which interchanges the two particles. Thus, the $2\pi^0$-state is even under both C and P and has CP-eigenvalue $+1$. For a $\pi^+\pi^-$-state, we must appeal to Bose symmetry. Since the total wave function, which consists of a product of a function of space coordinates and one of charge coordinates, is then symmetric under particle interchange, the eigenvalues of the C (charge interchange) operation and P (space inversion) operation must have the same sign, and hence the CP-parity is $+1$ again.

For a 3-pion state $\pi^+\pi^-\pi^0$, let us denote the total angular momentum as $J = l + l'$ where l is the relative orbital angular momentum of the $\pi^+\pi^-$ combination, and l' is that of the π^0 with respect to the dipion (Fig. 4.28). Since the kaon has spin zero (Section 7.1.2) we must have $l = -l'$ and therefore $l = l'$. The simplest possibility is $l = l' = 0$. From the preceding argument, the CP-parity of $\pi^+\pi^-$ is even. The π^0 has even C-parity, and space parity $P = (-1)(-1)^{l'} = -1$. Thus, the 3π-state of $l = l' = 0$ has CP-parity -1. For $l = l' = 1$, the CP-eigenvalue will be $+1$, but it can be shown that these modes are strongly inhibited by angular momentum barrier effects.

To summarize, therefore, and considering only the simplest 3π-state, the pion decay modes of the objects K_1 and K_2 will be:

$$K_1 \rightarrow 2\pi, \qquad CP = +1, \qquad \text{high } Q\text{-value, therefore short lifetime}$$
$$(\tau_1 = 0.87 \times 10^{-10} \text{ sec}),$$

$$K_2 \rightarrow 3\pi, \qquad CP = -1, \qquad \text{low } Q\text{-value, therefore long lifetime}$$
$$(\tau_2 = 0.53 \times 10^{-7} \text{ sec}).$$

The difference in lifetime is to be expected from the fact that the Q-value in the 2-pion decay (and hence the phase-space factor) is much greater than in the 3-pion decay (Fig. 4.29). The quoted lifetime τ_1 is the present value; actually it was known to be $\sim 10^{-10}$ sec from the early experiments, so that the Gell-Mann and Pais prediction was of a long-lived state K_2 of $\tau_2 \sim 10^{-7}$ sec. This was subsequently found in experiments at Brookhaven in 1956. Previously it had escaped discovery because K_2-decays would generally be a long way from the K^0-source; for example, a 1 GeV/c K_2-meson has a mean decay path of 30 m.

4.11.1 The K^0 Regeneration Phenomenon

The K_2 decay mode was not observed until some time after its prediction by Gell-Mann and Pais. However, Pais and Piccioni (1955) had observed that the $K_1 - K_2$ phenomenon would lead to the process of *regeneration*. Suppose we start with a pure K^0-beam, and let it coast *in vacuo* for the order of 100 K_1 mean lives, so that all the K_1-component has decayed and we are left with K_2 only. Now let the K_2-beam traverse a slab of material and interact. Immediately, the strong interactions will pick out the strangeness $+1$ and -1 components of the beam, i.e.

$$|K_2\rangle = \frac{1}{\sqrt{2}}(|K^0\rangle - |\overline{K^0}\rangle). \tag{4.57}$$

Thus, of the original K^0-beam intensity, 50 % has disappeared by K_1-decay. The remainder, called K_2, upon traversing a slab where its nuclear interactions can be observed, should consist of 50% K^0 and 50% $\overline{K^0}$. The existence of $\overline{K^0}$ ($S = -1$) a long way from the source of an originally pure K^0-beam ($S = +1$) was confirmed by the observation of production of hyperons by Fry *et al.* in 1956. Thus $\overline{K^0} + p \rightarrow \Lambda + \pi^+$, for example.

The K^0- and $\overline{K^0}$-components in (4.57) must be absorbed differently; K^0-particles can only undergo elastic and charge exchange scattering, while $\overline{K^0}$-particles can also be absorbed in nuclear collisions, giving hyperons.

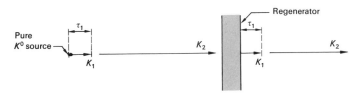

Fig. 4.30 Regeneration of short-lived K_1-mesons when a pure K_2-beam traverses a regenerator.

With more strong channels open, the $\overline{K^0}$ is therefore absorbed more strongly than K^0. After emerging from the slab, we shall therefore have a K^0-amplitude $f|K^0\rangle$ and a $\overline{K^0}$-amplitude $\bar{f}|\overline{K^0}\rangle$ say, where $\bar{f} < f < 1$. If we then ask what are the characteristics of the emergent beam with respect to decay, we must write, in place of (4.57):

$$\frac{1}{\sqrt{2}}(f|K^0\rangle - \bar{f}|\overline{K^0}\rangle) = \frac{(f + \bar{f})}{2\sqrt{2}}(|K^0\rangle - |\overline{K^0}\rangle) + \frac{(f - \bar{f})}{2\sqrt{2}}(|K^0\rangle + |\overline{K^0}\rangle)$$

$$= \tfrac{1}{2}(f + \bar{f})|K_2\rangle + \tfrac{1}{2}(f - \bar{f})|K_1\rangle. \qquad (4.58)$$

Since $f \neq \bar{f}$, it follows that some of the K_1-state has been regenerated (Fig. 4.30). This regeneration of short-lived K_1's in a long-lived K_2-beam was confirmed by experiment.

The regeneration phenomenon in K^0-decay, at first startling, is simply a consequence of the concepts of superposition and quantization in wave mechanics. The behavior of a proton or atomic beam in an inhomogeneous magnetic field provides a fairly close analogy. Referring to Fig. 4.31, suppose we have an initially unpolarized proton beam moving along the z-axis, which then traverses an inhomogeneous field H_y directed along the y-axis (representing the strong interaction in the K^0 case). The protons are then quantized in two spin eigenstates, $\sigma_y = +\tfrac{1}{2}$ and $\sigma_y = -\tfrac{1}{2}$, particles in the first state being deflected upward and those in the second downward. (These are analogous to the K^0 and $\overline{K^0}$ S-eigenstates.) We select the part of the beam deflected upward, and pass it through an inhomogeneous field along the x-axis, H_x (the analog of

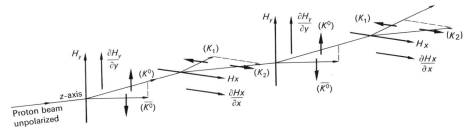

Fig. 4.31 Analogy between deflection of proton beam in two orthogonal planes, by means of inhomogeneous magnetic fields, and the K^0-decay and regeneration phenomena.

the weak interactions). The beam again splits into two components, 50% having $\sigma_x = -\frac{1}{2}$ and being deflected to the left, and 50% having $\sigma_x = +\frac{1}{2}$, deflected to the right. Note that *all* information about quantization along the y-axis has necessarily been lost. In turn, if we take the $\sigma_x = \frac{1}{2}$ component (corresponding to K_2), we may again pass it through a field along the y-axis, and recover the components $\sigma_y = \pm\frac{1}{2}$, and lose all knowledge of σ_x (analogous to regeneration of K^0 and $\overline{K^0}$ in an absorber). Finally, passage through a field H_x yields as eigenvalues $\sigma_x = \pm\frac{1}{2}$ (corresponding to reappearance of K_1). To make the analogy more exact, i.e. account for the different decay and nuclear absorption probabilities in the K^0 problem, one could arrange to absorb components of the proton beam deflected in two of the four directions.

The important feature of the Stern-Gerlach proton beam experiment is that it is impossible simultaneously to quantize the spin components of the beam along two orthogonal axes. This may also be stated by saying that the spin operators σ_x and σ_y do not commute; or, of the three matrix operators $\sigma_x, \sigma_y,$ and σ_z, only one may be diagonalized at a time (i.e. possess only diagonal elements, and have real, nonzero eigenvalues). Since an operator with real eigenvalues must commute with the Hamiltonian (energy) operator, the axis singled out in space is necessarily that defined by the magnetic field. In a similar fashion, the CP- and S-operators do not commute, so that one can have states which are eigenstates of CP or S, but not both.

4.11.2 Strangeness Oscillations

So far, we have combined K_1- and K_2-amplitudes without regard to their precise time dependence, in order to bring out the main features of the regeneration phenomena in the simplest way. To be more specific we should include phase factors with the amplitudes. The relative phase of K_1- and K_2-states of a given momentum will only be constant with time if these particles have identical masses. K_1 and K_2 are not charge conjugate states, having quite different decay modes and lifetimes, so that, in the same sense that the mass difference between neutron and proton can be attributed to differences in their electromagnetic coupling, a $K_1 - K_2$ mass difference—but a very much smaller one—is to be expected because of their different weak couplings.

The amplitude of the state K_1 at time t can be written as

$$a_1(t) = a_1(0)e^{-(iE_1/\hbar)t} \times e^{-\Gamma_1 t/2\hbar}, \tag{4.59}$$

where E_1 is the total energy of the particle, so that $E_1/\hbar$ is the circular frequency, ω_1, and $\Gamma_1 = \hbar/\tau_1$ is the width of the state, τ_1 being the mean lifetime in the frame in which the energy E_1 is defined. The second term must have the form shown, in accord with the law of radioactive decay of the intensity:

$$I(t) = a_1(t) \times a_1^*(t) = a_1(0)a_1^*(0)e^{-\Gamma_1 t/\hbar}$$
$$= I(0)e^{-t/\tau_1}.$$

Setting $\hbar = c = 1$ and measuring all times in the rest frame, so that τ_1 is the proper lifetime and $E_1 = m_1$, the particle rest mass, the K_1-amplitude is

$$a_1(t) = a_1(0)e^{-(\Gamma_1/2 + im_1)t}. \qquad (4.60a)$$

Similarly, for K_2,

$$a_2(t) = a_2(0)e^{-(\Gamma_2/2 + im_2)t}. \qquad (4.60b)$$

Now suppose that at $t = 0$, a beam of unit intensity consists of pure K^0. Then $a_1(0) = a_2(0) = 1/\sqrt{2}$. After a time t for free decay *in vacuo*, the K^0-intensity will be

$$I(K^0) = \frac{[a_1(t) + a_2(t)]}{\sqrt{2}} \frac{[a_1^*(t) + a_2^*(t)]}{\sqrt{2}}$$

$$= \tfrac{1}{4}[e^{-\Gamma_1 t} + e^{-\Gamma_2 t} + 2e^{-[(\Gamma_1 + \Gamma_2)/2]t} \cos \Delta mt], \qquad (4.61)$$

where $\Delta m = |m_2 - m_1|$. Similarly, the $\overline{K^0}$-intensity will be, writing the amplitude as $[a_1(t) - a_2(t)]/\sqrt{2}$,

$$I(\overline{K^0}) = \tfrac{1}{4}[e^{-\Gamma_1 t} + e^{-\Gamma_2 t} - 2e^{-[(\Gamma_1 + \Gamma_2)/2]t} \cos \Delta mt]. \qquad (4.62)$$

Thus, the K^0- and $\overline{K^0}$-intensities *oscillate* with the frequency Δm. Figure 4.32 shows the variation to be expected for $\Delta m = 0.5/\tau_1$. If one measures the

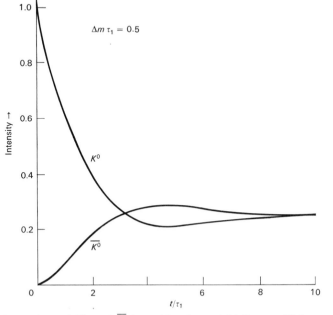

Fig. 4.32 Oscillations of K^0- and $\overline{K^0}$-intensities, for an initially pure K^0-beam, as calculated from Eqs. (4.61) and (4.62). A value $\Delta m \times \tau_1 = 0.5$ has been assumed.

number of $\overline{K^0}$ interaction events (i.e. the hyperon yield) as a function of position from the K^0-source, one can therefore deduce $|\Delta m|$, but not the sign. The first measurements (1962) of $|\Delta m|$ involved this method, and gave a value of order $1.5/\tau_1$, now known to be much too high. The currently accepted value is

$$\Delta m \times \tau_1 = 0.47 \pm 0.01, \tag{4.63}$$

where $\Delta m = m_2 - m_1$, and, as indicated in the next section, $m_2 > m_1$.

Using the value $\tau_1 = 0.86 \times 10^{-10}$ sec, (4.63) yields $(\Delta m)c^2 = 4 \times 10^{-6}$ eV, or a fractional $K_1 - K_2$ mass difference of

$$\frac{\Delta m}{m} = 10^{-14}. \tag{4.64}$$

The magnitude of the self-energy (i.e. contribution to the mass) of K_1 and K_2 can be estimated crudely by inspection of Fig. 4.33. (Incidentally, it is

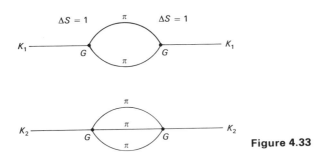

Figure 4.33

not known how to do a sophisticated calculation.) In the above diagrams the "weak" mass generated is clearly proportional to G^2 (all higher order terms being negligible in comparison), and thus to an effect of the second order in the weak coupling constant—the only one so far observed experimentally. Since $G = 10^{-5}/M_p^2$, we must introduce a mass in order to get the dimensions right. It is reasonable to choose this as the kaon mass. Since we have two strangeness-changing transitions, we should in fact replace G by $G \sin \theta_c$, θ_c being the Cabibbo angle. Hence we may estimate

$$\Delta m_K \text{ (for } \Delta S = 1 \text{ coupling)} \approx G^2 \times \sin^2 \theta_c \times m_K^5 = 10^{-13} m_K = 10^{-4} \text{ eV}, \tag{4.65}$$

within an order of magnitude of the observed value (4.64). By a similar argument, one can set very stringent limits on the strength of a possible direct $\Delta S = 2$ transition between K^0- and $\overline{K^0}$-states. If fG denotes the coupling strength, then

$$\Delta m_K \text{(for } \Delta S = 2 \text{ coupling)} \approx fGm_K^3 = 10^3 f \text{ eV}, \tag{4.66}$$

giving $f \leqslant 10^{-8}$ from the observed value of Δm. Such an extremely weak coupling (called *superweak*) has been postulated in connection with *CP*-violation (Section 4.12).

4.11.3 Transmission Regeneration: the Sign of Δm

A measurement of the mass difference Δm can also be made by observations on the regenerated K_1-intensity as a function of regenerator thickness. The regenerated amplitude originates from the different K^0 and $\overline{K^0}$ nuclear scattering amplitudes, and three types of scattering can be considered; scattering from individual nucleons, scattering from an entire nucleus, and finally, coherent scattering from all nuclei in the regenerator slab, within the beam width. The latter is sometimes called "transmission regeneration," since it means that the K_2-beam is attenuated in passing through matter and gives rise to a *parallel* beam of K_1. Just as in optical diffraction, the transmission-regenerated K_1-beam will have an extremely small angular spread, $\Delta\theta \sim \lambda/d$, where λ is the kaon wavelength and d the slab diameter (i.e. the "slit width"). For example, for a 1 GeV/c kaon beam, $\lambda = 4 \times 10^{-14}$ cm, so that for $d = 20$ cm (as commonly used), $\Delta\theta = 10^{-15}$. In contrast, the regeneration angular distribution from the incoherent processes of individual nuclear scattering will have $\Delta\theta \sim \lambda/R$, where R is the nuclear radius, and would be of order 10^{-1} rad. Figure 4.34 shows how these different regeneration amplitudes can be resolved.

The principle of one experiment to measure the mass difference Δm is illustrated in Fig. 4.35. A K^+-beam of momentum 0.99 GeV/c impinges on a charge exchange target of copper, generating a beam of K^0-mesons. After traveling some distance, in which the bulk of the K_1-component decays, the predominantly K_2-beam enters a regenerator of iron. The parallel beam of K_1 at the rear of the regenerator then consists of a superposition of two amplitudes; ψ_S due to K_1 directly from the target; and ψ_R, the transmission-regenerated K_1-component from the iron slab. The K_1-intensity, I, is measured by observing decays $K_1 \to \pi^+\pi^-$ in spark chambers. $I = |\psi_S + \psi_R|^2$, depending on both magnitudes and relative phases of ψ_S and ψ_R. The phase of ψ_S depends on the distance from source to detector, and the wavenumber $k_1 = p_1$ of the direct K_1-component. The phase of ψ_R depends on the wavenumbers of both the K_2 and regenerated K_1, denoted by p_2 and $p_{1'}$. These are not equal owing to the small mass difference; in fact, $\Delta p = p_1 - p_2 = m \times \Delta m/p$, where p and m are average values. The phase of ψ_R also depends on the inherent phase of the regenerated K_1-wave, relative to the incident K_2-wave, i.e. the phase of the quantity $f_{12} = (f - \bar{f})/2$ in (4.58). f_{12} can be found from data on $K^\pm p$ and $K^\pm n$ cross sections, and an optical model of the nucleus. Then the only unknown is Δm, which is measured by observing the change in the ψ_S/ψ_R interference as a function of distance L between source and regenerator. This is the principle of the method; the reader is referred to the original paper of Mehlhop *et al.* (1968) for the full algebraic detail.

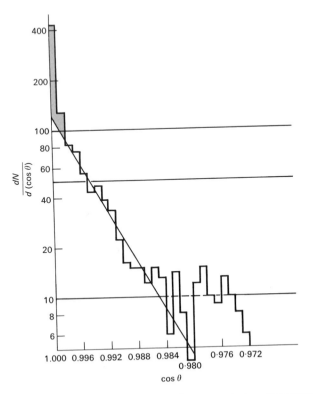

Fig. 4.34 Experimental results showing the transmission-regenerated peak (shaded) superimposed on the broader distribution due to nuclear (incoherent) regeneration (from experiment of Fig. 4.35).

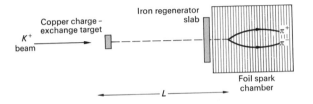

Fig. 4.35 Principle of experiment to determine the magnitude and sign of the $K_1 - K_2$ mass difference, by observing interference of the K_1-amplitude direct from a K^0-source, with that from regeneration by K_2 in an iron plate. (After Mehlhop *et al.*, 1968.)

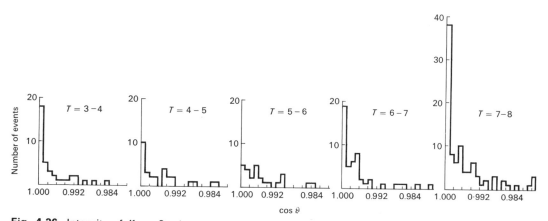

Fig. 4.36 Intensity of $K_1 \to 2\pi$ decay events plotted against cos θ, where θ is the angle between the K_1 momentum vector and the direction of the incident K_0-beam. As in Fig. 4.34, the forward peak (cos θ > 0.998) due to transmission regeneration is clearly visible. Note how the intensity of this peak varies with source-regenerator distance $T = L/\Lambda$ measured in units of K_1 mean lifetime, due to interference between direct and regenerated K_1-amplitudes.

Figure 4.36 shows how the forward intensity of 2π-decays fluctuates with L, expressed in terms of the proper time of flight $T = L/\Lambda$, where Λ is the mean decay length for K_1 of the beam momentum (4 cm in this case). The analysis of this data gave $\Delta m \times \tau_1 = 0.42 \pm 0.04$, and confirmed that $m_2 > m_1$ as in (4.63).

4.12 *CP*-VIOLATION IN K^0-DECAY

In 1964, the picture drawn above, of K^0 decay processes occurring through pure *CP*-eigenstates, K_1 and K_2, was changed by the observations by Christenson *et al.* and Abashian *et al.* that the long-lived K^0 could also decay to $\pi^+\pi^-$ with a branching ratio of order 10^{-3}. The experimental arrangement employed and results obtained are shown in Fig. 4.37. The nomenclature K_1 (for a state of $CP = +1$) and K_2 ($CP = -1$) has therefore been superseded by K_S (short-lived component) and K_L (long-lived component). The arguments quoted above on regeneration and mass difference, which follow from the superposition principle, remain essentially unchanged, although the formulae (4.56) need to be modified slightly if they are to describe K_S and K_L. The measure of the degree of *CP*-violation is usually quoted as the amplitude ratio

$$|\eta_{+-}| = \frac{\text{Ampl.}\ K_L \to \pi^+\pi^-}{\text{Ampl.}\ K_S \to \pi^+\pi^-} = (1.90 \pm 0.05) \times 10^{-3}. \tag{4.67}$$

The state K_S still consists principally of a $CP = +1$ amplitude, but with a little $CP = -1$, and K_L conversely. The conclusion, that in K^0-decay invariance under the operation CP was really violated, when it was conserved

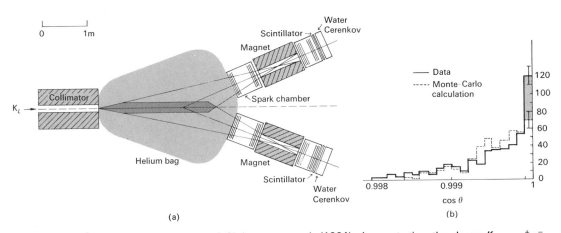

Fig. 4.37 (a) Experimental arrangement of Christenson *et al.* (1964) demonstrating the decay $K_L \rightarrow \pi^+ \pi^-$. The K^0-beam is incident from the left, and consists of K_L only, the K_S-component having died out. K_L-decays are observed in a helium bag, the charged products being analyzed by two spectrometers, consisting of bending magnets and spark chambers triggered by scintillators. The rare 2-pion decays are distinguished from the common 3-pion and leptonic decays on the basis of the invariant mass of the pair, and the direction, θ, of their resultant momentum vector relative to the incident beam. (b) Cos θ distribution of events of $490 < M_{\pi\pi} < 510$ Mev. The distribution is that expected for 3-body decays (dotted line), but with some fifty events (shaded) exactly collinear with the beam due to the $\pi^+ \pi^-$ decay mode.

in all other known interactions, was not finally accepted before exhausting all other possible explanations. Some of these were:

i)   The effect is due to a new long-range galactic field, in which the potential energy of K^0 and $\overline{K^0}$ would be slightly different. This has the effect of making K_L and K_S a mixture of CP-eigenstates in a region where matter preponderates over antimatter. The rate for the decay $K_L \rightarrow 2\pi$ then turns out to be proportional to γ^{2J}, where $\gamma = (1 - v^2/c^2)^{-1/2}$ is the Lorentz factor of the kaon relative to the field (essentially the laboratory system), and J is the spin of the field quantum ($J = 1$ for a vector field). Experiments at different K_L-momenta rule this explanation out completely—there is no observed γ-dependence (Fig. 4.38).

ii)   The argument $CP = +1$ for a $\pi^+ \pi^-$ state depends on the assumption of Bose symmetry for the pions. Perhaps pions do not always obey Bose statistics. This argument was invalidated when the decay $K_L \rightarrow 2\pi^0$ was also observed, since $CP = +1$ for two identical, spinless particles regardless of statistics.

iii)   The CP-eigenstate K_2 decays into K_1 plus an undetected particle S, lighter than the $K_L - K_S$ mass difference, and with $CP = -1$, i.e. $K_2 \rightarrow S + K_1 \rightarrow S + \pi^+ + \pi^-$. In such a situation, there could be no interference between the final state $\pi^+ \pi^- S$ of K_2-decay, and $\pi^+ \pi^-$ from decay of K_1 obtained by regeneration in a plate. As Fig. 4.39 indicates, interference between the K_L and K_S 2π-amplitudes is clearly demonstrated.

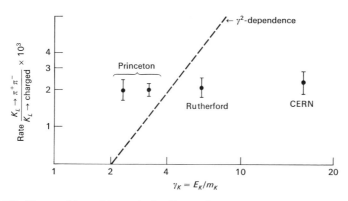

Fig. 4.38 Observed branching ratio for $K_L \to \pi^+\pi^-$ plotted as a function of K^0-momentum. Note the absence of any detectable momentum dependence.

Even more radical proposals, such as throwing doubt on the fundamental principle of superposition in wave mechanics, were advertised, but eventually one was left with the certain fact that *CP*-invariance really is violated in $K_L \to 2\pi$ decays. Subsequently, it was observed that the *leptonic* decay modes of K_L provided evidence for *CP*-noninvariance. These modes are

$$K_L \to e^+ \nu \pi^- \quad \text{or} \quad \mu^+ \nu \pi^-,$$

and

$$K_L \to e^- \bar{\nu} \pi^+ \quad \text{or} \quad \mu^- \bar{\nu} \pi^+, \tag{4.68}$$

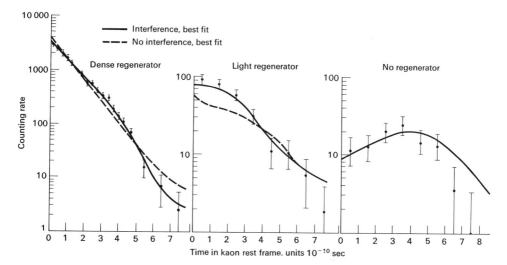

Fig. 4.39 Interference observed between the 2π final state from the *CP*-violating K_L-decay, and that from decay of K_S, the latter being produced by regeneration of the short-lived component in a plate. (After Alff-Steinberger *et al.*, 1966.)

where v stands for v_e or v_μ, whichever is appropriate. The state $ev\pi$ is *not* a CP-eigenstate, in fact it is better described in terms of the K^0 and $\overline{K^0}$ strangeness eigenstates. By the $\Delta S/\Delta Q = +1$ rule (Section 4.8.2), the allowed decays are $K^0 \rightarrow e^+ v\pi^-$ and $\overline{K^0} \rightarrow e^- \bar{v}\pi^+$. Nevertheless, the two decays (4.68) transform into one another under the CP-operation, and if CP-invariance is violated, one should obtain a small charge asymmetry. This has been detected, being

$$\delta = \frac{\text{Rate}\,(K_L \rightarrow e^+ v\pi^-) - \text{Rate}\,(K_L \rightarrow e^- \bar{v}\pi^+)}{\text{Rate}\,(K_L \rightarrow e^+ v\pi^-) + \text{Rate}\,(K_L \rightarrow e^- \bar{v}\pi^+)} = 2 \times 10^{-3}.$$

The source of the CP-violation is at present unknown. An important experiment in this connection is the precise determination of the ratio

$$\frac{K_L \rightarrow 2\pi^0}{K_L \rightarrow \pi^+\pi^-}. \tag{4.69}$$

For the short-lived component, the observed ratio

$$\frac{K_S \rightarrow 2\pi^0}{K_S \rightarrow \pi^+\pi^-} = 0.45 \pm 0.02. \tag{4.70}$$

This is nearly consistent with the value, $\frac{1}{2}$, expected from the $\Delta I = \frac{1}{2}$ rule. We recall that the two-pion state can have $I = 0$ or 2, and the $\Delta I = \frac{1}{2}$ rule allows $I = 0$ only. The Clebsch-Gordan coefficients for the addition of two isospin 1 particles then yield the above branching ratio. One method of accounting for the $K^+ \rightarrow \pi^+\pi^0$ decay ($I_{2\pi} = 2$) is to postulate an admixture of $\Delta I = \frac{3}{2}$ (or $\frac{5}{2}$) transitions (Section 4.8.1). For an $I = 2$ final state in $K^0 \rightarrow 2\pi$ decay, the expected value of the branching ratio (4.70) is 2, instead of $\frac{1}{2}$. Early measurements of the ratio (4.69) for the long-lived CP-violating decay suggested a value much greater than $\frac{1}{2}$. It was then plausible that the CP-violating amplitude was associated specifically with $\Delta I = \frac{3}{2}$ transitions. Later experiments have yielded lower values of the ratio; the average of these gives

$$|\eta_{00}| = \frac{\text{Ampl.}\ K_L \rightarrow 2\pi^0}{\text{Ampl.}\ K_S \rightarrow 2\pi^0} = (2.1 \pm 0.2) \times 10^{-3},$$

approximately equal to η_{+-} in (4.67). An example of a $K_L \rightarrow 2\pi^0$ decay event is shown in Fig. 4.40. It may be remarked that one theory predicts the ratios (4.69) and (4.70) to be identical. This is the *superweak theory* proposed by Wolfenstein (1964), in which the CP-violating amplitude arises exclusively from a new form of interaction obeying the selection rule $\Delta Q = -\Delta S$ and allowing $\Delta S = 2$. Comparing (4.65), (4.66) and inserting the observed CP-violating parameter η_{+-} (4.67), we see that the superweak coupling is of order 10^{-8} of the conventional weak coupling. Thus, there is probably no hope that it could be observed in any other physical process outside the extremely precise "interferometer" provided by the $K_L - K_S$ system.

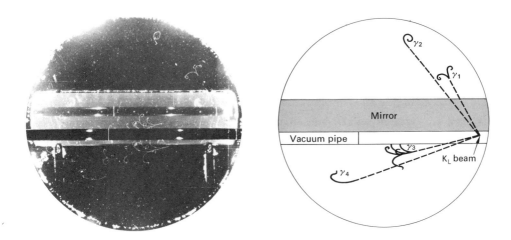

Fig. 4.40 Probable example of the decay $K_L \rightarrow 2\pi^0 \rightarrow 4\gamma$ obtained in an experiment to measure η_{00} in the CERN heavy liquid bubble chamber. The K_L-beam traverses a vacuum pipe through the chamber. γ-rays from the $2\pi^0$-mode are converted to e^+e^- pairs in the surrounding liquid (CF_3Br). Pairs occurring behind the pipe are rendered visible by a mirror at the rear of the chamber. (Courtesy, CERN Information Service.)

A general remark may be made in conclusion. Until the advent of *CP*-violation, there was no unambiguous way of defining left-handed and right-handed coordinate systems, or of differentiating matter from antimatter, on a cosmic scale. Thus, aligned Co^{60} nuclei emit negative electrons with forward-backward asymmetry and left-handed polarization (as we define left-handed). This is insufficient information to describe what we mean by a left-handed system, with the aid of light signals to an intelligent being in a distant part of the universe, unless one can also uniquely define negative and positive charge, or equivalently Co^{60} and anti-Co^{60}. *CP*-violation, however, provides an unambiguous definition. Positive charge is now defined as that of the lepton associated with the more abundant leptonic decay mode of the long-lived K_L-particle, $(K_L \rightarrow \pi^- e^+ v / K_L \rightarrow \pi^+ e^- \bar{v}) > 1$. This lepton has the same (or opposite) charge as the local atomic nuclei of matter (or antimatter).

BIBLIOGRAPHY

Feinberg, G., and L. Lederman, "Physics of muons and muon neutrinos," *Ann. Rev. Nucl. Science* **13,** 431 (1963).

Gasiorowicz, S., *Elementary Particle Physics*, John Wiley, New York, 1966.

Konopinski, E. J., "Experimental clarification of the laws of β-radioactivity," *Ann. Rev. Nucl. Science* **9,** 99 (1959).

Lederman, L., "Neutrino physics" in Burhop (ed.), *Pure and Applied Physics*, Academic Press, New York, 1967, Vol. 25-II.

Lee, T. D., and C. S. Wu, "Weak interactions," *Ann. Rev. Nucl. Science* **15**, 381 (1965).

Okun, L. B., *Weak Interactions of Elementary Particles*, Pergamon, London, 1965.

Prentki, J., "*CP* violation," and J. S. Bell and J. Steinberger, "Weak interactions of kaons," Proc. Oxford Conf. Elementary Particles 1965 (Rutherford High Energy Laboratory).

Reines, F., "Neutrino interactions," *Ann. Rev. Nucl. Science* **10**, 1 (1960).

Rubbia, C., "Weak interaction physics" in Burhop (ed.), *Pure and Applied Physics*, Academic Press, New York, 1968, Vol. 25-III.

PROBLEMS

4.1 Show that, if the neutrino in β-decay has mass m, the electron spectrum (4.2) has the form

$$N(p)\,dp = p^2(E_0 - E)^2 \sqrt{1 - \left(\frac{mc^2}{E_0 - E}\right)^2}\,dp.$$

4.2 O^{14} has a half-life of 71 sec and an end-point energy $E_0 = 1.8$ MeV. The pion normally decays in the mode $\pi \to \mu + \nu$ with a mean life of 2.6×10^{-8} sec. Find the branching ratio for the decay

$$\pi^+ \to \pi^0 + e^+ + \nu \qquad (J^P = 0^- \to 0^-),$$

given that

$$m_{\pi^+} = 140 \text{ MeV}, \qquad m_{\pi^0} = 135 \text{ MeV}, \qquad m_e = 0.5 \text{ MeV}.$$

Only a *rough* estimate is required, so crude approximations may be used. [The observed ratio $(\pi^+ \to \pi^0 e^+ \nu / \pi^+ \to \mu^+ \nu) = 1.02 \times 10^{-8}$; you should get within a factor of 3 of this number.]

4.3 The neutron has a mean lifetime $\tau_n = 930$ sec, and the muon $\tau_\mu = 2.2 \times 10^{-6}$ sec. Show that the couplings involved in the two cases are of the same order of magnitude, when account is taken of the phase-space factors.

4.4 Obtain an estimate of the branching ratio $(\Sigma^- \to \Lambda e^- \bar{\nu})/(\Sigma^- \to n\pi^-)$, assuming that the matrix element for the electron decay is the same as that of the neutron, and that baryon recoil may be neglected. (See the Table in the preface for neutron and Σ-decay data.)

4.5 Strangeness-changing decays do not conserve isospin but appear to obey the $\Delta I = \frac{1}{2}$ rule. Use this rule to compute the ratios

i) $K_s \to \pi^+\pi^- / K_s \to \pi^0\pi^0$,

ii) $\Xi^- \to \Lambda\pi^- / \Xi^0 \to \Lambda\pi^0$.

Compare your answers with the experimental values in the Table in the preface. [*Hint:* the $\Delta I = \frac{1}{2}$ rule may be applied by postulating that a hypothetical particle of $I = \frac{1}{2}$—called a "spurion"—is added to the left-hand side in a weak decay process, and then treating the decay as an isospin-conserving reaction.]

4.6 What is the expected ratio $K_L \to 2\pi^0 / K_L \to \pi^+\pi^-$ if the pions are in (i) an $I = 0$ state ($\Delta I = \frac{1}{2}$ rule), or (ii) an $I = 2$ state ($\Delta I = \frac{3}{2}$ or $\frac{5}{2}$)?

4.7 Using the $\Delta I = \frac{1}{2}$ rule, find a relation between the amplitudes for

$$\Sigma^+ \to n\pi^+, \qquad \text{say } a_+,$$

$$\Sigma^- \to n\pi^-, \qquad \text{say } a_-,$$

$$\Sigma^+ \to p\pi^0, \qquad \text{say } a_0.$$

You will need to introduce $I = \frac{3}{2}$ and $I = \frac{1}{2}$ amplitudes. The triangle relation you should obtain is

$$a_+ + \sqrt{2}\,a_0 = a_-.$$

Experimentally, $|a_+| \approx |a_0| \approx |a_-|$, so that a right-angled triangle results.

4.8 Given that the width of the $N^*(1236)$ π-p resonance is 150 MeV, estimate the branching ratio for β-decay

$$\frac{N^{*++} \to p + e^+ + \nu}{N^{*++} \to p + \pi^+}.$$

4.9 How can one produce experimentally a beam of pure, monoenergetic K^0-particles? If a short pulse of such particles travels through a vacuum, compute the intensity of K^0 and $\overline{K^0}$ as a function of proper time, assuming the mass difference Δm equal to (i) $0.5/\tau_1$, (ii) $2/\tau_1$. Display the phenomena on a graph.

4.10 An experiment in a gold mine in South Dakota has been carried out to detect solar neutrinos, using the reaction

$$\nu + Cl^{37} \to Ar^{37} + e^-.$$

The detector contained approximately 4×10^5 liters of tetrachlorethylene (C_2Cl_4). Estimate how many atoms of Ar^{37} per day would be produced, making the following assumptions:

a) solar constant $= 2$ cal cm^{-2} min^{-1};

b) 10% of thermonuclear energy of sun appears in neutrinos, of mean energy 1 MeV;

c) 1% of all neutrinos are energetic enough to induce the above reaction;

d) the cross section per Cl^{37} nucleus for "active" neutrinos is 10^{-45} cm^2;

e) Cl^{37} isotopic abundance is 25%;

f) density $C_2Cl_4 = 1.5$ g ml^{-1}.

Do you expect any difference between day rate and night rate? [For description of the experiment, see R. Davis, D. S. Harmer, and K. C. Hoffman, *Phys. Rev. Lett.* **20,** 1205 (1968).]

4.11 Energetic neutrinos may produce single pions in the following reactions on protons and neutrons:

i) $\nu + p \to \pi^+ + p + \mu^-,$

ii) $\nu + n \to \left. \begin{matrix} \pi^0 + p \\ \pi^+ + n \end{matrix} \right\} + \mu^-.$

Assume the process is dominated by the first pion-nucleon resonance $N^*(1236)$, so that the π-nucleon system has $I = \frac{3}{2}$ only. As in weak decay processes of $\Delta S = 0$, the isospin of the hadronic state then changes by one unit ($\Delta I = 1$ rule). Show that this rule predicts a rate for (i) three times that of (ii). Also show that, on the contrary, for a $\Delta I = 2$ transition (for which there is no present evidence), the rate for (ii) would be three times that for (i).

4.12 Use the $\Delta I = \frac{1}{2}$ rule to demonstrate the following relations between the rate of three-pion decays of charged and neutral kaons:

$$\Gamma(K_L \rightarrow 3\pi^0) = \tfrac{3}{2}\Gamma(K_L \rightarrow \pi^+\pi^-\pi^0),$$

$$\Gamma(K^+ \rightarrow \pi^+\pi^+\pi^-) = 4\Gamma(K^+ \rightarrow \pi^+\pi^0\pi^0),$$

$$\Gamma(K_L \rightarrow \pi^+\pi^-\pi^0) = 2\Gamma(K^+ \rightarrow \pi^+\pi^0\pi^0).$$

[Comment: These results depend on the (reasonable) assumption that the three pions are in a relative S-state. Then, any pair of pions must be in a symmetric isospin state, i.e. $I = 0$ and/or 2. The third pion ($I = 1$) must then be combined with the pair to yield a three-pion state of $I = 1, I_3 = 1$ for K^+, or $I = 1, I_3 = 0$ for K^0 (from the $\Delta I = \frac{1}{2}$ rule). It is necessary to write the three-pion wave function in a manner which is, as required for identical bosons, completely symmetric under pion label interchange. Thus, the $\pi^+\pi^+\pi^-$ state must be written

$$(++-) = \frac{1}{\sqrt{6}}(\pi_1^+\pi_2^+\pi_3^- + \pi_2^+\pi_1^+\pi_3^- + \pi_3^-\pi_2^+\pi_1^+ + \pi_3^-\pi_1^+\pi_2^+ + \pi_2^+\pi_3^-\pi_1^+ + \pi_1^+\pi_3^-\pi_2^+).$$

Each term in this expression will be the product of an isospin state of $I = 0$ or 2, for the first two pions, and one of $I = 1$, for the third pion, with appropriate Clebsch-Gordan coefficients.]

Chapter 5

Electromagnetic Interactions and Form Factors

5.1 GYROMAGNETIC RATIO OF ELECTRON AND MUON

We first consider the electromagnetic properties of the electron and the muon, which do not have any strong interactions. Each particle possesses unit electric charge, and a magnetic moment given to good approximation by the Dirac theory, which is suitable for describing relativistic particles of spin $s = \frac{1}{2}\hbar$. According to the Dirac theory, the wave function of an electron has four components, which can be interpreted in terms of two electron (e^-) states, eigenvalues $s_z = \pm\frac{1}{2}\hbar$ (spin up and down respectively), and two positron (e^+) states of spin up and down (a description which is rigorous only in the particle rest frame). Furthermore, in the interaction of such spin $\frac{1}{2}$ particles with a magnetic field, the energy eigenvalues are $E = \mu_z H$, corresponding to an intrinsic magnetic moment vector $\boldsymbol{\mu}$ with z-components

$$\mu_z = \pm\mu_B, \qquad \text{where} \qquad \mu_B = eh/2mc, \qquad (5.1)$$

m being the particle mass. μ_B is called the Bohr magneton. In general, for any particle the magnetic moment vector $\boldsymbol{\mu}$ is related to the spin vector s by the equation

$$\boldsymbol{\mu} = g\mu_B s, \qquad (5.2)$$

where g is called the Landé or g-factor, and $g\mu_B = \mu/s$ is the gyromagnetic ratio, i.e. the ratio of magnetic to mechanical moment of the particle. Thus, the Dirac theory predicts for particles of spin $\frac{1}{2}$,

$$g = 2. \qquad (5.3)$$

This result follows from the assumption of Dirac that the particle is localized relative to the field at a *point* in space-time—in other words it has no "spread" or structure. The g-values of electrons and muons have been determined experimentally with great precision, and are found to differ by a small amount from the value $g = 2$. Thus the Dirac picture of a point particle is not exact. Similar departures from the Dirac theory are encountered in other atomic phenomena; for example, the famous Lamb shift amounts to a 1 % correction to the $2P_{3/2} \rightarrow 2S_{1/2}$ level separation in the fine structure of hydrogen. These deviations are well accounted for in the more sophisticated treatment of quantum electrodynamics. Although this topic is really outside the scope of this book, it is not inappropriate to mention briefly some of the underlying ideas.

In the previous chapter, we remarked that the interaction between two charges may be viewed in terms of exchange of virtual photons. The spontaneous emission of a *real* photon by a stable charged particle is of course forbidden by conservation of energy. However, within the limits set by the uncertainty principle, the particle energy may differ by an amount ΔE from its time-averaged value during a short time Δt, where $\Delta E = \hbar/\Delta t$. The energy difference ΔE can be interpreted in terms of emission of one or more photons, which have a transient existence, being re-absorbed by the particle within the time Δt. Hence the term "virtual". Note that, although the energy of a particle is uncertain within a short interval, the electric charge is not. Charge is absolutely conserved at all times and in all processes, real or virtual. In classical physics, action at a distance, for example between two electric charges, is conveniently represented by inventing a fictitious quantity called electric field. In quantum mechanics, this concept may be replaced by that (equally fictitious) of the exchange of virtual photons between the charges. We may therefore view an electron as continually emitting and reabsorbing virtual photons, and virtual electron-positron pairs, and it is plausible that such processes will affect the ratio e/m and hence the gyromagnetic ratio. The amount of the correction will be proportional to the emission probability, determined by the characteristic coupling constant $\alpha = e^2/\hbar c$. The anomaly $(g - 2)/2$ in fact has the form of a perturbation series in α. A calculation gives a theoretical value for the *electron* of

$$\left(\frac{g - 2}{2}\right)_{\text{theory}} = \frac{\alpha}{2\pi} - 0.32848\,\frac{\alpha^2}{\pi^2}$$

$$+ 0.19\,\frac{\alpha^3}{\pi^3} + \cdots = (115964.1 \pm 0.3) \times 10^{-8}, \quad (5.4\text{a})$$

while the experimental value is (Wesley and Rich, 1970)

$$\left(\frac{g - 2}{2}\right)_{\text{expt.}} = (115964.4 \pm 0.7) \times 10^{-8}. \quad (5.4\text{b})$$

A similar calculation for the *muon* yields the result

$$\left(\frac{g - 2}{2}\right)_{\text{theory}} = \frac{\alpha}{2\pi} + 0.76578\,\frac{\alpha^2}{\pi^2} + 2.55\,\frac{\alpha^3}{\pi^3} + \cdots$$

$$= 116557.4 \times 10^{-8}, \quad (5.4\text{c})$$

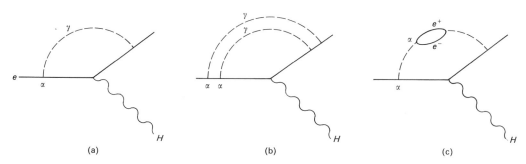

Fig. 5.1 Radiative corrections to the g-factor of electron and muon.

compared with the experimental value (Bailey *et al.*, 1968)

$$\left(\frac{g-2}{2}\right)_{\text{expt.}} = (116616 \pm 31) \times 10^{-8}. \tag{5.4d}$$

These results verify the theory of quantum electrodynamics to an astonishing degree of accuracy.

In Fig. 5.1 some of the diagrams indicating various corrections to the Dirac theory are shown. Diagram (a) shows the first order correction corresponding to the interaction of the electron with the magnetic field H when it is disassociated into a "bare" particle plus one photon. This has magnitude $\alpha/2\pi$ and is obviously the same for both electron and muon. In (b) we show one possible diagram contributing to 2-photon emission, and thus of magnitude α^2/π^2. Diagram (c) corresponds to virtual pair-production (sometimes called "vacuum polarization"), and, as we see, is also of order α^2. The electron and muon differ chiefly in the value of the vacuum polarization. This arises since the momenta of the particles in the intermediate virtual states scale in proportion to the parent particle mass. Further terms in (5.4a) and (5.4c) can be considered, but the accuracy required is so great that extremely small effects, coming from strong interactions, then need to be taken into account. We may remark that the agreement between experiment and theory is the more remarkable since quantum electrodynamics gives an infinite answer when used to calculate the self energy (physical mass) of the lepton itself.

The experiments on g-factors of electrons and muons have been carried out with a variety of extremely refined and cunning techniques. We shall just describe here the most recent experiments on the $(g - 2)$ value of the muon, which have been carried out at CERN using a "muon storage ring." The principle is as follows: A muon of velocity v normal to a uniform field H will describe a circular path with angular velocity ω_c given by the *cyclotron frequency:*

$$\omega_c = 2\pi\nu_c = \frac{eH}{mc\gamma}, \tag{5.5}$$

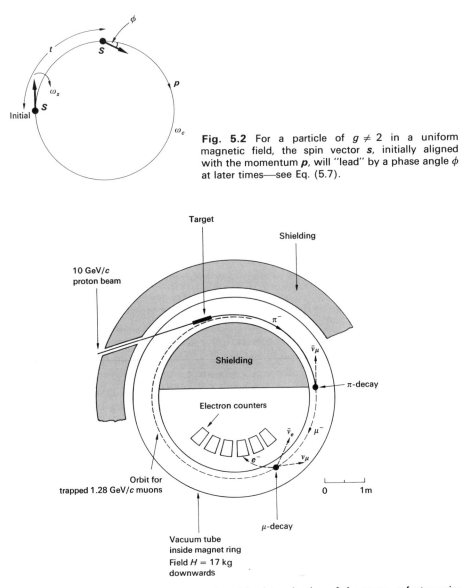

Fig. 5.2 For a particle of $g \neq 2$ in a uniform magnetic field, the spin vector s, initially aligned with the momentum p, will "lead" by a phase angle ϕ at later times—see Eq. (5.7).

Fig. 5.3 Experimental arrangement employed in determination of the muon g-factor using a "muon storage ring" (Bailey *et al.*, 1968).

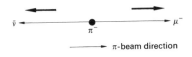

Fig. 5.4 Spin polarization sense of muon emitted in the "forward" direction in π-decay in flight.

where m is the muon rest mass, and $\gamma = (1 - v^2/c^2)^{-1/2}$. This formula follows from the usual expression for the radius of curvature ρ in terms of the momentum p, namely $p = He\rho$. Now, owing to its magnetic moment, it can be shown that the muon spin vector will *precess* about the field direction with a frequency*

$$\omega_s = \frac{eH}{mc\gamma}\left[1 + \left(\frac{g-2}{2}\right)\gamma\right]. \qquad (5.6)$$

Note that, in the muon rest frame ($\gamma = 1$) this formula reduces to the familiar expression for the Larmor precession frequency:

$$\omega_L = g \times \frac{eH}{2mc}.$$

Suppose now that the muon is initially polarized along the direction of motion. If $g = 2$ exactly, the momentum p and the spin vector s keep "in step." However, if $g \neq 2$, then after time t, s and p will not be parallel, but will make an angle

$$\phi = (\omega_s - \omega_c)t = \left(\frac{g-2}{2}\right)\frac{eH}{m}t, \qquad (5.7)$$

as in Fig. 5.2. The direction of the spin or magnetic moment vector at time t can be determined by observing the direction of the electrons produced by decay in flight of the muons (Section 4.6). The main stages in the experiment are as follows (see Fig. 5.3). Protons of momentum 10 GeV/c from a proton synchrotron are incident on a target placed inside a magnet ring. Secondary negative pions are produced in the target, and a small fraction of these, with the correct angle and momentum ~ 1.3 GeV/c, are "stored" for about $1\frac{1}{2}$ revolutions. These pions decay in flight. A muon emitted near to, but not in, the forward direction in the pion rest frame will have slightly less momentum than the parent pion, and can be trapped into an orbit which just clears the target in subsequent revolutions. Such muons are nearly fully (97%) longitudinally polarized, with spin vectors pointing initially along the direction of motion (see Fig. 5.4 and the discussion in Section 4.6). The detecting counters are biased to accept only the highest energy electrons from the muon decay, i.e. those projected forward in the muon rest frame. Thus, the electron angular distribution in the laboratory follows closely the precession of the muon spin, so that the counting rate of electrons is modulated by the frequency ($\omega_s - \omega_c$) which is a measure of $(g - 2)/2$—see Fig. 5.5. The formula (5.7) refers to frequency (or time t) measured in the *laboratory* frame. The lifetime of the muon, $\tau = 2.2$ μsec in its own rest frame, is however increased by a relativistic factor $\gamma \approx 12$ in the laboratory system (i.e. the muon lives on average 26 μsec). The precession

---

* Eqs. (5.5) and (5.6) refer to the laboratory frame. For a classical derivation of these formulae, see V. Bargmann, L. Michel, and V. L. Telegdi, *Phys. Rev. Lett.* **2**, 435 (1959).

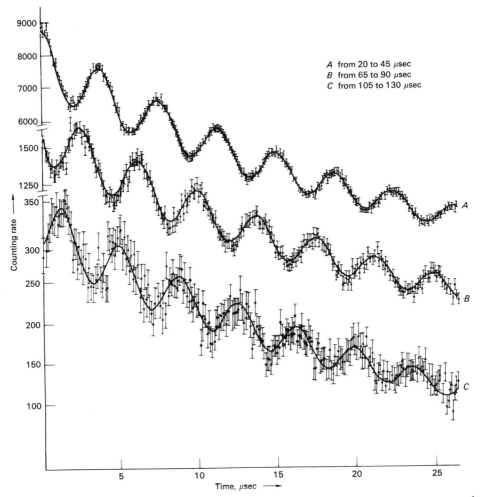

Fig. 5.5 Time dependence of the counting rate of electrons observed with the apparatus of Fig. 5.3. The general exponential decrease corresponds to a lifetime $\tau = 26\ \mu\text{sec}$ in the laboratory system. The rate is modulated by the frequency $(\omega_s - \omega_c)$, which measures $(g - 2)$.

period of the anomalous moment $2\pi/(\omega_s - \omega_c)$ is about 3.7 μsec, so that, by using high energy muons, the modulation can be followed out over some thirty periods and measured with great accuracy. The field H is calibrated by means of a proton precession magnetometer, so that one is essentially comparing two frequencies.

   An interesting by-product of the muon storage ring experiment is that it provides a confirmation of Einstein's clock hypothesis in the theory of relativity. The time dilation formula

$$\tau\ (\text{laboratory frame}) = \gamma \times \tau\ (\text{rest frame}) \tag{5.8}$$

is verified in this experiment to within 1%, for $\gamma \approx 12$, and indicates that, as predicted by Einstein, a clock (decaying muon) moving past with velocity v ticks more slowly, by a factor $\gamma = (1 - v^2/c^2)^{-1/2}$, than a similar clock at rest in the laboratory. In this case, the moving clock is not in an inertial frame, since if follows a circular path and suffers a perpendicular acceleration ($\sim 10^{19}$ g!), but nevertheless appears to obey the Einstein relation (5.8). Thus the muon is an "ideal clock," the rate depending only on the velocity and not the acceleration.

5.2 ANOMALOUS MAGNETIC MOMENT OF THE NUCLEON

If we apply the Dirac formula (5.1) to the neutron and proton, we obtain:

$$\text{Proton} \qquad \mu_{\text{Dirac}} = \frac{eh}{2Mc} = 1 \text{ nuclear magneton (n.m.)},$$

$$\text{Neutron} \qquad \mu_{\text{Dirac}} = 0.$$

Experimentally, it is found that

$$\mu_{\text{proton}} = +2.79 \text{ n.m.},$$

$$\mu_{\text{neutron}} = -1.91 \text{ n.m.}$$

Thus the deviations from the Dirac value are extremely large, the anomalous or Pauli moments being

$$\mu_p - 1 = +1.79 \text{ n.m.},$$

$$\mu_n = -1.91 \text{ n.m.},$$

of about the same magnitude, but opposite in sign. The existence of the anomalous moments demonstrates that, as for the electron and muon, the nucleon has structure. Crudely we can think of a nucleon as a "core" plus a circulating "meson cloud," like the photon cloud in the electron case. Now, however, the coupling is strong ($g^2 \sim 1$ instead of $e^2 \sim \frac{1}{137}$), so that the anomaly is much bigger (see Fig. 5.6). In the absence of a comprehensive theory of strong

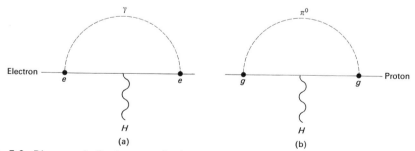

Fig. 5.6 Diagrams indicating contribution to anomalous magnetic moments observed in interaction with a field H. In (a) the first-order correction to the magnetic moment of the electron arises from single (virtual) photon emission, and is of magnitude $e^2/\hbar c = \alpha \sim 10^{-2}$. In the case of a proton, (b), with strong interaction, the anomaly arising from virtual pion emission will be of order $g^2/\hbar c \sim 1$.

interactions, we are, however, unable to calculate the expected value of $(g - 2)$.*
Therefore, the approach to the problem of nucleon structure has to be a
phenomenological one, describing the extent of the spatial distribution of
electric charge and magnetic moment in terms of structure or *form factors*.
These form factors have been determined by scattering experiments using
energetic electron beams which "probe" the charge distribution. To introduce
these ideas, let us first consider the simpler problem of scattering of electrons by
atomic nuclei.

5.3 SCATTERING OF SPINLESS CHARGED PARTICLES BY NUCLEI IN THE BORN APPROXIMATION

The Rutherford scattering of "spinless" electrons by nuclei can be derived from
first-order perturbation theory (and also classically). We employ formula
(3.16) for the transition probability:

$$W = \frac{2\pi}{\hbar} |M_{if}|^2 \, \rho_f. \tag{5.9}$$

For a perturbing central potential $V(r)$ provided by a stationary nucleus Ze,
the matrix element becomes the volume integral

$$M_{if} = \int \psi_f^* V(r) \psi_i \, d\tau, \tag{5.10}$$

where ψ_i, ψ_f are initial and final state wave functions of the scattered electron.
The Born approximation assumes the perturbation to be weak, so that only
single scattering is considered (this can be shown to be equivalent to the requirement $Z < 137$). We can therefore represent ψ_i and ψ_f as plane waves, before
and after scattering. Writing k_0 and k for the initial and final propagation
vectors, we obtain from (5.10)

$$M_{if} = \int e^{i(k_0 - k) \cdot r} V(r) \, d^3 r. \tag{5.11}$$

As in Section 4.10.1, the differential scattering cross section is W/v, where v is
the velocity of the incident beam relative to the scattering center. Setting the
density of states factor

$$\rho_f = \frac{p^2 \, d\Omega}{h^3} \frac{dp}{dE_f},$$

where $p = \hbar k$ is the momentum of the scattered electron and E_f is the total

---

* The *ratio* of μ_p/μ_n is, however, predicted fairly accurately by the quark model—see
Section 6.8.

energy in the final state, gives us

$$\frac{d\sigma}{d\Omega} = \frac{1}{(2\pi)^2\hbar^4} \frac{p^2}{v} \frac{dp}{dE_f} |M_{if}|^2. \tag{5.12}$$

So far, we have assumed the nucleus to be infinitely massive. In practice we have to consider the nuclear recoil. Let p', W, and M denote the momentum, total energy, and rest mass of the recoiling nucleus, and θ the angular deflection of the electron (see Fig. 5.8). Using units $\hbar = c = 1$, and assuming both incident and scattered electrons are extreme relativistic, we have

$$p_0 = k_0 = E_0, \qquad p = k = E, \qquad v \approx 1.$$

All quantities refer to the laboratory system. Applying energy/momentum conservation

$$E_i = p_0 + M = E_f = p + W; \qquad p_0 = p + p',$$

one finds

$$E_f = p + \sqrt{p'^2 + M^2} = p + \sqrt{p_0^2 + p^2 - 2pp_0 \cos\theta + M^2},$$

$$(E_f - p_0 \cos\theta) = Mp_0/p,$$

$$\frac{dp}{dE_f} = \frac{W}{(E_f - p_0 \cos\theta)} = \frac{W}{M} \frac{p}{p_0}, \tag{5.13}$$

and

$$\frac{p}{p_0} = \frac{1}{\left[1 + \dfrac{p_0}{M}(1 - \cos\theta)\right]}. \tag{5.14}$$

Equation (5.12) then becomes

$$\frac{d\sigma}{d\Omega} = \frac{1}{4\pi^2} p^2 \frac{W}{M} \frac{p}{p_0} \left| \int e^{i\mathbf{q}\cdot\mathbf{r}} V(\mathbf{r}) d^3r \right|^2, \tag{5.15}$$

where the momentum transfer

$$\mathbf{q} = \mathbf{p}_0 - \mathbf{p}.$$

Now let us represent the nucleus by a sphere of charge density $\rho(\mathbf{R})$ as shown in Fig. 5.7, normalized so that

$$\int_0^\infty \rho(\mathbf{R}) d^3R = 1.$$

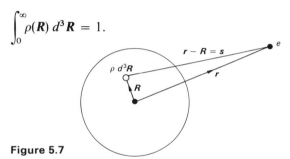

Figure 5.7

Then

$$V(r) = Ze^2 \int \frac{\rho(R)\, d^3R}{(r-R)},$$

and

$$M_{if} = Ze^2 \int\int \frac{\rho(R)e^{i\mathbf{q}\cdot\mathbf{R}}\, d^3R\, e^{i\mathbf{q}\cdot(\mathbf{r}-\mathbf{R})}\, d^3r}{(r-R)}$$

$$= Ze^2 \int \rho(R)e^{i\mathbf{q}\cdot\mathbf{R}}\, d^3R \int \frac{e^{iqs\cos\alpha}2\pi s^2\, ds\, d(\cos\alpha)}{s}, \qquad (5.16)$$

where $s = r - R$ and α is the polar angle between s and $\mathbf{q}$, the momentum transfer vector. We define the nuclear form factor by

$$F(q^2) = \int \rho(R)e^{i\mathbf{q}\cdot\mathbf{R}}\, d^3R. \qquad (5.17)$$

Then

$$M_{if} = 2\pi Ze^2 F(q^2) \int s\, ds \int e^{iqs\cos\alpha}\, d(\cos\alpha)$$

$$= 2\pi Ze^2 F(q^2) \int \frac{s\, ds(e^{iqs} - e^{-iqs})}{iqs}. \qquad (5.18)$$

This integral unfortunately diverges—as indeed it should, since the cross section for scattering of two particles via an inverse-square-law field is infinite. In fact, the nucleus is of course *screened* at large distances by atomic electrons. The trick is therefore to modify $V(r)$ by a factor $e^{-r/a}$, where a represents a typical atomic radius; afterwards we let $a \to \infty$ if we wish. Since $a \gg R$ (by a factor $\sim 10^4$), we can set $e^{-r/a} = e^{-s/a}$. Then (5.18) becomes

$$M_{if} = \frac{2\pi Ze^2 F(q^2)}{iq}\left\{ \int e^{-s(1/a - iq)}\, ds - \int e^{-s(1/a + iq)}\, ds \right\}$$

$$= \frac{2\pi Ze^2 F(q^2)}{iq}\left\{ \frac{1}{\left(\frac{1}{a} - iq\right)} - \frac{1}{\left(\frac{1}{a} + iq\right)} \right\} = \frac{4\pi Ze^2 F(q^2)}{\left(q^2 + \frac{1}{a^2}\right)}. \qquad (5.19)$$

Note that if $q < 1/a$ (i.e. very small momentum transfer), $M^2 \propto d\sigma/d\Omega \to$ constant. Now $a \sim 10^{-8}$ cm and thus $1/a \sim 1$ keV only; for the region of interest, $q \gg 1/a$ and we obtain for the differential cross section:

$$\frac{d\sigma}{d\Omega} = \frac{4Z^2 e^4}{q^4}p^2\frac{W}{M}\frac{p}{p_0}[F(q^2)]^2. \qquad (5.20)$$

For reasons that will appear later, $(W/M) - 1 = (q^2/2M^2) \ll 1$, so we can set $(W/M) = 1$ in all practical cases—the recoiling nucleus is nonrelativistic. Under the further assumption that the nuclear recoil momentum, $p' = q \ll p_0$, we can set $p = p_0$ and

$$q^2 = 2p_0^2 - 2p_0^2 \cos \theta = 4p_0^2 \sin^2 \frac{\theta}{2}, \tag{5.21}$$

so that

$$\frac{d\sigma}{d\Omega} = \frac{Z^2 e^4 [F(q^2)]^2}{4p_0^2 \sin^4 \dfrac{\theta}{2}}. \tag{5.22}$$

Note that, for a *point* nucleus, $F(q^2) = 1$ for all q^2; (5.22) then gives the *Rutherford scattering formula*.

5.4 4-MOMENTUM TRANSFER: MOTT SCATTERING

Equation (5.20) is expressed in terms of the 3-momentum transfer q in the laboratory system, between incident and target particle. In order to express the scattering in a form which is independent of the reference frame, it is better to consider q as the 4-*momentum transfer*. In Fig. 5.8, P_0, P, and Q represent the 4-momenta of the particles involved. Each 4-momentum has three space and one time component.

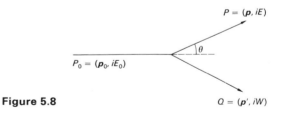

Figure 5.8

The square of the 4-momentum vector (of a real particle) is given by

$$P^2 = (\mathbf{p})^2 + (iE)^2 = p^2 - E^2 = -m^2,$$

and is thus invariant (see Appendix A). Similarly, the 4-momentum transfer squared, between the incident and emergent electron, is invariant, with a value

$$q^2 = (\mathbf{p}_0 - \mathbf{p})^2 - (E_0 - E)^2 = -2m^2 - 2pp_0 \cos \theta + 2EE_0$$

$$= 2pp_0(1 - \cos \theta) = 4pp_0 \sin^2 \frac{\theta}{2}, \tag{5.23}$$

if we neglect the electron mass ($m^2 \ll q^2$). Since the scattering angle is a real quantity ($-1 < \cos \theta < 1$), q^2 is positive. For exchange of a real particle, q^2 would be negative, i.e. with the same sign as the energy or time component $(iE)^2$. A value $q^2 > 0$ is sometimes referred to as a *space-like* momentum transfer, to distinguish it from $q^2 < 0$, which is called *time-like*. All scattering processes necessarily refer to the space-like region of q^2.

An alternative expression for q^2 is obtained by considering the transfer to the nucleus:

$$q^2 = (-p')^2 - (M - W)^2 = -2M^2 + 2MW = 2MT, \qquad (5.24)$$

where the kinetic energy acquired by the nucleus is $T = W - M$. Thus

$$\frac{W}{M} = 1 + \frac{q^2}{2M^2}. \qquad (5.25)$$

If the nucleus is to recoil coherently, it turns out that $q^2 \ll 2M^2$, so that $W/M \simeq 1$ in a practical case, as assumed in (5.22).

Although the values of q^2 in (5.23) and (5.24) refer to quantities measured in the laboratory frame, its numerical value is the same in all reference frames. The result of evaluating $|M_{if}|^2$ is as before, except that now, q being a 4-momentum transfer, (5.17) is strictly an integral over space-time, with qR as a scalar product of 4-vectors. However, if $W/M \simeq 1$, the energy transfer (time component) is small, and we can still interpret (5.17) as the integral over a spatial charge distribution.

If we describe electrons using Dirac (4-component) wave functions, i.e. we incorporate spin, then an extra term $\cos^2(\theta/2)$ appears in the cross section. (This arises essentially because, as indicated by the Dirac theory, relativistic electrons are longitudinally aligned; a 180° scattering would then involve a flip-over of the electron spin, which is forbidden by angular momentum conservation along the beam axis.) Inserting the expression (5.14) for p/p_0, we obtain the Mott formula for the scattering of relativistic spin $\frac{1}{2}$ electrons by spinless point-like nuclei:

$$\left(\frac{d\sigma}{d\Omega}\right)_{\text{Mott}} = \frac{Z^2 e^4 \cos^2 \dfrac{\theta}{2}}{4p_0^2 \sin^4 \dfrac{\theta}{2}\left[1 + \dfrac{2p_0}{M}\sin^2 \dfrac{\theta}{2}\right]}. \qquad (5.26)$$

Note that, for small scattering angles, this is identical with the Rutherford formula. The final term in the denominator allows for the nuclear recoil. It is important only for high energy electrons and large scattering angles.

5.5 THE FORM FACTOR OF THE NUCLEUS

In order to simplify the physics picture we shall revert to the case where the 4-momentum transfer is given essentially by the 3-momentum transfer, i.e. the energy transfer is small. The effect of the finite nuclear size is to introduce the term

$$F(q^2) = \int \rho(R)e^{iq \cdot R} \; d^3 R, \qquad (5.17)$$

so that

$$\frac{d\sigma}{d\Omega} = \left(\frac{d\sigma}{d\Omega}\right)_{\text{Mott}} |F(q^2)|^2. \tag{5.27}$$

Expression (5.17) is simply the Fourier transform of the nuclear charge density distribution $\rho(\mathbf{R})$, and $F(q^2)$ is termed the nuclear form factor. By integrating (5.17) over angles the reader should prove that

$$F(q^2) = \int \rho(R) \frac{\sin qR}{qR} 4\pi R^2 \, dR. \tag{5.28}$$

The interpretation of $F(q^2)$ as a (three-dimensional) Fourier transform of a charge distribution in space is strictly valid *only* when q is essentially equal to the 3-momentum transfer.

As an example, let us calculate $F(q^2)$ for a Yukawa-type charge distribution (Fig. 5.9):

$$\rho(R) = \rho_0 e^{-\alpha R}/R. \tag{5.29}$$

Then

$$F(q^2) = 4\pi\rho_0 \int \frac{e^{-\alpha R}}{2iqR^2} (e^{iqR} - e^{-iqR}) R^2 \, dR$$

$$= 4\pi\rho_0 \frac{1}{(\alpha^2 + q^2)} .$$

From normalization, one finds $\alpha^2 = 4\pi\rho_0$, hence

$$F(q^2) = \frac{1}{\left(1 + \dfrac{q^2}{\alpha^2}\right)} .$$

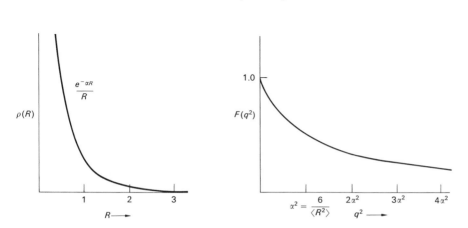

Fig. 5.9 Yukawa-type charge distribution, and corresponding form factor.

The rms radius of the charge distribution is given by

$$\langle R^2 \rangle = \frac{\int \rho(R)R^2 \times R^2 \, dR}{\int \rho(R)R^2 \, dR} = \frac{6}{\alpha^2}.$$

Hence

$$F(q^2) = \frac{1}{\left(1 + \dfrac{q^2 \langle R^2 \rangle}{6}\right)}. \tag{5.30}$$

For small values of $q^2 \langle R^2 \rangle$, *all* form factors reduce to the expression

$$F(q^2) = 1 - \frac{q^2 \langle R^2 \rangle}{6} + \cdots. \tag{5.31}$$

This may be demonstrated by expanding the exponential in (5.17) or the sine in (5.28) and retaining the first two terms.

5.5.1 Experiments on Nuclear Form Factors

Electron scattering experiments at 400 to 600 MeV have been employed to obtain fairly detailed charge distributions on many nuclei. From (5.28), $\rho(R)$ may be obtained from the inverse Fourier transform:

$$\rho(R) = \frac{1}{2\pi^2} \int F(q^2) \frac{\sin qR}{qR} q^2 \, dq. \tag{5.32}$$

The charge density observed in this way is found to vary slowly through the bulk of the nucleus and falls off sharply (by a factor 10) over an outer shell thickness of $\sim 2 \times 10^{-13}$ cm (see Fig. 5.10). The shape of the ρ-distribution in the

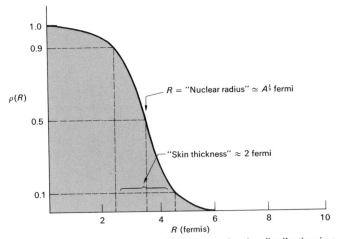

Fig. 5.10 Approximate radial variation of the charge density distribution in a medium-weight nucleus.

outer regions (the so-called "nuclear stratosphere") is not well known. For details, the reader is referred to the review article by R. Hofstadter in the bibliography. See also "Fundamentals of Modern Physics" by R. M. Eisberg, Chapter 16 (John Wiley, New York, 1961), or "Introduction to Nuclear Physics" by H. A. Enge (Addison-Wesley, Reading, Mass., 1966).

5.6 FORM FACTORS OF PROTON AND NEUTRON

The scattering of an electron by a point proton of spin $\frac{1}{2}$, magnetic moment $eh/2Mc$, is described by the Dirac theory, which gives

$$\left(\frac{d\sigma}{d\Omega}\right)_{\text{Dirac}} = \frac{e^4 \cos^2 \dfrac{\theta}{2}}{4p_0^2 \sin^4 \dfrac{\theta}{2}\left[1 + \dfrac{2p_0}{M}\sin^2\dfrac{\theta}{2}\right]}\left\{1 + \frac{q^2}{2M^2}\tan^2\frac{\theta}{2}\right\}. \qquad (5.33)$$

Note that this differs from the Mott formula in the inclusion of the second term in the curly brackets, due to magnetic scattering. This dominates the electrical scattering at large angle and high momentum transfer. Classically, one can see that the reason for this is that, for close collisions, the magnetic potential (varying as $1/r^2$) is larger than the electric potential ($\sim 1/r$).

For a real nucleon, we must take into account that the magnetic moment is anomalous, and that it is not a point particle, but has a form factor $G(q^2)$. Actually four independent form factors are involved, two for the proton and two for the neutron, corresponding to charge and magnetic moment distributions. We define these as

Electric form factor $\qquad G_E(q^2); \quad G_E^p(0) = 1; \quad G_E^n(0) = 0,$

Magnetic form factor $\qquad G_M(q^2); \quad G_M^p(0) = \mu_p = 2.79; \quad G_M^n(0) = \mu_n = -1.91,$

the units in the second line being nuclear magnetons. The cross section then has the form

$$\frac{d\sigma}{d\Omega} = \left(\frac{d\sigma}{d\Omega}\right)_{\text{Mott}}\left\{\left(\frac{G_E^2 + \dfrac{q^2}{4M^2}G_M^2}{1 + \dfrac{q^2}{4M^2}}\right) + \frac{q^2}{4M^2}\times 2G_M^2 \tan^2\frac{\theta}{2}\right\}. \qquad (5.34)$$

As $\theta \to 0$, $q^2 \to 0$ (see Eq. 5.21), and the curly bracket becomes $G_E(0)$, corresponding to the pure, point-like electrical scattering of the Mott formula.

Equation (5.34) is called the *Rosenbluth formula*. The essential feature is that

$$\left(\frac{d\sigma}{d\Omega}\right)\Big/\left(\frac{d\sigma}{d\Omega}\right)_{\text{Mott}} = A(q^2) + B(q^2)\tan^2\frac{\theta}{2}, \qquad (5.35)$$

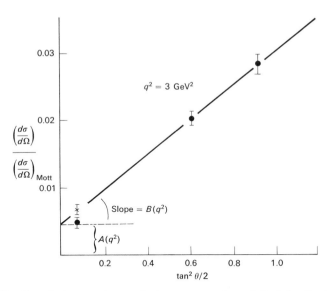

Fig. 5.11 The electron-proton scattering cross section plotted for fixed q^2 and different scattering angle θ (Rosenbluth plot). (After Weber, 1967.)

so that if one plots the cross section for different incident momenta p_0 and different scattering angle θ—such that q^2 is kept fixed—there should be a linear dependence on $\tan^2(\theta/2)$ (see Fig. 5.11). This is a direct consequence of the Born approximation, which assumes a single collision, or *single photon exchange*. The situation is depicted diagrammatically in Fig. 5.12(a), where we see the proton (or neutron) and electron coupled via an electromagnetic field, or virtual photon—a single one. Note that this photon carries the momentum transfer q^2, and has imaginary mass ($m^2 = -q^2$). Hence the term "virtual." The Rutherford scattering formula may be obtained by writing down the matrix element as the product of vertex functions and a factor $1/q^2$ for the virtual

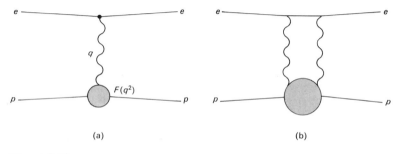

Figure 5.12

photon "propagator term":

$$M = \text{(coupling constant at top vertex)} \times \text{(propagator amplitude)}$$
$$\times \text{(coupling constant at bottom vertex)}$$

$$= e \times \frac{1}{q^2} \times e.$$

Thus

$$\frac{d\sigma}{dq^2} = |M|^2 \propto \frac{e^2}{q^4}, \tag{5.36}$$

which is the same as (5.22) if we note that

$$q^2 = 2p_0^2(1 - \cos\theta) \quad \text{and} \quad dq^2 = p_0^2 \, d\Omega/\pi.*$$

The quantity $1/q^4$ in the cross section simply gives the probability that the virtual photon shall carry the 4-momentum q; it corresponds to the classical physics statement that the electric field between the particles varies as $1/r^2$. The blob at the bottom of Fig. 5.12(a) is supposed to represent the nucleon form factor, left out in (5.36). Figure 5.12(b) represents double-photon exchange, which would not obviously give a straight line on the Rosenbluth plot. Actually the best test of single-photon exchange is the observed equality of e^+p- and e^-p-scattering, for which two-photon exchange introduces terms with opposite signs. All experiments support the single photon exchange picture.

The experimental determination of the proton form factor has been carried out by directing high energy (400 MeV to 16 GeV) electron beams at a hydrogen target, and making precision measurements of the momentum and angle of scattered electrons by means of magnetic spectrometers. Elastic events can be selected using the kinematic relation (5.14) between the energy and angle of the scattered electron (with M as the proton mass). For the neutron, scattering is observed with deuterium targets, and a subtraction procedure employed:

$$\frac{d\sigma}{d\Omega}(en) = \frac{d\sigma}{d\Omega}(ed) - \frac{d\sigma}{d\Omega}(ep) + \text{correction factor,}$$

where the correction factor involves nuclear physics of the deuteron. Because of this, the neutron data is less precise. The first experiments, demonstrating the deviation of the scattering from that expected for a point particle—and thus measuring the form factors—were carried out by Hofstadter and his collaborators in 1961, at Stanford, and have since been extended in numerous laboratories. The end result of the present experiments is that the form factors obey the

---

* The actual system of rules for writing down matrix elements by inspection from graphs was derived by Feynman. The interested reader cannot do better than consult Feynman's *Theory of Fundamental Processes*, Benjamin, New York, 1961.

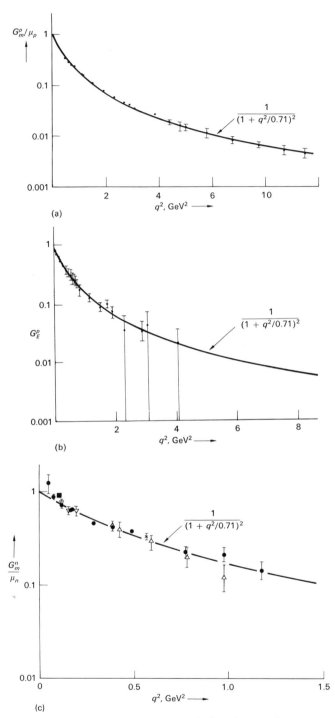

Fig. 5.13 Comparison of the magnetic and electric form factors of neutron and proton. They are consistent with the scaling law (5.37). (a) Proton magnetic form factor; (b) proton electric form factor; (c) neutron magnetic form factor. (After Weber, 1967.)

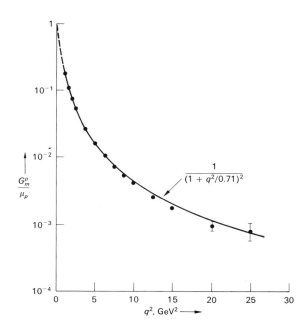

Fig. 5.14 The electromagnetic form factor of the proton, in the region of high q^2. The observations are fairly well fitted by the dipole formula. The data are principally from the SLAC laboratory. (After Panofsky, 1968.)

simple *scaling law*

$$G_E^p(q^2) = \frac{G_M^p(q^2)}{|\mu_p|} = \frac{G_M^n(q^2)}{|\mu_n|} = G(q^2), \qquad G_E^n(q^2) = 0, \qquad (5.37)$$

and the empirical *dipole formula*

$$G(q^2) = \left(1 + \frac{q^2}{M_V^2}\right)^{-2}, \qquad \text{with } M_V^2 = (0.84 \text{ GeV})^2. \qquad (5.38)$$

Relations (5.37) hold over the experimental range so far investigated, $q^2 = 0 - 2 \text{ GeV}^2$ (see Fig. 5.13). At higher q^2, only electron-proton scattering data exist, and, as (5.34) indicates, these refer essentially to the magnetic form factor $G_M^p(q^2)$, the electric contribution being small and unmeasurable. For this scattering, the dipole formula (5.38) fits to within 10 % accuracy up to $q^2 = 25$ GeV^2, as shown in Fig. 5.14. Over this range, the form factor squared falls by a factor of 1 million. Using Eq. (5.32), one finds that (5.38) can be interpreted in terms of an exponential charge/magnetic moment distribution of the proton of density

$$\rho(R) = \rho_0 \exp(-M_V R), \qquad (5.39a)$$

with a root mean square radius

$$R_{rms} = \frac{\sqrt{12}}{M_V} = 0.80 \text{ fermi.} \tag{5.39b}$$

5.7 INTERPRETATION OF NUCLEON FORM FACTORS

As we have mentioned, the elastic form factor, represented by the blob in Fig. 5.12(a), is a manifestation of the strong interaction properties of the nucleon. In other words, the form factor measures the probability that the target nucleon will "hold together" and recoil as a nucleon under the impact of the collision, rather than shake off one or more pions, for example. For very large momentum transfers, this probability is extremely small ($\approx 10^{-6}$ at $q^2 = 25 \text{ GeV}^2$).

Attempts have been made to understand the observations in terms of strong interaction dynamics, but these have met only moderate success. One of the first proposals was essentially to replace the blob of Fig. 5.12 by a single-particle exchange process, familiar in the Yukawa picture (Fig. 5.15).

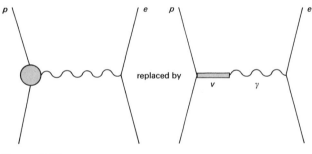

Figure 5.15

In the figure, V represents any particle which can mediate the interaction between the virtual photon and the nucleon (now treated as a structureless "point" particle). Obviously V must be neutral, and must have spin-parity 1^-, like the photon—in other words, it must be a neutral *vector* meson. Furthermore, it must have charge-conjugation parity $C = -1$, like the photon (Section 3.13). The only candidates of reasonably low mass are the ρ^0-, ω-, and ϕ-mesons (Table in preface). ω and ϕ have zero isospin, and ρ^0 has $I = 1$. Both $I = 0$ and $I = 1$ particles are required, for the following reasons. The form factors G_p and G_n (for both electric and magnetic parts) can be combined to form an *isoscalar* form factor

$$G_S = \tfrac{1}{2}(G_p + G_n), \qquad G_p = G_S + G_V,$$

and an *isovector* form factor

$$G_V = \tfrac{1}{2}(G_p - G_n), \qquad G_n = G_S - G_V. \tag{5.40}$$

G_S is the *same* for both proton and neutron, since it is a scalar in isospin space, and G_V is of *opposite sign* for proton and neutron, corresponding to the two isospin substates. Therefore, the vector meson responsible for the G_V part will have isospin $I = 1$, and that for G_S will have $I = 0$. We recall that the G-parity of a particle (of C-parity -1 in this case) is given by

$$G = C(-1)^I.$$

Thus the isovector meson has the properties of the ρ-meson:

$$J = 1, \quad I = 1, \quad G = (-1)(-1)^1 = +1, \quad \rho \rightarrow 2\pi,$$

and the isoscalar meson has the properties of the ω-meson:

$$J = 1, \quad I = 0, \quad G = (-1)(-1)^0 = -1, \quad \omega \rightarrow 3\pi,$$

recalling that $G(n\pi) = (-1)^n$. The ϕ-meson is also isoscalar.

As indicated in Section 7.8, the scattering amplitude which arises from a Yukawa potential, due to exchange of a single particle V of mass M as in Fig. 5.15, leads to a q^2-dependence (or propagator term) of the form

$$G(q^2) = \text{const.} \frac{1}{\left(1 + \dfrac{q^2}{M^2}\right)}, \tag{5.41}$$

or, if we sum over several single-particle exchanges,

$$G(q^2) = \sum_i \frac{g_i}{\left(1 + \dfrac{q^2}{M_i^2}\right)}, \tag{5.42}$$

where the g's are suitable parameters representing, in fact, the strong coupling constants of the appropriate mesons to the nucleon vertex. Choosing these parameters empirically, and including an additional constant term to ensure $G_E(q^2) \rightarrow 1$ and $G_M(q^2) \rightarrow \mu$ as $q^2 \rightarrow 0$, it is possible to fit the proton (and neutron) form factors in the region of fairly low q^2. This fit, however, requires the mass of the isovector meson to be about 550 MeV, compared with the observed ρ-mass of 765 MeV.

The historical development of the subject did not proceed in the way outlined above. First, an attempt was made to account for the form factors in terms of two-pion exchange, assuming *no* pion-pion interaction, since none was known at that time. When this failed, Frazer and Fulco (1959) postulated new vector mesons, i.e. *resonant* states of the 2π- and 3π-systems. They predicted the existence of the $\rho \rightarrow 2\pi$ and $\omega \rightarrow 3\pi$ states from the form factors, and, although the masses they deduced were too low, this triggered off an experimental search for pion resonances, the ρ and ω being first observed in 1961.

Expressions of the type (5.42), which we have introduced on a simple Yukawa picture, are expected from much more general considerations, using

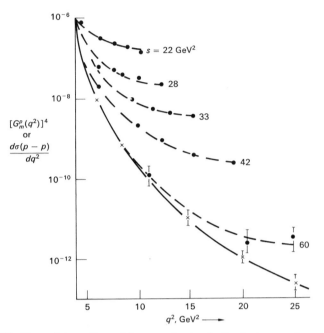

Fig. 5.16 The points on the solid curve refer to elastic electron-proton scattering, and those on the dashed curves to proton-proton elastic scattering, at different values of s, the square of the centre-of-mass energy. For p–p scattering, the ordinate is $[(d\sigma(q^2)/dq^2)/(d\sigma(0)/dq^2)]$, and for e p scattering, $[G_M(q^2)/G_M(0)]^4$, where G_M is the proton magnetic form factor. (After Abarbanel *et al.*, 1968.)

dispersion relations. It is therefore hard to understand why, at least in the region of high q^2, the data follow the unexpected and completely empirical "dipole formula" (5.38). A possible explanation may be that the above picture, in which just a few exchange particles are important—the so-called "vector dominance model"—is incorrect, and there are numerous background contributions from high-lying resonances, which combine in a complex way to give the observed q^2-dependence.

Not surprisingly, therefore, quite different approaches have been made to the interpretation of form factors. Wu and Yang (1965) made an interesting comparison between p–p and e–p scattering. The basic idea is that the spatial distribution of charge and magnetic moment in the proton is the same as that of the cloud of "strongly interacting matter." The electron is then a point probe, and e–p scattering can be considered as a point-cloud interaction. On the other hand, p–p scattering is a cloud-cloud interaction. The nature of the cloud is supposed to be the same in both cases. In other words, if the way a proton shatters under an impact q^2 (for example, the number of pions produced) is independent of what hits it, then so is the probability of it remaining intact. The elastic scattering cross section is then measured by $[G(q^2)]^2$ in e–p scattering, and by $[G(q^2)]^4$ in p–p scattering. The p–p scattering cross section, in fact, is not simply a function of q^2 alone, but of q^2 and s, the center-of-mass energy squared (see Section 7.9). We can see this qualitatively from the fact that the range of q^2 is from 0 to q^2_{max}, which depends on s. It turns out therefore that one ought to compare e–p and p–p scattering in the asymptotic limit $s \to \infty$, when the simple relation

$$\left[\frac{d\sigma(q^2)}{dq^2} \middle/ \frac{d\sigma(0)}{dq^2}\right]_{pp} = \left[\frac{d\sigma(q^2)}{dq^2} \middle/ \frac{d\sigma(0)}{dq^2}\right]^2_{ep} \tag{5.43}$$

should obtain. The results in Fig. 5.16 give quite striking support to this prediction.

In Chapter 6, a discussion is given of the quark model of strongly interacting particles, in which the hadrons are supposed to consist of very strongly bound combinations of more fundamental subunits of fractional charge, called quarks. According to this model, the structure of the nucleon will be determined by the internal dynamics of the quark-quark interaction, and R_{rms} of (5.39) would correspond, crudely, to the mean quark separation. If this is so, it is clear that the form factors, at least at high q^2, are not determined solely by the masses of vector mesons coupling singly to photons, as suggested by the vector dominance model, but are manifestations of a deeper, underlying substructure which we do not at present understand. The remarkable relation embodied in (5.43) supports this view. Our ignorance of what is really going on can be described in terms of an extra form factor $(1 + q^2)^{-1}$ with q^2 in GeV^2, to be multiplied into the terms of the vector dominance formula (5.42). This then reproduces the magic dipole formula reasonably well.

5.8 FORM FACTOR IN THE TIME-LIKE REGION: THE REACTION $p\bar{p} \to e^+e^-$

In the previous sections, we have considered form factors appearing in elastic scattering processes, i.e. with space-like momentum transfers $q^2 > 0$, as in Fig. 5.17(a).

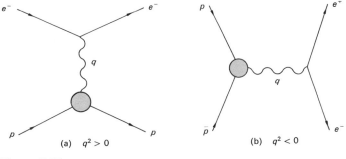

(a) $q^2 > 0$ (b) $q^2 < 0$

Figure 5.17

Now consider the proton-antiproton annihilation into an electron-positron pair. If we turn Fig. 5.17(a) through 90° and interchange outgoing particles for incoming antiparticles, we obtain Fig. 5.17(b), which describes exactly this process. In (a), the 4-momentum transfer squared, i.e. the invariant (mass)² of the intermediary photon, is

$$q^2 = (\boldsymbol{p}_1 - \boldsymbol{p}_2)^2 - (E_1 - E_2)^2 \approx 2p_1 p_2 (1 - \cos \theta) > 0,$$

where $\boldsymbol{p}_1$, E_1 and $\boldsymbol{p}_2$, E_2 are momenta and total energies of the incident and scattered electrons, θ the angle of scatter, and we neglect the electron mass. If in (b), $\boldsymbol{p}_1$, E_1 and $\boldsymbol{p}_2$, E_2 denote the same quantities for the electron and positron, we obtain

$$q^2 = (\boldsymbol{p}_1 + \boldsymbol{p}_2)^2 - (E_1 + E_2)^2,$$

or, in the center-of-mass frame, where $\boldsymbol{p}_1 = -\boldsymbol{p}_2$, $E_1 = E_2 = E$,

$$q^2 = -(2E)^2 = -(\text{CMS energy})^2 < 0. \tag{5.44}$$

In this case, q^2 corresponds to a real mass for the intermediary photon, i.e. the momentum transfer from nucleon vertex to electron vertex is *time-like*. Note that in *either* case the photon is virtual, since a real photon has $q^2 \equiv 0$. How can we describe the "blob" due to strong interaction structure in Fig. 5.17(b)? We can suppose that the form factor $G(q^2)$ is a continuous function of q^2, existing for both positive and negative values of q^2. As shown in Fig. 5.18, the *physical* region for the reaction $p\bar{p} \to e^+ e^-$ starts at $q^2 = -4M^2$, where M is the nucleon mass; this corresponds to $\cos \theta > +1$, or the *unphysical* region of the elastic scattering process, Fig. 5.17(a). Experiments on the annihilation $p\bar{p} \to e^+ e^-$ at rest or in flight are exceedingly difficult, and the observations to date only set upper limits for the cross section. These show that, at $q^2 \approx -7(\text{GeV})^2$, the region of the experiments, $G(q^2)$ has fallen off at high negative q^2 at least as fast as at high positive q^2. (Conversi *et al*, 1965; Hartill *et al*, 1969.)

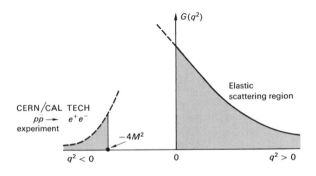

Figure 5.18

5.9 THE REACTIONS $e^+e^- \to \pi^+\pi^-$ AND $\pi^+\pi^-\pi^\circ$

From (5.42), we expect that singularities or *poles* should occur whenever $q^2 = -M_i^2$, when $G(q^2) \to \infty$. Actually, since vector meson resonances are quite broad (the ρ-meson has a width of ~ 120 MeV), these poles will appear as strong maxima and minima in the cross section, considered as a function of the center-of-mass energy. If we think in terms of the ρ-, ω-, and ϕ-mesons, it is clear that the antinucleon annihilation process described above will not detect them, since $|q^2|$ is much too large in the accessible region. Therefore, it is better to start off with e^+ and e^- as the incident particles, and this can be done in a "clashing beam" experiment. Figure 5.21(a) represents the reaction $e^+e^- \to \pi^+\pi^-$. Comparison with Fig. 5.17(b) shows that this in fact measures the electromagnetic form factor of the pion for time-like q^2.

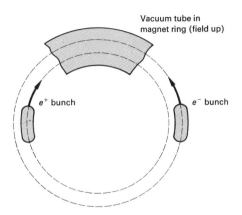

Fig. 5.19 Principle of e^+e^- clashing beam experiment.

In a special storage ring, consisting of a vacuum tube inside a magnet ring, electrons of some hundreds of MeV energy are constrained in a circular path. They circulate in the ring with a radiofrequency bunch structure. For the same field, injected positrons of equal momentum will circulate in the opposite direction. As in Fig. 5.19, the two bunches meet head-on twice per revolution. Currents of order 100 mA are attained, the useful rate for $e^+e^- \to \pi^+\pi^-$ being a few events per hour. It is necessary to obtain very high vacua in the rings in order to reduce the background from scattering of electrons or gas atoms. The pioneer experiments were carried out at Novosibirsk (USSR) and Orsay (Paris) during 1967 and 1968. With $2E$ the center-of-mass energy as in (5.44), the cross section for $e^+e^- \to \pi^+\pi^-$ in the region of the ρ-meson mass (Fig. 5.20) has the Breit-Wigner form, as in (7.31):

$$\sigma = \frac{\pi\lambda^2}{4} \frac{(2J+1)\Gamma^2 B}{(2E-M_\rho)^2 + \Gamma^2/4}, \tag{5.45}$$

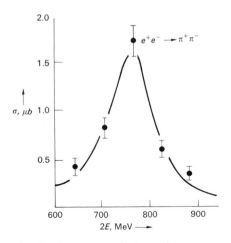

Fig. 5.20 Cross section for the process $e^+e^- \rightarrow \pi^+\pi^-$ as measured in the electron-positron colliding beam experiments at Orsay (Augustin *et al.*, 1969). The dominant feature is the production of the ρ-meson resonance. The curve is from the Breit-Wigner formula, using best-fit parameters; central mass $M_\rho = 765$ MeV, width $\Gamma_\rho = 104$ MeV.

where Γ is the total width of the ρ-meson (~ 104 MeV in these experiments), B is the branching ratio for the decay $\rho \rightarrow e^+e^-$, M_ρ is the central value for the ρ-meson mass (765 MeV), and $J = 1$ is the ρ-spin. $\lambda = \hbar/E$ is the CMS de Broglie wavelength of either of the colliding particles. The factor 4 appears from the average over the initial spin states of the electron and positron, to provide $J = 1$ in the final state. The measurements yield $B = (6.0 \pm 0.5) \times 10^{-5}$. This is the order of magnitude expected from purely qualitative arguments. In Fig. 5.21(a) we note that the electromagnetic coupling constant $\sqrt{\alpha} = e$ (in units $\hbar = c = 1$) appears twice, coupling the photon to the e^+e^- pair and the ρ-meson respectively. Thus, we expect that, in order of magnitude, $B \sim (\sqrt{\alpha}\sqrt{\alpha})^2 = (137)^{-2} \sim 10^{-4}$. In the clashing beam experiments, the

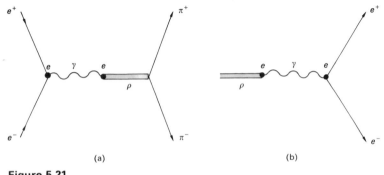

(a)   (b)

Figure 5.21

normalization of the cross section is made by reference to the Bhabha scattering $e^+e^- \rightarrow e^-e^+$, for which the cross section is calculable from quantum electro-dynamics.

It is also possible to produce ρ-mesons, either in strong interactions or in the inelastic collisions of energetic electron or photon beams with nucleons, and to observe the decay into (instead of formation from) an e^+e^- pair, as in Fig. 5.21(b). The value of B obtained is consistent with that above.

In the colliding beam experiments, the processes

$$e^+e^- \rightarrow \omega \rightarrow \pi^+\pi^-\pi^\circ, \qquad e^+e^- \rightarrow \phi \rightarrow K\bar{K} \text{ or } \pi^+\pi^-\pi^0$$

have also been observed. Figure 5.22 shows data on ω-production. Since the total width for the ω-decay (~ 16 MeV) is only one-tenth of that of the ρ, the observed branching ratio $B_\omega = 6 \times 10^{-4}$ for leptonic decay is correspondingly larger.

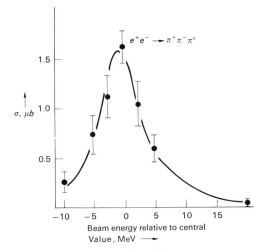

Fig. 5.22 Cross section for $e^+e^- \rightarrow \pi^+\pi^-\pi^0$ as a function of beam energy. The curve corresponds to a Breit-Wigner resonance of width $\Gamma_\omega = 16$ MeV.

In the vector dominance model described above, the coupling of hadrons to the electromagnetic field is supposed to be dominated by the vector mesons ρ, ω, and ϕ of low-lying mass. As we have seen, this accounts for the nucleon form factors at least in the region of low q^2. Further, it allows the unification of many electromagnetic processes in a single picture. The basic assumption is that the coupling of a particular vector meson, say the ρ^0, to a photon is the same for all values of the photon mass, $-q^2$. To be quantitative, the intrinsic coupling constant is specified, not by e alone as in Fig. 5.21, but by ef_i where f_i is some constant which has specific values (to be determined by experiment) for the ρ, ω, and ϕ respectively. As examples, we instance the processes in Fig. 5.23.

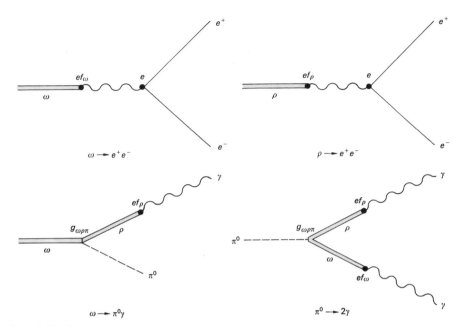

Fig. 5.23 Diagrams illustrating the electromagnetic decays of mesons, as described by the vector dominance model.

Since f_ρ and f_ω are known from the leptonic decay rates, the width for the $\pi^0 \to 2\gamma$ decay can be expressed in terms of the partial width for $\omega \to \pi^0\gamma$, plus phase-space factors (see Problem 5.4). It is also possible to obtain relations between f_ρ, f_ω, and f_ϕ from considerations of unitary symmetry, treating the photon as a U-spin scalar (Chapter 6). These predictions are in tolerably good agreement with experiment.

5.10 INELASTIC ELECTRON-PROTON SCATTERING

In addition to elastic scattering, inelastic processes occur in high energy electron-nucleon collisions. These result in production of one or more pions or other mesons. One of the characteristic features of these collisions is the production of pion–nucleon resonances, which are discussed in Section 7.4 and one of which, the $N^*(1236)$, we have already met with in Section 3.11. Such resonant states can be identified on the basis of the observed mass M^* of the recoiling hadronic system, using only the measured energy and direction of the scattered electron. Thus in Fig. 5.24, the energy E' and angle θ of the recoil electron define q^2 and the energy transfer, v:

$$v = E - E',$$
$$q^2 = 2EE'(1 - \cos \theta),$$

(5.46)

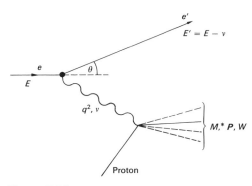

Figure 5.24

where the electron rest mass is neglected. In the laboratory frame, the target proton has energy M (its rest mass) and the final state a rest mass M^*, momentum P, and total energy $W = M + v$. Then another expression for q^2 is

$$q^2 = (0 - P)^2 - (M - W)^2$$
$$= -M^2 - (W^2 - P^2) + 2MW$$
$$= M^2 - M^{*2} + 2Mv. \qquad (5.47)$$

Thus q^2 and v together define M^*. Note that $q^2 = 2Mv$ for the elastic reaction, $M^* = M$, as in (5.24). Figure 5.25 shows a typical excitation curve, in which the elastic peak and several peaks due to formation of pion–nucleon resonances are visible. For excitation of a particular resonance, the q^2-dependence of the cross section can again be defined by structure factors. Since the resonances do not have a unique mass, however, the cross section is now double differential, $d^2\sigma/dq^2\, dE'$, which again can be expressed by a Rosenbluth-type formula, (5.35), with A and B now functions of both q^2 and v, as in (5.48). Integration of E' or v, performed over the resonance width (for fixed q^2), then yields the q^2-dependence. Figure 5.26 shows a comparison of this dependence with that of the elastic process, for three pion–nucleon resonances. After some threshold growth at small q^2 (depending on the resonance spin) it is seen that the cross sections are very similar to the elastic cross section in their q^2-behavior.

In our discussion so far, the final state of strongly interacting particles has had well-defined quantum numbers and mass; thus, for the elastic reaction

$$e^- + p \rightarrow e^- + p$$

the final state proton has spin-parity $J^P = \frac{1}{2}^+$ and isospin $I = \frac{1}{2}$. For the quasi-elastic reaction

$$e^- + p \rightarrow e^- + N^{*+}(1236)$$
$$ \hookrightarrow \pi^+ + n, \text{ or } \pi^0 + p,$$

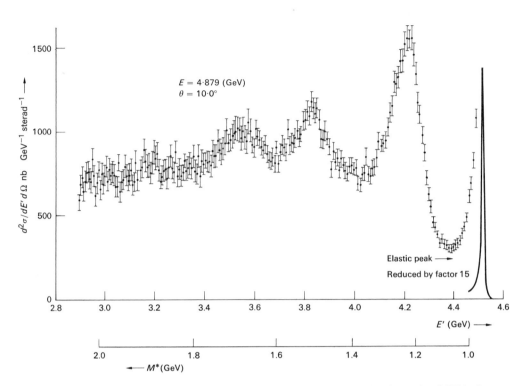

Fig. 5.25 Excitation curve of inelastic $e-p$ scattering, obtained at the DESY electron accelerator (Bartel *et al.*, 1968). E and E' are the energies of incident and scattered electron, and M^* is the mass of the recoiling hadronic state. The peaks due to the pion-nucleon resonances of masses 1.24, 1.51, and 1.69 GeV are clearly visible. After corrections are made for other radiative processes in the target (bremsstrahlung), the strong elastic peak would disappear.

the final state has a well-defined central mass (1236 MeV), spin-parity $J^P = \frac{3}{2}^+$, and isospin $I = \frac{3}{2}$. Feynman has called such reactions *exclusive* reactions. Suppose, however, that we measure $d^2\sigma/dq^2\,dv$ without regard to details of the final state, i.e. q^2 and v are allowed to act as independent variables, and we integrate over the continuum of final-state masses—what Feynman calls *inclusive* reactions. Then, remarkably enough, when q^2 and v are large, we get a quite different cross section dependence. Figure 5.27 shows the results of such measurements at the SLAC laboratory. The elastic and quasi-elastic peaks (corresponding to resonance production) are visible for low energy transfer, but at higher values, $d^2\sigma/dq^2\,dv$ shows no such structure. In other words, the initial nucleon, instead of undergoing transitions to specific, well-defined levels, is being excited into the continuum. This continuum cross section is seen to be large and, for a given energy transfer v, more or less independent of q^2.

This behavior may be contrasted with that of the form factors in the elastic cross section, varying as q^{-8} at large q^2, [Eq. (5.38)]. The inelastic cross section

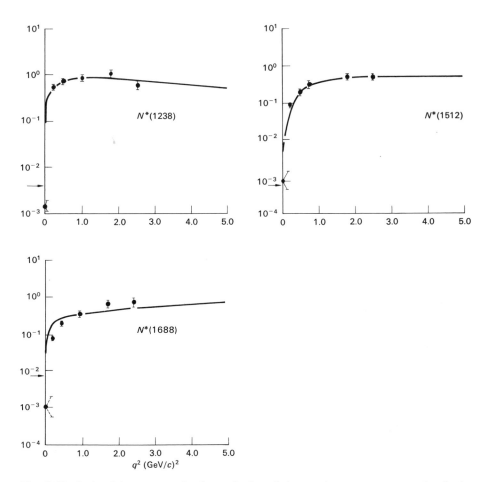

Fig. 5.26 Ratio of the cross section for excitation of pion-nucleon resonances, to the elastic cross section, as measured in high energy electron-proton scattering experiments at the Stanford linear accelerator. At high q^2, the ratio tends to a constant value for all three resonances. (After Panofsky, 1968.)

is often expressed in terms of two structure factors [analogous to A and B of (5.35)], defined by

$$\frac{d^2\sigma}{dq^2\,dv} = \frac{E'}{E}\frac{4\pi\alpha^2}{q^4}\left[W_2(q^2, v)\cos^2\frac{\theta}{2} + 2W_1(q^2, v)\sin^2\frac{\theta}{2}\right]. \quad (5.48)$$

The data for Fig. 5.27 were taken at only one angle ($\theta = 6°$) of the scattered electron, so that W_2 and W_1 were not separated. However, we expect the W_2-term to dominate at small angles, so that it is this one structure factor which the cross section effectively measures. The data can also be presented by plotting

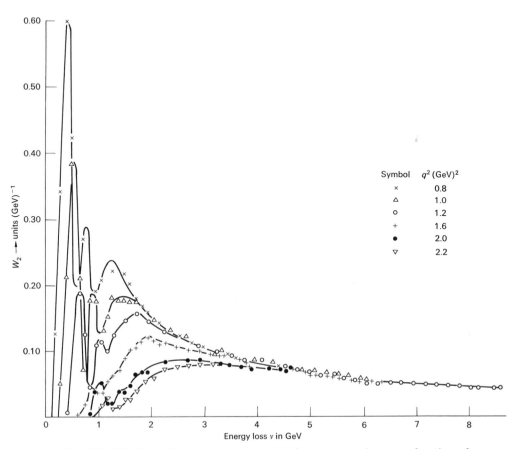

Fig. 5.27 The inelastic electron-proton scattering cross section as a function of energy transfer v, for different values of q^2, the invariant (4-momentum transfer)^2. For small q^2, the data at low v-values is dominated by the elastic and resonance peaks. Note that, at high q^2, these have largely disappeared. On the other hand, for $v > 3$ GeV, the continuum excitation falls off slowly and smoothly with v and has an extremely weak q^2-dependence. (SLAC data, after Panofsky, 1968.)

vW_2, found in this way, against the quantity

$$\omega = \frac{Mv}{q^2}, \tag{5.49}$$

as shown in Fig. 5.28. For $\omega > 2$, vW_2 is about the same for all values of v and q^2, with the points falling on a universal curve. For large values of ω, the W_1-term cannot be neglected, and the data is plotted twice for extreme assumptions, $R = 0$ and $R = \infty$, on the value of a ratio R defined by $W_1/W_2 = (1 + v^2/q^2)/(1 + R)$. More recent measurements show that R is small (of order 0.2), corresponding to the lower curve in the diagram. Thus vW_2 tends to a rather constant value for $\omega > 2$.

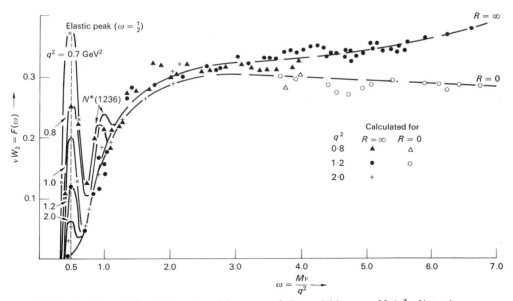

Fig. 5.28 The data of Fig. 5.27 replotted in terms of the variable $\omega = Mv/q^2$. Note the resonance peaks at small ω, and the continuum excitation which rises to a fairly constant value for $\omega > 2$. The points for all q^2 and v then appear to lie on a universal curve—the so-called scale-invariant behavior.

5.11 SCALE INVARIANCE AND PARTONS

The experiments thus show that the quantity $vW_2(q^2, v) = F(\omega)$ is a function of the *dimensionless* variable ω, in the region of continuum excitation, where q^2 and v are large. This means there is no *scale* to the scattering phenomenon; if there were, the q^2-dependence would be characterized by a unique mass or length, as in (5.38), in the elastic, or exclusive, reaction. One of several hypotheses put forward to account for this interesting behavior is the parton model, due originally to Feynman (1969) (see also Bjorken and Paschos, 1969, where further references are given). It is supposed that the nucleon consists of a number of elementary, point-like constituents or *partons*. These are supposedly massive but very strongly bound. Nevertheless, for high energy-momentum transfers, large compared with the nucleon mass, an individual parton can be regarded as effectively independent of the rest (just as an individual nucleon bound in a nucleus can be regarded as quasi-free under a large external impulse). Since the parton, by definition, has no structure, the electron-parton collision is elastic, so that in analogy with (5.24)

$$q^2 \approx 2mv,$$

where m is the effective parton mass (the difference between the mass of the free particle and the binding energy). Thus $F(\omega) = F(Mv/q^2) = F(M/2m)$ is a universal parameter measuring the parton effective mass distribution in the nucleon. On this basis, the elastic electron-proton scattering, or any other exclusive process, would be regarded as the coherent recoil of all constituents together, the form factor $G(q^2)$ then measuring the Fourier transform of the spatial distribution of partons (in complete analogy with the nuclear case).

In terms of such a model, one would hope to describe the multitude of inelastic channels, summing to the total cross section, in terms of simple properties of the constituents. For example, the observed ratio W_1/W_2, which measures the relative amounts of magnetic and electric scattering, indicates spin $\frac{1}{2}$ for the partons. It is therefore tempting to identify them with the quarks and antiquarks proposed as the basis states in the unitary symmetry schemes for classifying hadrons (see Chapter 6). If one makes the rather natural assumption that each type of quark carries, on average, the same fraction of the proton mass, then it turns out that the quantity $\frac{1}{2}\int vW_2(\omega)\,d\omega/\omega^2$ should equal the mean square quark charge. From the data in Fig. 5.28, the integral is found to be somewhat smaller than the value expected if the constituents have the fractional charge values $2e/3$ and $e/3$ usually ascribed to the quarks. Furthermore such assignments are unable to account for the observed difference in inelastic cross sections on neutrons and protons. Thus the simple constituent models have to be modified and in this process, lose some of their appeal.

Analogous studies of inclusive hadronic processes at large, timelike q^2 are possible with electron-positron clashing beam machines, and in their general features support the electron scattering results. Recent data from Frascati (Table 2.2) on the total cross-section for the inclusive reaction $e^+e^- \rightarrow$ hadrons $(2\pi, 3\pi \cdots)$ at 3 GeV centre-of-mass energy, give a value of about 25 μb, approximately equal to that for the purely leptonic "pointlike" process $e^+e^- \rightarrow \mu^+\mu^-$.

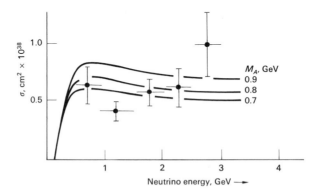

Fig. 5.29 Cross section for the reaction $v_\mu + n \rightarrow \mu^- + p$ as measured in CERN experiments, employing a heavy-liquid bubble chamber (see Fig. 4.25). The curves show the expected variation for different values of the axial-vector form factor. This was parameterized in terms of M_A, the mass in the dipole formula (5.38).

The real test of any constituent model of the nucleon is obviously to produce the constituents in a free state. As discussed in the following chapter, quarks have not so far been observed as free particles, possibly because they are very massive and strongly bound. Therefore, it is more correct to describe the features of inelastic electron scattering by the statement that hadronic matter behaves *as if* it were composed of elementary (quark and antiquark) constituents.

5.12 WEAK FORM FACTORS

The form factors discussed so far have been used to describe the q^2-dependence of the coupling of the hadrons to the electromagnetic current. (Although we have discussed only electron scattering, similar experiments are possible with muon beams and yield quite similar results.) We can also expect to describe the *weak* coupling in terms of form factors, corresponding, crudely speaking, to the spatial distribution of the weak or Fermi charge, rather than the electric charge. In weak interactions there are both vector (V) and axial-vector (A) currents,

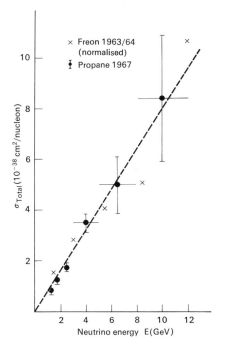

Fig. 5.30 Total cross section for the collision of neutrinos on nucleons, in the CERN bubble chamber experiments. A linear dependence would be expected if the appropriate inelastic form factors are scale invariant, so that the interaction is point-like (Budagov *et al.*, 1969).

hence two sets of form factors. According to the conserved vector current (CVC) hypothesis, it is proposed that the q^2-dependence of the vector form factor should be the same as in the electromagnetic case, leaving the axial-vector form factor $G_A(q^2)$ as the unknown. For the elastic process

$$v_\mu + n \rightarrow \mu^- + p \tag{5.50}$$

the q^2-dependence of the cross section has been investigated in high energy neutrino experiments, which suggest that G_A has approximately the same form as the vector form factor (5.38), i.e. with $M_A \sim 0.8$ GeV again. The existence of weak form factors implies that the cross section for (5.50) should reach a constant value at high energy, as indicated in Fig. 5.29. By way of contrast, the total neutrino cross section—the inclusive process—appears to rise linearly with neutrino energy (Fig. 5.30) as would be expected from scale-invariance, and Eq. (4.47) for point-like scattering.

BIBLIOGRAPHY

Feynman, R. P., *Theory of Fundamental Processes*, Benjamin, New York, 1961.

Farley, F. J., "Electromagnetic properties of the muon," *Prog. Nucl. Phys.* **9**, 257 (1964).

Griffy, T. A., and L. I. Schiff, "Electromagnetic form factors," *Pure and Applied Phys.* **25**, (1), 341 (1967).

Haissinski, J., "Physics with Electron-Positron Colliding Beams," Proc. CERN School, Leysin, 1969, p. 97 (CERN 69-29).

Hofstadter, R., "Nuclear and nucleon scattering of high-energy electrons," *Ann. Rev. Nucl. Science* **7**, 231 (1957).

Kendall, H. W., and W. K. Panofsky, "The Structure of the Proton and the Neutron", Scient. American, **224**, 60 (1971).

Minten, A., "Electron Scattering, Form Factors, Vector Mesons," Lectures in Academic Training Programme of CERN 1968–1969 (CERN 69-22).

PROBLEMS

5.1 In an electron-positron clashing beam experiment, the cross section at the ω-peak for the reaction $e^+e^- \rightarrow \pi^+\pi^-\pi^0$ is 1.5 μb. If the ring radius is 10 m, and the electron and positron beams are each of 10 mA, with a cross-sectional area of 0.1 cm^2, what would be the rate per hour for the above reaction? Assume the e^+ and e^- bunches meet head-on twice per revolution.

5.2 Show that an exponential charge-density distribution, of the form (5.39a), leads to a form-factor corresponding to the dipole formula (5.38) and verify that, for $M_V = 0.84$ GeV, the rms radius is 0.8 F.

5.3 From Fig. 5.21 it is seen that the process $e^+e^- \rightarrow \pi^+\pi^-$ measures the electromagnetic form-factor of the pion in the time-like region of q^2. If this process is dominated by the ρ intermediate state, show that the rms radius of the pion is 0.64 F (for $M_\rho = 765$ MeV).

5.4 The partial width for the decay $\omega \to \pi^0\gamma \approx 1$ MeV. Use the diagrams of Fig. 5.23 to estimate the lifetime for the decay $\pi^0 \to 2\gamma$. Assume the coupling of ω to γ to be characterized by a dimensionless constant $f_\omega = 0.1$. (The $\omega\rho\pi$ and $\rho\gamma$ couplings will cancel in the calculation.) Show that, under these very crude assumptions, one obtains

$$\frac{\pi^0 \to 2\gamma}{\omega \to \pi^0\gamma} = \frac{e^2}{\hbar c}\frac{f_\omega^2}{2}\left(\frac{m_\pi m_\omega}{m_\omega^2 - m_\pi^2}\right)^2,$$

and thus determine τ_{π^0}.

5.5 A 10 GeV electron collides with a proton and emerges from the collision with a $10°$ deflection and an energy of 7 GeV. Calculate the rest mass M^* of the recoiling hadronic state.

5.6 By considering the scalar product of the four-momentum transfer q with the four-momentum of the target nucleon, show that the quantity ω of Eq. (5.49) is Lorentz-invariant, provided ν is measured in the rest frame of the target.

5.7 A pencil electron beam of energy 15 GeV and intensity 10^{14} particles sec^{-1} impinges on a liquid hydrogen target of length 1 m parallel to the beam and of cross section sufficient to cover the beam. Estimate the number of electrons per second scattered elastically through 0.1 rad and into a solid angle of 10^{-4} sr, for (i) point-like spinless protons, (ii) protons with the form-factors of Eq. (5.23). (Hydrogen density $= 0.06$.)

Chapter 6

Strong Interactions I — Unitary Symmetry and the Quark Model

6.1 UNITARY SYMMETRY — TERMINOLOGY AND INTRODUCTION

So far, we have classified the strongly interacting particles (hadrons) in an empirical way by assigning them quantum numbers such as spin, parity, isospin, strangeness, baryon number, etc. In making these assignments, symmetry principles have frequently played an essential role. It is not surprising, therefore, that numerous attempts have been made over the years to guess at new symmetries which would embrace the ever-growing list of particle states and quantum numbers. Undoubtedly the most successful of the schemes put forward has been the "Eightfold Way" proposed independently by Gell-Mann (1962) and Ne'eman (1961), which, together with its subsequent development, is usually referred to as unitary symmetry.

The Gell-Mann-Ne'eman scheme is based on the symmetry group SU3. To discuss this, it is best to start with the groups called SU2, or R3, which arise in the treatment of angular momentum or isospin. We have seen that the conservation of isospin in strong interactions is equivalent to invariance of the interaction under rotations of the coordinate axes in a three-dimensional "isospin space." Such rotations can be accomplished with rotation operators, as in (3.10). Thus, if χ represents the two-component isospin function of a nucleon, it is transformed under rotation through θ about any axis $\mathbf{n}$ (unit vector) to

$$\chi' = Q\chi, \tag{6.1}$$

where

$$Q = \exp(i\boldsymbol{\tau} \cdot \mathbf{n}\theta), \tag{6.2}$$

and $\boldsymbol{\tau}$ is the Pauli operator (3.38). Since θ can have any value, (6.1) describes an infinite set of transformations generated by $\boldsymbol{\tau}$ in a three-dimensional, real (i.e. not complex) space. These transformations form the so-called rotation group,

R3. Of course, as far as strong interactions are concerned, isospin is exactly conserved and all the states (6.1) are degenerate.

It is not essential to describe isospin invariance in terms of rotations. The operator Q can also be written as a 2×2 matrix with complex elements, generating transformations in a complex, two-dimensional space. It then has the form

$$Q = \begin{vmatrix} \alpha & \beta \\ -\beta^* & \alpha^* \end{vmatrix},$$ (6.3)

where

$$\det Q = \alpha\alpha^* + \beta\beta^* = 1, \qquad \alpha = \cos\left(\frac{\theta}{2}\right) + in_3 \sin\left(\frac{\theta}{2}\right),$$

and

$$\beta = i(n_1 - in_3) \sin\left(\frac{\theta}{2}\right).$$

This result is easily proved by expanding (6.2) in a series in $\theta/2$. The matrix Q has the property $Q_{jk}Q_{kj}^* = 1$ (the unit matrix), and is therefore said to be *unitary* (i.e. it preserves the norm of the wave function χ). $Qe^{i\gamma}$ would also be unitary, but since we are not interested in unobservable phases, we consider only the *unimodular* matrix with det $Q = +1$. The transformations produced by the 2×2 unitary, unimodular matrix Q form the symmetry group SU2. The "2" stands for the dimension of the matrix, which in this case has arisen because we started off with the nucleon isospin doublet.

One of the tenets of group theory is that, if a and b are elements of a group, $c = ab$ is also an element of the group. Thus, we can combine two $\frac{1}{2}$-spin objects, each from the fundamental "2" *representation* of SU2, to form another representation. We did this in Section 3.10, where we combined proton and neutron states in pairs, and found four combinations which could be broken down or reduced into an $I = 1$ triplet (or "3" representation) and $I = 0$ singlet ("1" representation). Symbolically, this is written as

$$2 \otimes 2 = 1 \oplus 3.$$ (6.4)

All this mathematical juggling really adds nothing to our knowledge of spin or isospin. It does, however, stress the fact that it *is* possible to build up multiplets of arbitrary complexity from the basic doublet, and that therefore any angular momentum or isospin state is a representation of SU2. As it happens, nature has decreed that both $J = \frac{1}{2}$ and $I = \frac{1}{2}$ basic states actually exist.

In the real world, of course, one does not have exact SU2 symmetry, with all members of a multiplet having identical properties. Electromagnetic effects break the isospin symmetry, but are sufficiently small for the symmetry to be easily recognized. As an example, the components of the Σ-multiplet have closely similar masses:

	I_3	Mass, MeV
Σ^-	-1	1197.4
Σ^0	0	1192.6
Σ^+	$+1$	1189.4.

SU3 is simply an extension of SU2 to three dimensions (in a complex space). One starts off with a basic triplet, instead of doublet. This means we incorporate a further quantum number. Besides isospin I, we can have strangeness S. Combining three triplets (or "3" representations of SU3), one will get a "9" representation, reducible to a singlet and octet. The "8" representation or multiplet gave rise to the name "Eightfold Way." Singlets and octets, incorporating particle states into isospin and strangeness multiplets, are indeed observed, whereas the basic triplets, so far, are not. The SU3 symmetry is badly broken by interactions which differentiate between different values of S as well as I_3, but is still readily recognizable. The mathematical construction of SU3 multiplets is best achieved using tensor algebra. In order, however, to keep the discussion as simple as possible, and to have a "physical" picture of what is going on, we shall assume the existence of particles corresponding to the basic triplet, and combine them using simple symmetry arguments. This is usually called the "quark model." First, however, we shall revert to SU2, as exemplified by the Fermi model of pions.

6.2 PIONS AS NUCLEON-ANTINUCLEON COMBINATIONS

One of the first attempts to classify particle states as combinations of more "elementary" particles was made by Fermi and Yang in 1949. They considered pions as combinations of nucleon-antinucleon pairs (some seven years before antinucleons were observed experimentally). It will be recalled from Section 3.10 that nucleon-nucleon pairs can form an $I = 1$ triplet and an $I = 0$ singlet. By applying G-conjugation to one of the nucleons, one can form the corresponding $N\bar{N}$ isospin states. As in (3.65), the G-operation is defined as $G = CR = C \exp(i\pi\tau_2)$ where τ_2 is the y-component Pauli operator (3.38). Then, expanding the exponential as a power series in $\pi\tau_2$, and using the fact that $(2\tau_2)^2$ is a unit matrix, it is found that the nucleon state, with I_3 eigenfunctions p or n, transforms as

$$\begin{pmatrix} p \\ n \end{pmatrix} \xrightarrow{R} \left(\cos\frac{\pi}{2} + 2i\tau_2 \sin\frac{\pi}{2} \right) \begin{pmatrix} p \\ n \end{pmatrix} = \begin{pmatrix} 0 & 1 \\ -1 & 0 \end{pmatrix} \begin{pmatrix} p \\ n \end{pmatrix} = \begin{pmatrix} n \\ -p \end{pmatrix},$$

so that

$$\begin{pmatrix} p \\ n \end{pmatrix} \xrightarrow{G = CR} \begin{pmatrix} \bar{n} \\ -\bar{p} \end{pmatrix}, \tag{6.5}$$

where we have denoted the charge-conjugate (antiparticle) states of neutron and proton by $\bar{n}$ and $\bar{p}$. This choice of phase is a natural one, since nucleon and antinucleon then transform in the same way under isospin rotations. Note the change in sign of the antiproton state, relative to the antineutron, in (6.5). This arises from the way in which R is defined, as a rotation about the y-axis. Had we defined $R = \exp(i\pi\tau_1)$, which is an equally valid choice, no such change of sign would have arisen, although a common phase factor $\exp(i\pi/2)$ would have appeared. The definition $R = \exp(i\pi\tau_2)$ is adopted in this text because of its common use, but more particularly because the singlet $I = 0$ nucleon-antinucleon state is then symmetric in the particle/antiparticle labels, as shown below. This feature is useful later in spotting SU3 singlet states of baryon number zero.

Written out specifically in terms of isospin components, we thus have:

TABLE 6.1

	$B = +1$	$B = -1$
$I_3 = +\frac{1}{2}$	p	$\bar{n}$
$I_3 = -\frac{1}{2}$	n	$-\bar{p}$

Referring to (3.43) we may now write out the NN, and, by G-conjugation, the corresponding $N\bar{N}$, isospin wave functions:

TABLE 6.2

		Nucleon-nucleon ($B = 2$)	Nucleon-antinucleon ($B = 0$)	J^{PG}
$I = 1$	$I_3 = +1$	pp	$p\bar{n} \leftrightarrow \pi^+$	
	$I_3 = -1$	nn	$-n\bar{p} \leftrightarrow \pi^-$	0^{--}
	$I_3 = 0$	$\dfrac{np + pn}{\sqrt{2}}$	$\dfrac{n\bar{n} - p\bar{p}}{\sqrt{2}} \leftrightarrow \pi^0$	
$I = 0$	$I_3 = 0$	$\dfrac{np - pn}{\sqrt{2}}$	$\dfrac{n\bar{n} + p\bar{p}}{\sqrt{2}} \leftrightarrow \eta$	0^{-+}

The first three $N\bar{N}$-states form an isospin triplet, and the last forms an isospin singlet. Take the case where the two-nucleon system has $l = 0$ and spins antiparallel. The corresponding $I = 1 N\bar{N}$ states have $J = 0$, and odd intrinsic parity and G-parity, while the $I = 0$ state has $J = 0$, $P = -1$, and $G = +1$ (see Eq. 3.67). Thus, the triplet has the quantum numbers of the pion, which could be regarded as a very tightly bound combination of nucleon and antinucleon. The singlet state was a bit of a problem 20 years ago. Nowadays, it would be identified with the η-meson.

The combination of two doublet states $N\overline{N}$ gives us four independent states, which form an $I = 1$ triplet and an $I = 0$ singlet; the $I = 0$ state is the only one which is symmetric under interchange of particle/antiparticle labels. It is therefore quite unique. The first three states can be transformed one into the other by isospin rotations, for example by using the operators $I_{\pm}$ (Appendix C), but there is no operation which one can perform on the $I = 0$ state to transform it into $I = 1$. As we have seen, in the language of group theory the nucleon is a two-component, or simply a "2" representation of the symmetry group SU2 and the antinucleon a "$\overline{2}$" representation. The combination $2 \times \overline{2}$ is reducible into the $I = 0$ (singlet, or "1") representation of SU2 and the $I = 1$ (triplet, or "3") representation. Symbolically,

$$2 \otimes \overline{2} = 1 \oplus 3. \tag{6.4}$$

As emphasized above, the transformation properties of the $I = 1$ and $I = 0$ multiplets can be expressed in terms of the appropriate operators $I_{\pm}, I_3, I^2$, etc. (with $I = 1$ or 0), *independent* of the real existence of the fundamental $I = \frac{1}{2}$ doublet—just as the description of states of integral angular momentum can be made independently of the existence of states of half-integral spin.

6.3 THE SAKATA MODEL

An obvious defect of the Fermi model is that it does not incorporate strange particles. In 1956, Sakata proposed to remedy this by enlarging the fundamental nucleon doublet to a triplet, consisting of p, n, and Λ, with, of course, their antiparticles $\bar{p}, \bar{n}$, and $\overline{\Lambda}$. Let us now see how we can build up meson states using combinations of these fundamental "bricks."

We denote the Sakata states by equal distances along three orthogonal axes (Fig. 6.1). We can make a symmetrical two-dimensional plot by projecting these states onto a plane making equal angles with all three axes. Let the dotted equilateral triangle have sides of length $2/\sqrt{3}$ units. Plot I_3 along the x-axis, and strangeness S along the y-axis. Then the projections give us

$$\text{isotopic doublet} \begin{cases} n, & I_3 = -\tfrac{1}{2}, & S = 0 \\ p, & I_3 = +\tfrac{1}{2}, & S = 0 \end{cases}$$

$$\text{isosinglet} \quad \Lambda, \quad I_3 = 0, \qquad S = -1.$$

Now suppose we combine p, n, Λ and $\bar{p}$, $\bar{n}$, $\overline{\Lambda}$ in baryon-antibaryon pairs— giving nine possible states. Instead of S, we shall in fact plot the hypercharge $Y = B + S$ along the y-axis; since $B = 0$ for baryon-antibaryon combinations, Y and S are equal in this case.

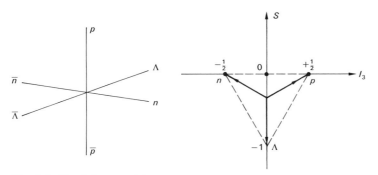

Fig. 6.1 The Sakata model.

Thus, the state $p\bar{\Lambda}$ is obtained by going along the p-axis at 30°, and then along the $\bar{\Lambda}$-axis upward, thus giving a point $I_3 = +\frac{1}{2}$, $Y = 1$ (Fig. 6.2). The other five "outer" states are obtained similarly. We are then left with three combinations of $I_3 = Y = 0$, made up of $p\bar{p}$, $n\bar{n}$, and $\Lambda\bar{\Lambda}$. In our previous discussion we said $(n\bar{n} - p\bar{p})$ corresponds to the π^0, and $(n\bar{n} + p\bar{p})$ to η. The latter was a separate singlet as far as isospin was concerned, but now we have an extra quantum number (S) and we must include $\Lambda\bar{\Lambda}$-combinations. As before, there is one combination, $(n\bar{n} + p\bar{p} + \Lambda\bar{\Lambda})$, which is symmetric under particle/antiparticle label interchange and is thus an SU3 singlet, which cannot transform into the other eight SU3 states. We call it η'. The remaining two combinations may be formed in more than one way. To keep as close to our previous discussion as possible, we choose one of these to be the π^0, i.e. the combination

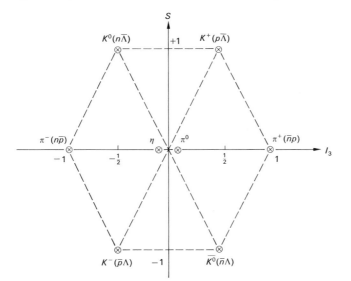

Fig. 6.2 Meson states in Sakata model.

$(n\bar{n} - p\bar{p})$, which transforms to the π^+ and π^- states by isospin rotations (i.e. an SU2 transformation). The third combination is then the isospin singlet, the η. It has to be orthogonal to the other two. With the correct normalization, the three states are then found to be:

$$\pi^0 = (n\bar{n} - p\bar{p})/\sqrt{2}, \qquad I = 1, I_3 = Y = 0$$

$$\left. \begin{array}{l} \\ \eta = (n\bar{n} + p\bar{p} - 2\Lambda\bar{\Lambda})/\sqrt{6}, \qquad I = Y = 0 \end{array} \right\} \text{ central members of octet}$$

$$\eta' = (n\bar{n} + p\bar{p} + \Lambda\bar{\Lambda})/\sqrt{3}, \qquad I = Y = 0 \qquad \text{singlet (symmetric).}$$

Thus, η, π^0, and η' are orthonormal, with $\langle \eta \,|\, \eta' \rangle = 0$, $\langle \eta \,|\, \eta \rangle = 1$, etc. While the state η, as we have defined it, cannot transform into the other states π^+, π^-, and π^0 along the I_3-axis of the figure, it (or rather a linear combination of π^0 and η) *can* transform into the other neutral states K^0 and $\bar{K}^0$ along the U-spin axis at 10 o'clock. We return to this point later. The essential point is that eight states can transform one into the other, while the ninth, η', is an SU3 singlet. We can write this:

$$3 \otimes \bar{3} = 8 \oplus 1. \tag{6.6}$$

What we have drawn out is the *octet* of pseudoscalar mesons, $J^P = 0^-$:

$$\pi^+, \pi^-, \pi^0 \qquad \text{mass} \sim 140 \text{ MeV,}$$

$$K^+, K^-, K^0, \bar{K}^0 \qquad \text{mass} \sim 490 \text{ MeV,}$$

$$\eta \qquad \text{mass} \sim 540 \text{ MeV,}$$

plus a pseudoscalar *singlet*, usually ascribed to the meson η' of mass 960 MeV.

There is as yet no real indication here that these states form a supermultiplet, and that, if we were to turn off the symmetry-breaking interactions, which split the states of different I_3 and S, we would end up with eight degenerate 0^- states, of a unique mass, as required by perfect SU3 symmetry. Also, the allocation $J^P = 0^-$ has not been justified. However, if we made the assumption that the p, n, Λ and $\bar{p}$, $\bar{n}$, $\bar{\Lambda}$ combinations are singlet s-states (i.e. spin antiparallel, $l = 0$), we would expect them to have $J = 0$ and negative parity—since the particle-antiparticle states have intrinsic parity -1. Equally, one would expect $l = 0$ states with spins parallel, i.e. $J^P = 1^-$, and indeed there *is* a multiplet of vector mesons with these properties (see Section 6.7).

6.4 BARYON STATES AND THE QUARK MODEL

If we try to apply the Sakata scheme to baryons, we run into difficulties. Combinations such as $pn\Lambda$ give $B = 3$, and such states are not observed. Thus we must restrict ourselves to combinations like $p\bar{n}\Lambda$, with $B = 1$. But then we cannot obviously exclude the combination $pn\bar{\Lambda}$, with strangeness $S = +1$,

TABLE 6.3 Quark quantum numbers

	Symbol	B	I_3	S	q/e
p-type quark	Q_1	$\frac{1}{3}$	$+\frac{1}{2}$	0	$+\frac{2}{3}$
n-type quark	Q_2	$\frac{1}{3}$	$-\frac{1}{2}$	0	$-\frac{1}{3}$
Λ-type quark	Q_3	$\frac{1}{3}$	0	-1	$-\frac{1}{3}$

which also is not observed in nature. The alternative is to assume the basic triplet is not p, n, Λ, but other particles with *fractional* baryon number. If one considers combinations of up to three such particles, the simplest assumption is that each has $B = \frac{1}{3}$. Such objects have been called *quarks* by Gell-Mann (1964) and Zweig (1964). They showed that the "Eightfold Way" of Gell-Mann and Ne'eman, based on SU3 and a basic octet of particles, could be thought of in terms of combinations of a basic triplet of quarks. At this point, we do not wish to discuss whether such objects exist as free particles, but to see the logical consequences of building up particle states out of quark combinations. As before, the basic triplet will consist of a doublet, of $I_3 = \pm\frac{1}{2}$, $S = 0$, and a singlet, $I = 0$, $S = -1$. From the formula (3.36), the particle charge is given by

$$\frac{q}{e} = \frac{B+S}{2} + I_3,$$

which must hold for quarks if we are to build the observed particle states from them. Thus, the quark quantum numbers are as in Table 6.3, and involve fractional electric charges. Antiquarks $\bar{Q}_1$, $\bar{Q}_2$, $\bar{Q}_3$ have opposite values for I_3, B, and S and thus opposite charges also.

Going back to the meson states of Fig. 6.2 with $Q_1\bar{Q}_2$ replacing $p\bar{n}$, and so on, we find no change. Thus K^0 corresponds to $Q_2\bar{Q}_3$, π^- to $Q_2\bar{Q}_1$, etc. Now consider three-quark combinations. First consider the nine QQ pairs one may form from three quarks. These can be grouped into a *sextet* whose members are symmetric under quark label interchange, and a *triplet* which is antisymmetric:

$$
\begin{array}{ccc}
Q_1Q_1 & & Q_1Q_2 - Q_2Q_1 \\
Q_2Q_2 & & Q_1Q_3 - Q_3Q_1 \\
Q_3Q_3 & \text{and} & Q_2Q_3 - Q_3Q_2 \\
Q_1Q_2 + Q_2Q_1 & & \\
Q_1Q_3 + Q_3Q_1 & & \\
Q_2Q_3 + Q_3Q_2 & &
\end{array}
$$

A under quark label interchange

S under quark label interchange

Symbolically this is written as

$$3 \otimes 3 = 6 \oplus 3$$
$$\begin{array}{cc} \uparrow & \uparrow \\ S & A. \end{array}$$

(6.7)

There are no known candidates for these representations, nor do we expect any, since $B = \frac{2}{3}$. The next step is to combine a further quark with these pairs. There are a total of $3^3 = 27$ different QQQ combinations, which will break down into four distinct representations, with the following properties:

Number of states	Symmetry of QQ pair under label interchange	Symmetry of third Q relative to pair
10	S	S
8	S	A
8	A	S
1	A	A

The first row of ten states (decuplet) which is completely symmetric under interchange of all quark labels is easily written down. Writing only the subscripts, these are

$$111, \quad 112 + 211 + 121, \quad 113 + 311 + 131,$$
$$222, \quad 221 + 122 + 212, \quad 223 + 322 + 232, \quad (6.8)$$
$$333, \quad 331 + 133 + 313, \quad 332 + 233 + 323,$$

and
$$123 + 231 + 312 + 213 + 321 + 132.$$

Of the total of 27 states, there are $6 \times 3 = 18$ where the first QQ pair is symmetric under label interchange [from the sextet of (6.7)], and, thus, there must be $18 - 10 = 8$ states where the pair is S and the third quark is A with respect to the pair (i.e. they have no particular symmetry). There are $3 \times 3 = 9$ states where the QQ system is A, and in one of these, the third quark is also A with respect to the first two. This completely antisymmetric state is

$$123 + 231 + 312 - 213 - 321 - 132. \quad (6.8a)$$

This leaves, finally, 8 states where the QQQ system is A and the third quark is S—for example, (121-211) would be a member. We will not write out all the combinations in full. The main point is that the 27 three-quark combinations subdivide into a decuplet which is totally S, a singlet which is totally A, and two octets of mixed symmetry. Symbolically,

$$3 \otimes 3 \otimes 3 = 27 = 1 \oplus 8 \oplus 8 \oplus 10.$$
$$\begin{array}{cccc} & \uparrow & \underbrace{} & \uparrow \\ & A & \text{mixed} & S \end{array}$$

(6.9)

No further simplification is possible for combinations of our basic quark triplet, three at a time. The singlet, octets, and decuplet are said to be irreducible representations of SU3. They cannot be transformed one into another by rotations in "unitary spin space." We see that the quark interchange symmetry determines the SU3 representation; just as the nucleon-nucleon exchange symmetry determined the SU2 representation, i.e. $I = 1$ ("3," S) or $I = 0$ ("1," A). But now, instead of symmetry under isospin rotations (transforming the I_3 substates one into the other) we are involved in a higher degree of symmetry, interchanging quarks of different I_3 and S. Depending on whether these interchanges are symmetric, antisymmetric, or of mixed symmetry, one obtains the different representations.

The completely antisymmetric state (6.8a) is a singlet, in the same way that the nucleon-nucleon combination $n(1)p(2) - n(2)p(1)$ of (3.43d), which we would denote (12-21) in the above notation, was an isospin (SU2) singlet. Note that the three-quark singlet is completely antisymmetric under interchange of quark labels, while the singlet quark-antiquark state appearing in the previous discussion of the meson multiplets was symmetric under quark-antiquark interchange. This difference is just one of convention, arising from the way we have defined phases in the operation of charge conjugation. In either SU2 or SU3, the singlet state is unique and distinct in that, under rotations in isospin space or "unitary spin" space, it can transform only into itself.

In Chapter 1 we observed that pairs of identical particles have definite exchange symmetry (S or A). The occurrence of octet combinations of *mixed* symmetry arises because we are considering three particles at a time, and because we have not considered what happens under spin (or space) coordinate interchange. When we incorporate quark spin later, it is found that the proper

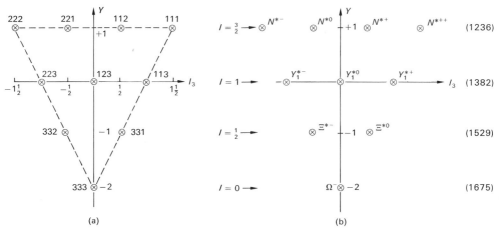

Fig. 6.3 (a) Quark label assignments in the baryon decuplet. Except for the states at the apices, each site label corresponds to the first term only of (6.8). (b) The observed decuplet of baryon states of spin-parity $\frac{3}{2}^+$. The mean mass of each isospin multiplet is given in brackets.

combinations do, in fact, have a definite symmetry; the three-quark wave-function is symmetric under interchange of any pair of quarks and their spin.

The predicted baryon decuplet is drawn in Fig. 6.3(a), together with the quark labeling. The coordinates are I_3 along the x-axis and hypercharge $Y = B + S$ along the y-axis. Figure 6.3(b) shows observed baryon states, of $J^P = \frac{3}{2}^+$, consisting of the $N^*(1236)$ isospin quartet, $Y_1^*(1382)$ triplet, $\Xi^*(1530)$ doublet, and $\Omega^-(1675)$, plotted in the same way. Note the consecutive mass differences, which are large, but remarkably equal. Figure 6.4 shows the baryon octet of $J^P = \frac{1}{2}^+$.

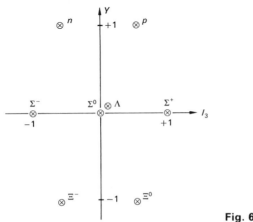

Fig. 6.4 Baryon octet of spin-parity $\frac{1}{2}^+$.

The first predictions of the quark model, that (i) baryons consist of QQQ combinations, and should occur as singlets, octets, and decuplets, and (ii) mesons consist of $Q\bar{Q}$ combinations, giving rise to singlets and octets, are therefore verified. Note that we affix a given value of J^P to a particular multiplet—we have not considered quark spin, but obviously in all our discussion so far on exchange symmetry, it was tacitly assumed that the quark spins were not interfered with—hence all members of a multiplet should have the same J^P.

6.5 U-, V-, AND I-SPIN.
SYMMETRY BREAKING AND MASS FORMULAS

It is useful, instead of plotting the quark coordinates on three orthogonal axes, as in Fig. 6.1, to rotate the axes by 30°, then plot along them three equivalent quantities, which we call U-, V-, and I-spin—U- and V-spin not being defined as yet—and then to project the third components U_3, V_3, and I_3 onto the Y-I_3 plane (Fig. 6.5). Note that, for any state, $U + I + V = 0$ from geometry, so one of the three variables is redundant—we just use U and I. U, I, and V are called the *unitary spin*. Instead of building up our meson and baryon states out

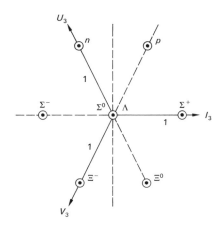

Figure 6.5

of "physical" quarks, we can disregard quarks altogether and think of the supermultiplets as arising from simultaneous transformations of *both* U- and I-spin—i.e. the symmetry group SU3—this symmetry being broken by U- and I-spin breaking interactions, producing the observed states of definite Y, I_3 eigenvalues.

Therefore, along the I_3-axis, the isospin states are split up into I_3-components of slightly different masses, for example $\Sigma^+, \Sigma^0, \Sigma^-$, associated with their different electric charges. Equally, along the U_3-axis, the mass splitting of n, Λ, Σ^0, Ξ^0 can be considered as due to different "U-charges." In fact, if we go to the decuplet of Fig. 6.3, we note that the four $q = -1$ states—members of a $U = \frac{3}{2}$ quartet—have the masses listed in Table 6.4.

There is an accurate linear dependence of mass M on U_3. In fact, this equal spacing rule for the decuplet led to the prediction of the Ω^- particle before it was discovered experimentally. A particle of $S = -3$ and $M = 1675$ MeV, cannot decay by strong or electromagnetic interactions (the strong decay mode $\Omega^- \to K^- + \Xi^0$ requires a minimum mass of 1805 MeV). So it must decay weakly ($\Delta S = 1$) by one of the modes $\Omega^- \to \pi\Xi, K^-\Sigma$, or $K^-\Lambda$. Figure 6.6 shows the first Ω^- event, observed by Barnes and 32 co-authors in 1964 in a hydrogen bubble chamber irradiated with a beam of 5 GeV/c negative kaons.

TABLE 6.4 Mass splitting in $J^P = \frac{3}{2}^+$ decuplet

U_3 State	$\frac{3}{2}$ $N^{*-}(1236)$	$\frac{1}{2}$ $Y_1^{*-}(1382)$	$-\frac{1}{2}$ $\Xi^{*-}(1529)$	$-\frac{3}{2}$ $\Omega^-(1675)$
ΔM(MeV)		146	147	146
Quark assignment	222	223	332	333

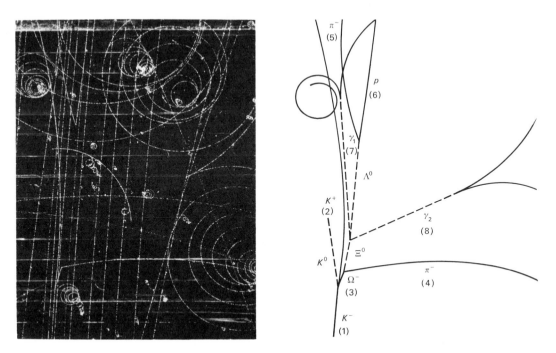

Fig. 6.6 The first Ω^- event (Barnes *et al.*, 1964). (Courtesy, Brookhaven National Laboratory.)

It depicts the following chain of events:

$$K^- + p \rightarrow \Omega^- + K^+ + K^0$$

$$\hookrightarrow \Xi^0 + \pi^- \ (\Delta S = 1 \text{ weak decay})$$

$$\hookrightarrow \pi^0 + \Lambda \ (\Delta S = 1 \text{ weak decay})$$

$$\hookrightarrow \pi^- + p \ (\Delta S = 1 \text{ weak decay})$$

$$\hookrightarrow \gamma + \gamma \ (\text{E.m. decay})$$

$$\downarrow \quad \downarrow$$

$$e^+ e^- \quad e^+ e^-.$$

The equal spacing rule tells us that the symmetry-breaking interaction has a simple form; the mass

$$M = a + bU_3, \tag{6.10}$$

i.e. the interaction is a combination of scalar and vector in U-spin space. Note that mass splittings for a U-spin multiplet are of order 100 MeV, much larger than the electromagnetic splittings. The symmetry-breaking interaction is for this reason called "medium strong." On the quark model, one way of accounting for the splitting is to suppose the quarks have different masses. In the limit of exact SU3 symmetry, all three quarks would have identical masses, so that their combinations are then SU3 degenerate states. Since the equal spacing rule holds to an accuracy of order 1%, it is presumed that the quark mass is very large (many GeV), so that the symmetry breaking follows the linear rule (6.10). This means that, to yield the observed baryon masses, the three quarks must be bound together in a very deep potential well, by a _superstrong_ interaction. When the symmetry-breaking (i.e. medium strong) interactions are turned on, the three quarks separate into strange (Q_3) and nonstrange (Q_1 and Q_2) with masses differing by ~ 100 MeV. Q_1 and Q_2 are further separated along the I_3-axis (~ 1 MeV) by the electromagnetic interactions.

We now turn to the mass relation for the $\frac{1}{2}^+$ baryon octet, where the center position is occupied by both Λ and Σ^0 (Fig. 6.4). In order to apply the mass formula, one must express these not as I-spin but as U-spin states. Consider the $U = 1$ triplet:

$$n, \quad \alpha\Sigma^0 + \beta\Lambda, \quad \Xi^0$$
$$U_3: \quad +1 \qquad \quad 0 \qquad \quad -1,$$

where we write the middle state as a linear combination of the I-spin states Σ^0 and Λ. All states in an irreducible representation have unit weight, so that $\alpha^2 + \beta^2 = 1$. Now apply the Clebsch-Gordan shift operators on the U_3 substrates. From Appendix C (Eq. C5), we have, in complete analogy with angular momentum and isospin,

$$U_-\psi(U, U_3) = \sqrt{U(U + 1) - U_3(U_3 - 1)}\,\psi(U, U_3 - 1),$$
$$U_+\psi(U, U_3) = \sqrt{U(U + 1) - U_3(U_3 + 1)}\,\psi(U, U_3 + 1). \tag{6.11}$$

Let us work our way from the neutron state to the Σ^+ state (Fig. 6.7) by steps 1 and 2. Step 1, using U_- on the neutron wave function, denoted by $|n\rangle$,

$$U_- |n\rangle = U_-\psi(1, 1) = \sqrt{2}\,\psi(1, 0) = \sqrt{2}(\alpha\,|\Sigma^0\rangle + \beta\,|\Lambda\rangle). \tag{6.12}$$

Step 2, using the isospin raising operator I_+,

$$I_+U_- |n\rangle = I_+\sqrt{2}\alpha\,|\Sigma^0\rangle = \sqrt{2}\,\alpha\sqrt{2}\,|\Sigma^+\rangle, \tag{6.13}$$

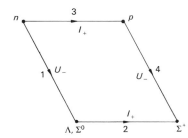

Figure 6.7

where, since $I_\Lambda = 0$, the I_+ operator acts only on the Σ^0 part of the state (6.12). One can also get to the Σ^+ state by steps 3 and 4. Step 3 gives

$$I_+ |n\rangle = I_+ \psi(\tfrac{1}{2}, -\tfrac{1}{2}) = \psi(\tfrac{1}{2}, \tfrac{1}{2}) = |p\rangle.$$

Step 4 gives

$$U_- I_+ |n\rangle = U_- |p\rangle = U_- \psi(\tfrac{1}{2}, \tfrac{1}{2}) = \psi(\tfrac{1}{2}, -\tfrac{1}{2}) = |\Sigma^+\rangle. \qquad (6.14)$$

Thus $I_+ U_- |n\rangle = 2\alpha |\Sigma^+\rangle$ and $U_- I_+ |n\rangle = |\Sigma^+\rangle$. These two operations are equivalent, since I and U must commute (they are both "good" quantum numbers in the limit of exact SU3 symmetry). So, $\alpha = \tfrac{1}{2}$ and since $\alpha^2 + \beta^2 = 1$, we have $\beta = \sqrt{3}/2$. The $U = 1$ triplet in the baryon octet is then

$$n, \quad \Sigma_u^0 = \tfrac{1}{2}[\Sigma^0 + \sqrt{3}\,\Lambda], \quad \Xi^0$$
$$U_3: \quad +1 \qquad\qquad 0 \qquad\qquad\qquad -1, \qquad (6.15)$$

where the state $U = 1$, $U_3 = 0$ is denoted Σ_u^0. The $U = 0$, $U_3 = 0$ singlet state (which must exist in precise analogy with the isospin singlet and triplet) must be orthogonal to Σ_u^0 and therefore has the form

$$\Lambda_u = \tfrac{1}{2}[\sqrt{3}\,\Sigma^0 - \Lambda]. \qquad (6.16)$$

If the states (6.15) obey the spacing rule for the decuplet (6.10), then squaring the amplitudes and using the fact that the isospin states Σ^0 and Λ are orthogonal, we expect that the following masses are equally spaced:

$$M_n, \quad \underbrace{\tfrac{1}{4}[M_{\Sigma^0} + 3M_\Lambda]}, \quad M_{\Xi^0}$$
$$\Delta M: \quad 195\ \text{MeV} \qquad\quad 180\ \text{MeV}. \qquad (6.17)$$

Usually this relation is quoted as

$$\frac{(M_n + M_{\Xi^0})}{2} = \frac{(M_{\Sigma^0} + 3M_\Lambda)}{4}, \qquad (6.18)$$
$$\qquad\uparrow \qquad\qquad\qquad \uparrow$$
$$1127\ \text{MeV} \qquad\quad 1135\ \text{MeV}$$

which looks much better! Equation (6.18) is an example of the octet mass formula of Okubo and Gell-Mann.

For *mesons*, the linear mass formula, at least of the above type, does not work. If we consider the 0^- meson octet (Fig. 6.2), we would predict, in analogy with (6.18),

$$\frac{(M_{K^0} + M_{\overline{K^0}})}{2} = \frac{(M_{\pi^0} + 3M_\eta)}{4},$$

where

$$M_{K^0} \equiv M_{\overline{K^0}} \qquad \text{by } CPT.$$

Hence

$$M_\eta = \frac{4M_{K^0} - M_{\pi^0}}{3} \sim 620\ \text{MeV}.$$

The η-meson, which is the only likely $I = 0$, $J^P = 0^-$ candidate available, was found after this prediction, but with $M_\eta = 549$ MeV. Subsequently it was discovered, for no clearly convincing reason,* that the meson mass formula worked better if one used the (mass)^2:

$$\frac{(M_{K^0}^2 + M_{\bar{K}^0}^2)}{2} = M_{K^0}^2 \approx (M_{\pi^0}^2 + 3M_\eta^2)/4.$$

$$\underset{0.247 \text{ GeV}^2}{\uparrow} \qquad \underset{0.231 \text{ GeV}^2}{\uparrow} \qquad (6.19)$$

The concept of U-spin invariance of the SU3 symmetry can also be used to predict branching ratios for the decay of baryon resonances, exactly as isospin invariance was exploited in Section 3.11. Consider, for example, the two processes

$$Y_1^{*-} \rightarrow \Lambda + \pi^-, \qquad (6.20a)$$

and

$$N_{3/2}^{*-} \rightarrow n + \pi^-. \qquad (6.20b)$$

In each case, the initial state has $U = \frac{3}{2}$, being a member of a U-spin quartet. Further, as seen in Fig. 6.2, the π^- has $U = \frac{1}{2}$, $U_3 = +\frac{1}{2}$. The Λ-state, on the other hand, is a U-spin mixture, which as we have seen has the form

$$|\Lambda\rangle = -\tfrac{1}{2}\psi(U = 0, U_3 = 0) + \frac{\sqrt{3}}{2}\psi(U = 1, U_3 = 0).$$

In the decay $Y_1^* \rightarrow \Lambda + \pi$, $U = \frac{3}{2}$, so only the $U = 1$ part of the Λ can contribute. Now, from the Clebsch-Gordan coefficients for adding $J_1 = 1$ and $J_2 = \frac{1}{2}$, to give $J = \frac{3}{2}$ or $\frac{1}{2}$, one finds

$$\psi(1, 0) \times \psi(\tfrac{1}{2}, \tfrac{1}{2}) = \sqrt{\tfrac{2}{3}}\,\psi(\tfrac{3}{2}, \tfrac{1}{2}) + \sqrt{\tfrac{1}{3}}\,\psi(\tfrac{1}{2}, \tfrac{1}{2}).$$

Thus, the matrix element for (6.20a) will be

$$M(Y_1^{*-} \rightarrow \Lambda + \pi^-) = \sqrt{\frac{2}{3}}\frac{\sqrt{3}}{2}M_1 = \frac{M_1}{\sqrt{2}}.$$

Now, for reaction (6.20b) the neutron has $U = 1$, $U_3 = 1$, so that $(n + \pi^-)$ is a pure $U_3 = \frac{3}{2}$, $U = \frac{3}{2}$ state. Therefore

$$M(N_{3/2}^{*-} \rightarrow n + \pi^-) = M_1,$$

so that the ratio of decay rates (a)/(b) $= \frac{1}{2}$, times the appropriate phase-space factors, which in this case reduce the expected ratio to about the observed value of $\Gamma(Y_1^* \rightarrow \Lambda\pi^-)/(N_{3/2}^* \rightarrow n\pi^-) \approx \frac{1}{4}$. In general, decay rates and cross section ratios calculated using U-spin invariance do not agree with experiment to the level of accuracy of the mass formulae or the magnetic moment estimates, described in the next section.

---

* In Appendix B, it is observed that fermions (baryons) obey the Dirac equation of motion, containing the mass to first order; while bosons (mesons) obey the Klein-Gordon equation. of second order in the mass.

6.6 ELECTROMAGNETIC PROPERTIES IN SU3

In Figs. 6.3 and 6.4, it is observed that members of a U-spin multiplet have the same electric charge. It is therefore plausible that the electromagnetic interaction (i.e. the photon) is a U-spin *scalar*. This is to be contrasted with the behavior of the electromagnetic field under rotations in isospin space; here the electromagnetic operator acts as a combination of scalar and vector giving rise to the selection rule $\Delta I = 0$ or 1 (examples being $\eta \to 2\gamma$ and $\Sigma^0 \to \Lambda + \gamma$ respectively), and producing the mass splitting between members of an isospin multiplet.

Let us denote by Δm the difference between the actual physical mass of a particle and its "bare" mass with the electromagnetic interactions "turned off." According to the above assumption, the value of Δm should be the same for all members of a U-spin multiplet, and for the baryon octet,

$$\Delta m_p = \Delta m_{\Sigma^+},$$
$$\Delta m_{\Sigma^-} = \Delta m_{\Xi^-},$$
$$\Delta m_{\Xi^0} = \Delta m_n.$$

Adding the "bare" masses to each side, and summing these equations, gives

$$m_p + m_{\Sigma^-} + m_{\Xi^0} = m_{\Sigma^+} + m_{\Xi^-} + m_n,$$

or

$$(m_p - m_n) = (m_{\Sigma^+} - m_{\Sigma^-}) + (m_{\Xi^-} - m_{\Xi^0}).$$
$$\uparrow \qquad\qquad \uparrow \qquad\qquad\qquad \uparrow$$
$$-1.3 \text{ MeV} \qquad -8.0 \text{ MeV} \qquad\qquad +6.6 \text{ MeV}$$

$$\underbrace{\qquad\qquad\qquad}$$
$$-1.4(\pm 0.7) \text{ MeV}$$

This formula was due originally to Coleman and Glashow, and is well verified within the experimental errors.

The hypothesis that the electromagnetic interaction is a U-spin scalar allows one to make predictions about *baryon magnetic moments*.

One can arrive at the following relations:

i) Consider first the Σ isospin triplet. The electromagnetic operator, O, having the form $O = a + bI$, will give an equal-spacing rule for the magnetic moments, which are eigenvalues of the operator I_3. Therefore

$$\mu_{\Sigma^+} + \mu_{\Sigma^-} = 2\mu_{\Sigma^0}. \tag{6.21}$$

ii) From U-spin invariance of the electromagnetic interactions,

$$\mu_p = \mu_{\Sigma^+}, \tag{6.22}$$
$$\mu_{\Xi^-} = \mu_{\Sigma^-}, \tag{6.23}$$
$$\mu_n = \mu_{\Xi^0} = \mu_{\Sigma^0_u}. \tag{6.24}$$

iii) Recalling (6.15) and (6.16), we may write

$$\Lambda = \tfrac{1}{2}[\sqrt{3}\,\Sigma_u^0 - \Lambda_u],$$
$$\Sigma^0 = \tfrac{1}{2}[\Sigma_u^0 + \sqrt{3}\,\Lambda_u],$$

(6.25)

so that

$$\mu_\Lambda = \langle\Lambda|\,O\,|\Lambda\rangle = \tfrac{3}{4}\mu_{\Sigma_u^0} + \tfrac{1}{4}\mu_{\Lambda_u},$$

and

$$\mu_{\Sigma^0} = \tfrac{1}{4}\mu_{\Sigma_u^0} + \tfrac{3}{4}\mu_{\Lambda_u},$$

where we have used the fact that the electromagnetic operator O has zero matrix element between the $U = 1$ state (Σ_u^0) and $U = 0$ state (Λ_u), as it is a U-spin scalar. From the last two equations we get

$$\mu_{\Sigma_u^0} = \tfrac{1}{2}[3\mu_\Lambda - \mu_{\Sigma^0}].$$

(6.26)

iv) In the limit of exact SU3, all three quarks have the same mass, zero charge, and zero strangeness. Thus their combination, in this limit, must have zero magnetic moment. The symmetry-breaking interactions split the degeneracy *symmetrically* along the U- and I-spin axes, so that the net magnetic moment of the baryon octet is still zero. (Similarly, the net value of the magnetic quantum number, J_z, of the $(2J + 1)$ substates belonging to the angular momentum J—a representation of SU2—is zero.) Therefore we obtain

$$\mu_p + \mu_n + \mu_{\Sigma^+} + \mu_{\Sigma^0} + \mu_{\Sigma^-} + \mu_\Lambda + \mu_{\Xi^-} + \mu_{\Xi^0} = 0.$$

(6.27)

The six equations (6.21) to (6.24), (6.26), and (6.27) give predicted values of the magnetic moments of the $\tfrac{1}{2}^+$ baryon octet states in terms of any two of them. If these are chosen to be the proton and neutron, straightforward algebra gives

$$\left.\begin{aligned}
\mu_{\Sigma^+} &= \mu_p, \\
\mu_{\Sigma^-} &= \mu_{\Xi^-} = -(\mu_n + \mu_p), \\
\mu_{\Xi^0} &= \mu_n, \\
\mu_\Lambda &= \frac{\mu_n}{2}, \\
\mu_{\Sigma^0} &= -\frac{\mu_n}{2}.
\end{aligned}\right\}$$

(6.28)

The only magnetic moments observed experimentally to date are

$$\mu_p = 2.79, \qquad \mu_n = -1.91,$$
$$\mu_{\Sigma^+} = 2.4 \pm 0.6, \qquad \mu_\Lambda = -0.70 \pm 0.07,$$

(6.29)

in units of the nuclear magneton, $e\hbar/2Mc$, where M is the nucleon mass. Since SU3 symmetry is broken, the octet members do not have the same mass, and it is more reasonable to take for M the actual particle masses. One then obtains $\mu_\Lambda/\mu_n = 0.44 \pm 0.04$, and $\mu_{\Sigma^+}/\mu_p = 1.1 \pm 0.25$, which are consistent with the SU3 predictions.

A word may be added here about the experimental difficulties in measuring magnetic moments of the short-lived hyperon states. We take the Λ-hyperon as an example. The usual technique is to generate Λ's in π-p collisions at just above 1 GeV/c incident momentum:

$$\pi^- + p \rightarrow \Lambda + K^0.$$

As described in Section 4.5, the Λ's are polarized with spins normal to the production plane. The degree of polarization is a maximum at this incident momentum; its value averaged over all CMS production angles is approximately 0.7. The sense of polarization can be determined by observing the up-down asymmetry in the subsequent decay $\Lambda \rightarrow p + \pi^-$. Suppose now that the Λ traverses a strong magnetic field B before decaying. The Λ-spin will precess with the Larmor frequency $v_L = \mu_\Lambda B/h$. In a time t it will rotate through angle $\theta = \mu_\Lambda Bt/\hbar$ rad. Setting $\mu_\Lambda \sim e\hbar/2Mc$, $t = \tau_\Lambda = 3 \times 10^{-10}$ sec, $B = 2 \times 10^5$ G, gives $\theta = 0.3$ rad. Thus the typical precession angle is quite small, even with the largest practicable magnetic fields. Advantage is here taken of the fact that, in a strong-focusing proton synchrotron, the proton beam (yielding pions as a secondary beam from a target) may be extracted from the machine in a very short burst (~ 1 μsec), thus allowing the use of pulsed magnetic fields generated by short circuiting a large condenser bank through an air-cored coil. Figure 6.8 shows a sketch of the arrangement employed in the CERN experiments.

In this particular experiment, the field B is directed normal to the line-of-flight of the Λ-hyperons, and in the production plane. The Λ-spin s_Λ precesses about B as shown (for μ_Λ negative). The advantage of using a transverse field is that background due to charged pions and protons from the target is partially swept away. The $\Lambda \rightarrow \pi^- + p$ decays are detected in a stack of nuclear emulsions. For each event, the angle of the decay pion in the hyperon rest frame, relative to the normal to the production plane, and also the proper time-of-flight of the hyperon, are recorded. From the combined data, the precession angle per unit time in the Λ rest frame, and thus μ_Λ, is computed.

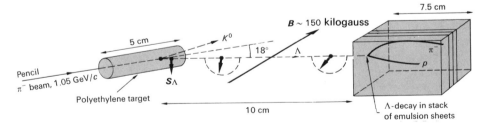

Fig. 6.8 Experimental arrangement employed in the CERN experiments to determine the magnetic moment of the Λ-hyperon. Λ's are selected at 18° to the pion beam direction. This is near the maximum angle of emission for the Λ in the reaction $\pi^- + p \rightarrow \Lambda + K^0$, and there the transverse polarization is greatest. Thick arrows show the Λ spin vector. (After Charrière *et al.*, 1965.)

6.7 THE VECTOR MESON NONET; SU6, QUARK SPIN, AND φ − ω MIXING

The meson multiplet so far considered was the pseudoscalar octet, consisting of $Q\bar{Q}$-states. Clearly, if we assume antiparallel quark spins of $\frac{1}{2}$, and odd intrinsic parity of a $Q\bar{Q}$-pair, we can account for this octet and singlet as $J = 0$ bound states. Equally, for quark spins parallel, and $l = 0$, we might expect an octet and singlet of vector mesons, of $J^P = 1^-$:

$$l = 0 \quad \overset{\downarrow\uparrow}{Q\bar{Q}} \quad J^P = 0^- \qquad \text{pseudoscalar octet and singlet,}$$

$$l = 0 \quad \overset{\uparrow\uparrow}{Q\bar{Q}} \quad J^P = 1^- \qquad \text{vector octet and singlet.}$$

(6.30)

In fact, a total of nine 1^- states (meson resonances) have been observed, with the following properties (Fig. 6.9 and Table 6.5).

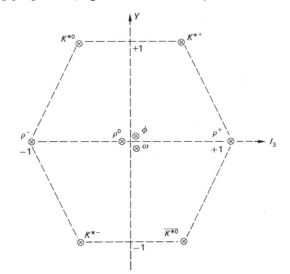

Fig. 6.9 The nonet of vector mesons ($J^P = 1^-$).

The Gell-Mann-Okubo formula (6.19) predicts for the $Y = I = 0$ member of the octet, a mass given by

$$(M_8)^2 = \frac{4M_{K*}^2 - M_\rho^2}{3} = (930 \text{ MeV})^2,$$

(6.31)

which fits neither the ω nor ϕ candidates. We must therefore assume that there is some "mixing" between the eighth member of the octet and the singlet. This phenomenon is not too surprising, since only in the case of exact SU3 symmetry would the octet and singlet be completely distinct states. We have

TABLE 6.5 Vector meson nonet

State	I	Y	Mass (MeV)	Dominant decay mode
ρ	1	0	765	$\rho \to 2\pi$
K^*	$\frac{1}{2}$	± 1	890	$K^* \to K\pi$
ω	0	0	783	$\omega \to 3\pi$
ϕ	0	0	1019	$\phi \to K\bar{K}$

already mentioned that a similar phenomenon occurs in isospin (SU2). Re-actions which are forbidden, by conservation of isospin, to proceed as strong reactions, can take place via electromagnetic coupling (provided of course they conserve charge or I_3). As a result, states of $I = 1$ and $I = 0$ may be mixed, although they belong to different representations of SU2, because of electromagnetic interaction. In strong processes, this is a small effect, because the fine structure constant α is small. However, in SU3 the symmetry breaking, as judged by the mass differences in the multiplets, is relatively much larger, so that a considerable degree of octet/singlet mixing could and does occur.

In connection with the mixing in the vector meson octet and singlet, it is of interest to remark that, by including the quark spin, we have enlarged the symmetry. Instead of three basic quark states, we now have six, since each quark can have spin up or down; i.e. we are dealing with the symmetry group SU6. We can transform one quark into another in SU3 space, and one spin configuration into another in SU2 space. Combined, we have SU6 symmetry; invariance under SU6 transformations means that the binding energy of a $Q\bar{Q}$- or QQQ-state is independent of the type of quarks involved ($Q_1, Q_2,$ or Q_3) *and* of their spin orientations. (We do run into difficulty about statistics—this is discussed below.) Then, for the $Q\bar{Q}$-combinations, we have, instead of $\bar{3} \otimes 3 = 8 \oplus 1$ states, a total of $\bar{6} \otimes 6 = 36$ states $= 35 \oplus 1$ (since, as always, we can find one state which is again completely symmetric in all SU6 variables). These 35 states are made up as follows:

$Q\bar{Q}$-spins	SU3 part	SU2 (spin) part	Number of states	
Parallel	8	3 (triplet)	$8 \times 3 = 24$	$\left.\right\}1^-$ vector nonet
Parallel	1	3 (triplet)	$1 \times 3 = 3$	
Antiparallel	8	1 (singlet)	$8 \times 1 = 8$	0^- octet
Antiparallel	1	1 (singlet)	$1 \times 1 = 1$	0^- singlet

We recall that the 0^- octet mass formula seems to work well, and this must mean that any octet-singlet (η-η') mixing must be small. As (6.31) indicates, the case is otherwise for the vector meson members of the 35-plet, and for this reason it is better to group them all into one nonet, rather than an octet and singlet.

Let us now discuss the mixing in the vector nonet. The ϕ- and ω-mesons actually observed will be linear combinations of the eighth $(Y = I = 0)$ octet member—which we can call ϕ_8—with the SU3 singlet member, denoted ϕ_0. The mixing angle θ is defined as follows:

$$\phi = \phi_0 \sin\theta - \phi_8 \cos\theta, \tag{6.32a}$$

$$\omega = \phi_8 \sin\theta + \phi_0 \cos\theta, \tag{6.32b}$$

so as to satisfy the conditions of orthogonality, namely

$$\langle\phi\,|\,\omega\rangle = \langle\phi_0\,|\,\phi_8\rangle = 0.$$

Put in terms of the particle masses, it may be shown that

$$\tan^2\theta = \left(\frac{M_\phi^2 - M_8^2}{M_8^2 - M_\omega^2}\right), \tag{6.32c}$$

where substituting the values in Table 6.5 and (6.31), one finds $\theta \approx 40°$. Now, recalling the discussion of the pseudoscalar octet, we write the $I = 1, I_3 = Y = 0$ (ρ^0-meson), the $I = Y = 0$ octet member (ϕ_8), and the SU3 singlet (ϕ_0) in terms of the following quark-antiquark combinations:

$$\rho_0 = \frac{1}{\sqrt{2}}(1\bar{1} - 2\bar{2}),$$

$$\phi_8 = \frac{1}{\sqrt{6}}[1\bar{1} + 2\bar{2} - 2(3\bar{3})], \tag{6.33}$$

$$\phi_0 = \frac{1}{\sqrt{3}}(1\bar{1} + 2\bar{2} + 3\bar{3}).$$

These satisfy the conditions for normalization of the wave functions, and are such that ϕ_8 and ϕ_0 are orthogonal as required. Consider the special case when $\cos = \sqrt{\frac{2}{3}}$, $\sin\theta = 1/\sqrt{3}$, $\theta = 35°$; (6.33) and (6.32a and b) then give

$$\phi = 3\bar{3},$$
$$\omega = \frac{(1\bar{1} + 2\bar{2})}{\sqrt{2}}. \tag{6.34}$$

In this case of *ideal mixing*, we see that the physical state ϕ is composed only of Λ-type quarks, and the state ω is composed only of p- and n-type quarks. From the baryon decuplet, we know that $\Omega^-(333)$ is heavier than $N^{*++}(111)$, so that the symmetry breaking is such as to make the strange quark (Q_3) heavier than the members of the nonstrange isodoublet $(Q_1$ and $Q_2)$. Thus we expect ϕ to be heavier than ω, which in turn should have a mass equal to that of the ρ—as observed. Since (6.32c) shows θ is somewhat greater than 35°, the mixing is not in fact quite ideal.

From the existence of mixing we can also say something about the decay modes of ϕ and ω. If one considers the dominant decay mode $\phi \to K\bar{K}$, the two zero spin kaons must be in an $l = 1$ state, i.e. a state antisymmetric under space exchange. But ϕ_0 is completely symmetric, so that only the ϕ_8 component of ϕ can go to $K\bar{K}$. In analogy with (6.25) it has the same form, in terms of U-spin wave functions, as the Λ-hyperon:

$$\phi_8 = \frac{\sqrt{3}}{2} \psi \, (U = 1, U_3 = 0) - \tfrac{1}{2}\psi \, (U = 0, U_3 = 0). \qquad (6.35)$$

Consider just the $\psi(1, 0)$ part of (6.35). The decay $\phi_8 \to K\bar{K}$ (that is, $\psi \, (U = 1, U_3 = 0) \to 2$ pseudoscalar mesons of $U = 1, U_3 = \pm 1$) will have the *same* matrix element as the decay $\rho^0 \to \pi^+\pi^-$ (that is, $\chi \, (I = 1, I_3 = 0) \to 2$ pseudoscalar mesons of $I = 1, I_3 = \pm 1$)—just because in SU3 we can interchange U- and I-spin as we please. Hence, from (6.32) and (6.35) the decay rates are related by

$$\frac{\Gamma(\phi \to K\bar{K})}{\Gamma(\rho^0 \to \pi^+\pi^-)} = \tfrac{3}{4} \cos^2 \theta. \qquad (6.36)$$

Because of symmetry breaking, the masses of parent and daughter particles in numerator and denominator are not the same, so this expression needs to be corrected by phase-space factors, using the actual particle masses. Assuming $\theta = 40°$ and a ρ-width $\Gamma = 125$ MeV, we find that (6.36) gives

$$\Gamma(\phi \to K\bar{K}) \approx 2.6 \text{ MeV (predicted)},$$

compared with the experimental value

$$\Gamma(\phi \to K\bar{K}) = 2.9 \pm 0.7 \text{ MeV (observed)}.$$

We see therefore that the particle ϕ has predominantly the properties of an octet member—decaying with 90% probability to the spatially antisymmetric state $K\bar{K}$. On the other hand, the ω-meson decays with 90% probability in the mode $\omega \to 3\pi$ which, as shown in Section 7.1.3, is spatially symmetric; the ω is "more singlet than octet."

6.8 MAGNETIC MOMENTS OF NEUTRON AND PROTON

SU3, or the spinless quark model, gave a relation between the magnetic moments of the $\frac{1}{2}^+$ baryon octet in terms of those of neutron and proton. SU6, or the quark model with spin, predicts one more number, μ_n/μ_p, a ratio which is very accurately measured.

We recall that the decuplet of Fig. 6.3 is composed of three-quark states completely symmetric under quark label interchange. The $J^P = \frac{3}{2}^+$ decuplet members are assumed to be built from quark triplets bound in an S-state with parallel spins; it follows that they are in an SU6 representation which is completely symmetric in all six quark variables. This representation is a 56-plet which arises from the combination $6 \otimes 6 \otimes 6 = 20 \oplus 70 \oplus 70 \oplus 56$. The 56-plet consists of the SU3 multiplets:

$$56 = \underset{\underset{\frac{1}{2}^+ \text{ octet}}{\uparrow}}{(2, 8)} + \underset{\underset{\frac{3}{2}^+ \text{ decuplet}}{\uparrow}}{(4, 10),} \qquad (6.37)$$

where the first number in rounded brackets denotes the spin multiplicity $(2J + 1)$, the second the SU3 representation or multiplicity. In Section 6.4, we found *two* baryon octets, of mixed symmetry with respect to quark label interchange (without spin). Given parallel quark spins (i.e. $J = \frac{3}{2}$), neither of these could form a symmetric SU6 configuration. However, for two spins parallel and one antiparallel ($J = \frac{1}{2}$) one can get one SU6 representation which is totally symmetric, as indicated by (6.37). Once this is decided, the rest is pure algebra.

The states at the corners of the octet in Fig. 6.4 have the following SU3 quark compositions, which are obtained by considering charge and strangeness quantum numbers:

$$n = 221 \qquad p = 112$$

$$\Sigma^- = 223 \qquad\qquad\qquad \Sigma^+ = 113$$

$$\Xi^- = 332 \qquad \Xi^0 = 331.$$

Thus, a proton consists of two p-type quarks (Q_1) of charge $+\frac{2}{3}$, and one n-type quark (Q_2) of charge $-\frac{1}{3}$. The two like quarks in each state must have spins parallel, in order to ensure symmetry under interchange of these quarks (SU3 label *and* spin). Therefore the two like quarks must be in a triplet spin state, denoted by the spin function χ ($J = 1$; $m = 0, \pm 1$), while the third, unlike quark is denoted by Φ ($J = \frac{1}{2}$; $m = \frac{1}{2}$). The total spin function of all three quarks must be $\psi(\frac{1}{2}, \frac{1}{2})$ since the proton has spin $\frac{1}{2}$. To combine χ and Φ to form ψ we use the Clebsch-Gordan coefficients (Appendix C, or Table 3.4) to obtain

$$\psi(\tfrac{1}{2}, \tfrac{1}{2}) = \sqrt{\tfrac{2}{3}}\chi(1, 1)\Phi(\tfrac{1}{2}, -\tfrac{1}{2}) - \sqrt{\tfrac{1}{3}}\chi(1, 0)\Phi(\tfrac{1}{2}, +\tfrac{1}{2}). \qquad (6.38)$$

Let μ_A denote the magnetic moment of the pair of like quarks, μ_B that of the unlike quark. In the first term on the right in (6.38), these moments are antiparallel, so the combination has moment $(\mu_A - \mu_B)$. In the second term, the pair has $m = 0$, and thus no contribution, and the moment is simply $+\mu_B$. Since, on squaring (6.38) the intensities of the terms are in the ratio $\frac{2}{3}:\frac{1}{3}$, the total moment will be

$$\mu_{AB} = \tfrac{2}{3}(\mu_A - \mu_B) + \tfrac{1}{3}\mu_B = \tfrac{2}{3}\mu_A - \tfrac{1}{3}\mu_B. \qquad (6.39)$$

If μ denotes the characteristic scale magnetic moment of a quark (i.e. "quark magneton"), the charge assignments of Table 6.3 give

$$\mu_1 = +\tfrac{2}{3}\mu, \qquad \mu_2 = \mu_3 = -\tfrac{1}{3}\mu. \tag{6.40}$$

This result also follows since, from (6.27), $\mu_1 + \mu_2 + \mu_3 = 0$, and further, as Q_2 and Q_3 form a U-spin doublet and the electromagnetic interaction is a U-spin scalar, $\mu_3 = \mu_2$. From the quark compositions for the outside members of the $\tfrac{1}{2}^+$ baryon octet displayed above, one obtains from (6.39) and (6.40):

$$\mu_p = [\tfrac{2}{3}(\tfrac{2}{3} + \tfrac{2}{3}) - \tfrac{1}{3}(-\tfrac{1}{3})]\mu = \mu; \qquad \mu_n = -\tfrac{2}{3}\mu;$$

$$\mu_{\Sigma^+} = \mu; \qquad \mu_{\Sigma^-} = -\tfrac{1}{3}\mu; \qquad \mu_{\Xi^0} = -\tfrac{2}{3}\mu; \qquad \mu_{\Xi^-} = -\tfrac{1}{3}\mu.$$

The Λ and Σ^0 moments, not considered here, have already been given in (6.28). The important prediction here is that

$$\mu_n/\mu_p = -0.67, \qquad \mu_n/\mu_p = -\frac{1.913}{2.793} = -0.68, \tag{6.41}$$

$$\text{SU6 and quark model} \qquad\qquad \text{Experiment}$$

which is a fantastically good agreement.

6.9 FURTHER CONSIDERATIONS ON THE QUARK MODEL. FREE QUARKS?

Ever since quarks were proposed by Gell-Mann and Zweig an intensive search has been made for evidence of their existence as free particles. At existing high energy accelerators, the complete absence of heavy, fractionally charged particles has now been established beyond reasonable doubt, even under the most pessimistic assumptions regarding their production mechanism. This allows one to set a lower limit $M_Q > 5$ GeV on the quark mass. Searches have also been made for quarks liberated in very high energy collisions of primary cosmic rays in the atmosphere. The flux of any fractionally charged particles near sea level is $<10^{-10}$ cm^{-2} sr^{-1} sec^{-1}, compared with the flux of cosmic ray muons of 10^{-2} cm^{-2} sr^{-1} sec^{-1}. Quite recently, the attention of cosmic ray physicists has turned to detailed examination of particles in extensive air showers appearing at sea level as a result of collisions of primary cosmic ray nuclei of 10^5 GeV or more. Such collisions might lead to production of quarks even if their rest mass is very large (say 50 GeV). As yet, however, there appears to be no decisive evidence for their existence (1971).

Even if quarks are never observed, this is not a condemnation of the quark model of hadrons, or SU3 and SU6. As emphasized at the beginning of the chapter, the basis states, or quarks, may be simply convenient mathematical entities out of which to build up the octets and other multiplets of real particle states. If taken literally as a dynamical model of hadrons, the quark model possesses some extremely curious features as well as some resounding successes. Consider the difficulties first.

In atomic physics, the mechanical spin is always coupled to orbital angular momentum. In high energy processes too, for example in $\rho \to 2\pi$, the spin of the ρ-meson is transformed into orbital angular momentum of the pions. Spin is not an invariant of strong interactions. It is therefore rather strange that the quark model relies on invariance with respect to spin alone. However, it has been argued that such invariance, and the ability to quantize separately spin σ and orbital motion l, follows because the model is nonrelativistic. The quarks are supposed extremely massive, and bound in a deep potential well W by the superstrong forces, so as to yield the physical masses; thus $M_p = 2M_{Q_1} + M_{Q_2} - W$. On the other hand, the mean quark momentum, $k \approx R^{-1}$, where R is the well radius. If we take R as the "size" of a proton, of order 1 F, $k \sim 200$ MeV $\ll M_Q$. So the quarks are in nonrelativistic motion. Another problem relates to quark statistics. The decuplet of $J^P = \frac{3}{2}^+$ is assumed to be built from three quarks in an S-state with spins parallel, for example $N^{*++}(1238)$ has the configuration $Q_1\uparrow Q_1\uparrow Q_1\uparrow$. Thus quarks may not obey Fermi statistics, even though they have spin $\frac{1}{2}$, since the Pauli principle forbids this configuration. Therefore, it is necessary to invent a new kind of statistics for quarks.

Another difficulty is that, while several clear, fully-occupied baryon octets and decuplets are certainly observed, almost half the known baryon states do not fit into obvious octet or decuplet patterns. In other words, they have to be classed as singlets, far outstripping the octets and decuplets in number.

These objections are really a statement of our lack of understanding. Any theoretical model has to be judged by what it predicts, rather than just whether it is intuitively comprehensible according to presently conceived ideas. So let us list a few of the successes of the model:

1. All observed particle multiplets can be accounted for as combinations of either $Q\bar{Q}$ (for bosons) or QQQ (for baryons). In particular, no strangeness $+1$ baryons appear to be observed (which would require $QQQ\bar{Q}Q$ combinations, with the transformation properties of the SU3 multiplet $(\overline{10})$); nor are doubly charged boson states, of the form $Q\bar{Q}Q\bar{Q}$. While there is no way of excluding such "exotic" states from unitary symmetry alone, the appeal of the quark model is that it can account for all known states, and exclude all unknown ones, in terms of two, and only two, types of quark combination. This argues strongly that the systematics of the observed particle states may be determined by the detailed dynamical features of the quark-quark interaction.

2. While, in this chapter, we have only considered particles built from quarks in a relative S-state ($l = 0$), it also appears possible to account for the particles of high mass and spin angular momentum (see, for example, Fig. 7.33) in terms of orbital and radial excitations of the $Q\bar{Q}$ or QQQ system.

3. Quite independently of the spectroscopy of hadronic states, the study of deep inelastic lepton scattering by nucleons has led to the model, described in Chapter 5, of hadrons as composed of many point-like constituents of spin $\frac{1}{2}$. The idea is that, just like the chemically inert closed shells of electrons in an atom, these constituents $(Q\bar{Q})$ are mostly "paired-off" so as to have zero net spin, baryon number, isospin, and strangeness, in other words, to be SU3 scalars. This leaves two or three unpaired "valence quarks," which are manifested in the unitary symmetry schemes. Such a simple model, while not able to account for the detailed features of the electron scattering data, does at least give a natural interpretation of scale invariance and predict gross cross-sections in rough accord with experiment.

BIBLIOGRAPHY

Carruthers, P., *Introduction to Unitary Symmetry*, Interscience, 1966.

de Swart, J. J., "Symmetries of strong interactions," Proc. 1966 CERN School of Physics, Vol. II (CERN 66-29).

Gell-Mann, M., and Y. Ne'eman, *The Eightfold Way*, Benjamin, New York, 1964.

Kokkedee, J. J., *The Quark Model*, Benjamin, New York, 1969.

Lipkin, H., *Lie Groups for Pedestrians*, North-Holland, Amsterdam, 1966.

Matthews, P. T., "Unitary symmetry," *Pure and Applied Phys.* **25**, 1 (1967).

Van Hove, L., "SU$_3$ and SU$_6$ symmetry of strong interactions," Proc. 1965 CERN Easter School, Vol. III (CERN 65-24).

PROBLEMS

6.1 Verify that the quark wave functions for the π^0, η, and η' states on p. 230 are mutually orthogonal and correctly normalized.

6.2 Estimate the expected lifetime of the Ω^- particle, given that of the Λ as 2.5×10^{-10} sec.

6.3 Calculate the magnetic moments of the Ω^- particle and the $N^{*++}(1238)$ resonance, on the basis of the quark model.

6.4 Plan an experiment to measure the Σ^+ magnetic moment.

6.5 Using Clebsch-Gordan coefficients (in U-spin), deduce the ratios of the following amplitudes:

$$\pi^- + p \rightarrow K^+ + Y_1^{*-}, \qquad K^- + p \rightarrow K^+ + \Xi^{-*},$$
$$\pi^- + p \rightarrow \pi^+ + N^{*-}, \qquad K^- + p \rightarrow \pi^+ + Y_1^{*-}.$$

6.6 If the photon is treated as a U-spin scalar, which of the following decays are allowed by SU3 symmetry?

i) $\Xi^{*-} \rightarrow \Xi^- + \gamma$,

ii) $Y_1^{*-} \rightarrow \Sigma^- + \gamma$,

iii) $N^{*+} \rightarrow p + \gamma$,

iv) $Y_1^{*+} \rightarrow \Sigma^+ + \gamma$.

6.7 Compare the amplitudes predicted by SU3 for the following transitions (not all are allowed energetically):

i) $N^{*-} \rightarrow \Sigma^- + K^0$,

ii) $Y_1^{*-} \rightarrow \Sigma^- + \pi^0$,

iii) $\Xi^{*-} \rightarrow \Xi^- + \eta$,

iv) $\Xi^{*-} \rightarrow \Xi^- + \pi^0$.

6.8 Comment on the relative magnitudes of the p-p and π-p total cross sections at high energy (Fig. 7.24) in terms of a very simple quark model.

Chapter 7

Strong Interactions II—Dynamical Features

Thus far, our discussion of strong interactions has emphasized symmetry and invariance principles, without any detailed consideration of the actual dynamics of the interaction. However, until one has some formalism for describing the interaction in detail, one cannot make calculations of cross sections and angular distributions, and thus a real confrontation with the mass of experimental data from collision processes. As indicated in Chapter 5 the analogous quantum mechanical treatment of the interaction between electrons and photons, known as quantum electrodynamics, has been resoundingly successful. This stems essentially from the fact that the coupling constant $e^2/\hbar c = \frac{1}{137}$ is small, and perturbation theory can therefore be employed. There is at present no corresponding, comprehensive treatment available for strong interactions. This is so not simply because the interaction is strong. In electromagnetic processes, the associated quanta are identified as photons. On the contrary, in strong interactions there are numerous particles which can play the role of "quanta" of the field. Any one type of particle therefore both generates, and is generated by, the interaction of all the other types of particle—the so-called "bootstrap" mechanism.

In this chapter, we discuss some of the important dynamical features of strong interactions. First we deal with the determination of the quantum numbers of particles and resonances. This is followed by a section on the subject of scattering amplitudes and phase shifts, and the definition of a resonance. The final sections introduce some of the salient features of high energy cross sections, and their attempted interpretation in terms of models such as the one-particle-exchange model, and Regge poles.

7.1 DALITZ PLOTS: EXAMPLES OF DETERMINATION OF QUANTUM NUMBERS

In a reaction such as

$$\pi + p \to \pi + \pi + p$$

one wants to investigate possible final state resonances (πp, $\pi\pi$), and thus how the observed distributions in secondary momentum, angle, or invariant mass depart from the situation where the matrix element is constant (i.e. there is no strong final state interaction). In the latter case, the distributions will be determined simply from the phase-space factors. A vital technique in assessing whether, and how, the distributions depart from phase-space is the *Dalitz plot*. Historically this was first used to determine the spin and parity of the final state in $K \rightarrow 3\pi$ decay (*à propos* of the "$\tau - \theta$ puzzle" discussed in Chapter 4).

7.1.1 Three-Body Phase Space

The number of states available per particle per unit normalization volume with momentum $p \rightarrow p + dp$, inside the solid angle $d\Omega$, is

$$\frac{p^2 \, dp \, d\Omega}{h^3},$$

provided p and Ω are independent variables [see (3.17)]. For three particles of momenta p_1, p_2, and p_3 in the final state, and all quantities defined in the center-of-momentum system (CMS) of the particles, the number of states per unit volume will be

$$dN = \text{const.} \, p_1^2 \, dp_1 p_2^2 \, dp_2 \, d(\cos \theta), \tag{7.1}$$

since $p_3 = -(p_1 + p_2)$ is fixed, and the overall orientation of the configuration in space has no significance (provided the initial state has zero spin or is unpolarized). Here, θ is the angle between p_1 and p_2, and, for p_1 fixed, $2\pi \, d(\cos \theta)$ is the element of solid angle available to particle 2.

In writing down the matrix element M for the interaction, a factor $1/\sqrt{E}$ occurs in the wave function normalization for each particle, where E is the particle energy. When squared, this accounts essentially for the Lorentz factor $\gamma^{-1} = m/E$ in the particle density, which has first to be taken out to make M independent of the reference frame. It is customary to take these factors for the *final-state particles* over into the density-of-states term. This then acquires the relativistically invariant form (see Appendix D)

$$\rho = \frac{d}{dE_f} \frac{\int p_1^2 \, dp_1 \int p_2^2 \, dp_2 \int d(\cos \theta)}{E_1 E_2 E_3}, \tag{7.2}$$

where constants like 2π and h have been omitted, and the integration proceeds from right to left. The total energy $E_f = E_1 + E_2 + E_3$. Rewriting (7.2) as

$$\rho = \frac{d}{dE_f} \frac{\int p_1 \, dp_1 \int p_2 \, dp_2 \int p_1 p_2 \, d(\cos \theta)}{E_1 E_2 E_3},$$

and using the relations

$$E_1^2 = p_1^2 + m_1^2, \qquad E_2^2 = p_2^2 + m_2^2,$$

$$E_3^2 = p_3^2 + m_3^2 = p_1^2 + p_2^2 + m_3^2 + 2p_1 p_2 \cos \theta,$$

$$E_1 \, dE_1 = p_1 \, dp_1, \qquad E_2 \, dE_2 = p_2 \, dp_2,$$

$$(E_3 \, dE_3)_{p_1, p_2 \text{ fixed}} = p_1 p_2 \, d(\cos \theta),$$

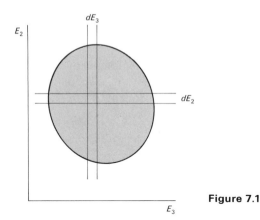

Figure 7.1

one obtains

$$\rho = \frac{d}{dE_f} \frac{\int E_1 \, dE_1 \int E_2 \, dE_2 \int E_3 \, dE_3}{E_1 E_2 E_3}. \qquad (7.3)$$

Here, $d/dE_f = d/dE_1$ suppresses the final integration. If we assume the matrix element is a constant, (7.3) also gives the total event rate. The partial rate, giving particle 2 in dE_2 and particle 3 in dE_3, is simply

$$d^2\rho = \text{const. } dE_2 \, dE_3. \qquad (7.4)$$

Thus, three-body phase space (and nothing else) predicts that if we plot E_2 and E_3 along orthogonal axes, the density of events per unit area (i.e. per unit of $dE_2 \, dE_3$) should be uniform (Fig. 7.1). In the sketch the curve indicates the kinematically allowed region. Such a plot is called a Dalitz plot. Clearly, if $|M|^2$ depends on the momenta and angles of the particles, then one will not get a uniform Dalitz plot;

$$d^2\rho = |M|^2 \, dE_2 \, dE_3, \qquad (7.5)$$

so that the density at any point is a measure of the square of the matrix element.

7.1.2 $K_{\pi 3}$-Decay

The original Dalitz plot was made for the decay

$$K^+ \to \pi^+ + \pi^+ + \pi^-, \qquad Q = 75 \text{ MeV} = \text{total kinetic energy.}$$

In this case, the three particles have equal mass and are nonrelativistic (or very nearly so). Let p_i and T_i be the momenta and kinetic energies of the pions ($i = 1, 2, 3$). One can then make the construction shown in Fig. 7.2. ABC is an equilateral triangle of height Q. Then, if one plots T_1, T_2, T_3 along axes normal to the sides of the triangle, all events (represented by points P) will lie

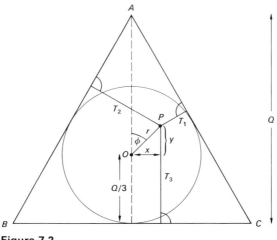

Figure 7.2

inside the triangle, since

$$Q = T_1 + T_2 + T_3.$$

Furthermore, since $p_1 + p_2 + p_3 = 0$, the point P is restricted to lie inside the inscribed circle, center O and radius $Q/3$. In order to prove this, one can express T_1, T_2, T_3 as a function $f(r, \phi, Q)$, where r and ϕ are indicated in Fig. 7.2. Eliminating ϕ, one gets

$$r^2 = \tfrac{4}{9}[Q^2 - 3(T_1T_2 + T_2T_3 + T_1T_3)].$$

Also,

$$|p_1 - p_2| < p_3 < |p_1 + p_2|.$$

Then, using the nonrelativistic relation $p^2 = 2mT$, one obtains

$$|T_3 - T_1 - T_2| < 2\sqrt{T_1T_2}, \qquad \text{etc.,}$$

and, substituting in the expression for r^2, one then finds

$$r^2 < Q^2/9,$$

i.e. P must lie within the inscribed circle. This is only true *nonrelativistically*. Even in $K_{\pi3}$-decay there is a slight correction, so that C becomes distorted to C'. In the *extreme relativistic case* ($p \sim E \sim T$), the boundary becomes the inscribed triangle C'' (Fig. 7.3).

Returning to Fig. 7.2, if the Cartesian coordinates of P relative to O are denoted by x and y, then

$$y = T_3 - \frac{Q}{3}, \qquad \frac{dT_3}{dy} = \frac{dE_3}{dy} = -1,$$

$$x = \frac{(T_2 - T_1)}{\sqrt{3}}, \qquad \left(\frac{dT_2}{dx}\right)_{T_1} = \frac{dE_2}{dx} = \sqrt{3}.$$

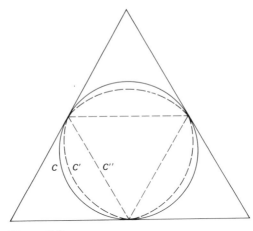

Figure 7.3

Then from (7.4) it follows that again the density of points in Fig. 7.2 will be uniform if the matrix element is a constant. The display of the *kinetic energies* along three axes at 120° thus results in a simple circular Dalitz plot, for the case of three final state particles with equal masses and nonrelativistic velocities in the CMS.

We may now ask what are the departures from phase space for various spin-parity assignments to the $\pi^+\pi^+\pi^-$ combination in $K_{\pi 3}$-decay. Let l_+ be the angular momentum of the two *like* pions in the CMS, and let l_- be the angular momentum of the π^- relative to this dipion system. If J is the total angular momentum, then $|l_+ - l_-| \leqslant J \leqslant |l_+ + l_-|$, since the pions are spinless. Now by symmetry, the orbital angular momentum l_+ of two identical pions must be even. We therefore obtain the assignments shown in Table 7.1 for the spin-parity J^P of the three-pion state, for l_+ or $l_- \leqslant 2$.

If $J_K \neq 0$, then l_+ and/or l_- must be nonzero. We may consider the following possibilities:

i) $l_+ \geqslant 2$. Here we expect the matrix element to vanish when the two positive pions are mutually at rest, since they cannot then possess relative angular momentum. This situation occurs when T_- is a maximum, the two positive pions recoiling in the same direction with equal velocity. Thus, for any

Table 7.1

l_- \ l_+	0	2
0	0^-	2^-
1	1^+	$3^+, 2^+, 1^+$
2	2^-	$4^-, 3^-, 2^-, 1^-, 0^-$

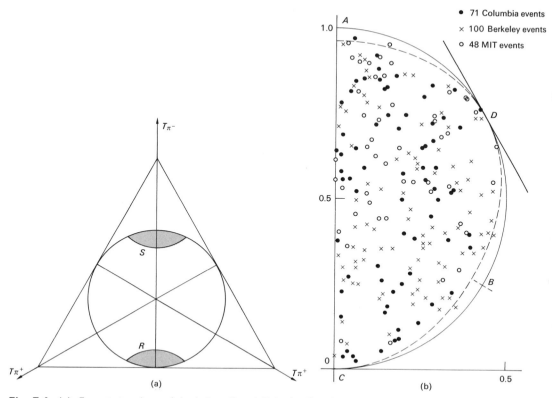

Fig. 7.4 (a) Expected regions of depletion, R and S, in the $K \to 3\pi$ Dalitz plot, for a nonzero kaon spin, J_K. (b) Observed distribution of events in $K \to 3\pi$, showing uniform population, and thus $J_K = 0$. All events are plotted in one semicircle. The dotted curve is the boundary using relativistic kinematics, and departs slightly from a circle. (After Orear *et al.*, 1956.)

assignment $l_+ > 0$, we expect a depletion of events in the region S of Fig. 7.4.

ii) $l_- \geqslant 1$. In this case, there should be a depletion in density in the region R, corresponding to very low kinetic energy of the negative pion, relative to the dipion pair.

The observed Dalitz plot in $K \to 3\pi$ decay (Fig. 7.4) is uniform to a high degree of accuracy, proving that $l_+ = l_- = 0$ and $J_K = 0$ is the only possible assignment.

It must be emphasized that the above analysis gives no information about the kaon *parity*. Three pions in a relative S-state have parity $(-1)^3 = -1$; since, however, parity is not conserved in the weak decay $K \to 3\pi$, this tells us nothing about the initial state. The kaon parity may be determined by the study of *hypernuclei*. These are nuclei in which one neutron is replaced by a bound Λ-hyperon. For example, when negative kaons are brought to rest in helium, and undergo capture from an atomic S-state, a few per cent of the events give rise to bound, rather than free, Λ-hyperons, according to the reaction

$$K^- + He^4 \to {}_\Lambda H^4 + \pi^0,$$

where the hypernucleus ${}_\Lambda H^4$ consists of a triton (H^3) plus a bound Λ-particle. Detailed measurements of its (weak) decay modes establish that $J_{{}_\Lambda H^4} = 0$. Thus $J = l = 0$ on both sides of the equation. If the Λ-parity is defined to be positive, like the nucleon, the kaon must then have $J^P = 0^-$, like the pion.

7.1.3 ω-Decay

The ω-meson was first observed in 1961 in hydrogen bubble chamber experiments at the Berkeley Bevatron, carried out by Maglic *et al.* An analysis was

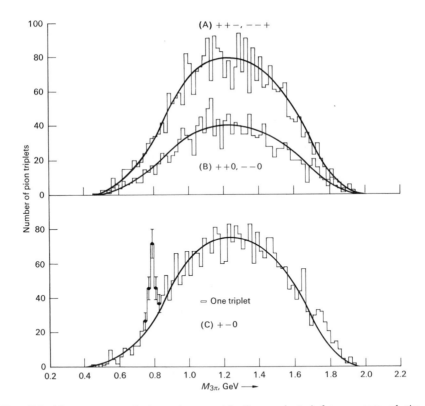

Fig. 7.5 Mass spectra of three-pion combinations selected from events of the type $p + \bar{p} \to \pi^+ + \pi^+ + \pi^- + \pi^- + \pi^0$. (A) Combinations of charge ± 1; (B) combinations of charge ± 2; (C) neutral combinations. The curves refer to the phase-space distribution, i.e. the spectrum expected if there is no strong pion-pion interaction in the final state. (After Maglic *et al.*, 1961.)

made of antiproton annihilations in the chamber, of the type

$$p + \bar{p} \to \pi^+ + \pi^+ + \pi^- + \pi^- + \pi^0.$$

In this reaction, one can measure the magnetic curvatures, and hence momenta, of all the charged particles. The neutral pion decays to two γ-rays, which are not normally observed since the conversion length in hydrogen is some 15 m. However, one can compute the missing energy and momentum, and see if it is consistent with the π^0-mass, and in this way some 800 examples of the above reaction were uniquely identified. The invariant masses of all possible three-pion combinations were plotted, as shown in Fig. 7.5. For the $\pi^+\pi^+\pi^-$, $\pi^-\pi^-\pi^+$, $\pi^+\pi^+\pi^0$, and $\pi^-\pi^-\pi^0$ combinations, the invariant three-pion mass spectrum is consistent with a phase-space distribution. For the $\pi^+\pi^-\pi^0$ events, however, there is a strong peak at $M_{3\pi} = 790$ MeV, with a width determined in later, more precise experiments to be $\Gamma = 12$ MeV. This departure from phase-space proves the existence of a three-pion resonance ω, decaying by strong interactions, so that the reaction is $p\bar{p} \to \pi^+\pi^-\omega$; $\omega \to \pi^+\pi^-\pi^0$. Since there is no peak in the singly or doubly charged triplets, the isospin $I_\omega = 0$. Also, since $\omega \to 3\pi$, its G-parity is $(-1)^3 = -1$.

The full spin-parity analysis of the ω-decay Dalitz plot is more complicated than for $K_{\pi3}$-decay. In the latter case, one could capitalize on the fact that there are two identical π^+-particles in the final state. The first point to remark on is that the Dalitz plot (Fig. 7.6) is not uniformly populated, so that the matrix element describing the process is not constant, and depends on the energies or momenta of the pions. The dependence of the spatial part of the matrix element on these quantities can be denoted by the amplitude

$$A = A(\mathbf{p}_i, E_i), \tag{7.6}$$

where $\mathbf{p}_i$ and E_i refer to the ith pion ($i = 1, 2, 3$), and are of course measured in the ω rest frame.

The second point is that, since $I_\omega = 0$ and $I_\pi = 1$, any pair of pions must be in a state $I = 1$, the third having $I = 1$ to give a resultant of zero. As discussed in Chapter 3, a pair of like bosons with $I = 1$ must have an antisymmetric space wave function. Thus the amplitude A must change sign under interchange of any two of the indices i in (7.6). Thirdly, we consider the properties of the initial (ω) and final (3π) states under space inversion. Since $\omega \to 3\pi$ is a strong decay, the parity of initial and final states must be the same. Thus, if the space amplitude is even under inversion, that is $A(-\mathbf{p}_i, E_i) = +A(\mathbf{p}_i, E_i)$, the final state parity will be $(+1)(-1)^3 = -1$, where the second factor comes from the intrinsic parity of each pion. On the other hand, if A is odd, the final state parity—and hence that of the ω-meson—must be even.

We now consider in turn various possible assignments J^P for the spin-parity of the ω-meson, and what they imply for the form of $A(\mathbf{p}_i, E_i)$, which in turn determines the Dalitz plot population.

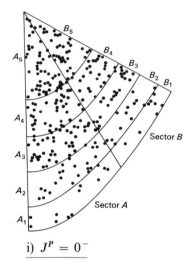

Fig. 7.6 Observed Dalitz plot distribution for the $\omega \rightarrow 3\pi$ decay. Since the plot has sextant symmetry, all events have been displayed in one sextant. (After Stevenson *et al.*, 1961.)

i) $J^P = 0^-$

A particle of zero spin and odd parity is called *pseudoscalar*, since it has the spatial transformation properties of such a quantity. From the above discussion, it follows that A must be a scalar quantity, that is, even under space inversions. The simplest expression with the required properties is

$$A = (E_1 - E_2)(E_2 - E_3)(E_3 - E_1). \tag{7.7}$$

This is clearly a scalar, involves the energies of all three pions, and changes sign under interchange of any two pion labels. It is obviously not the most general form one can write. For example, one could multiply (7.7) by some symmetric scalar function of the energies, or raise it to any odd power. However, one always starts out by trying the simplest matrix elements, in the hope that one of them will work. In any case we are not attempting in the first instance to predict the exact Dalitz plot population, but rather enquiring about general features, such as where the density is zero or has a maximum value.

The most obvious prediction from (7.7) is that the population should vanish at the center of the plot, when $E_1 = E_2 = E_3$, as well as along radial lines separated by $60°$ (when $E_1 = E_2$, $E_2 = E_3$, or $E_1 = E_3$). After some algebra the population is found to be

$$A^2 = [E_1 - E_2)(E_2 - E_3)(E_3 - E_1]^2 \propto r^6 \sin^2 3\phi, \tag{7.8}$$

where the coordinates r and ϕ are defined in Fig. 7.2. The population has a maximum at the boundary, and has the distribution shown in Fig. 7.7(a). The observed distribution, Fig. 7.6, shows these predictions to be incorrect, so that the ω-meson cannot be pseudoscalar.

ii) $J^P = 0^+$

In this case, the ω-meson would be *scalar*, so that A must be odd under inversion and hence a pseudoscalar. The only simple way to form a pseudoscalar from

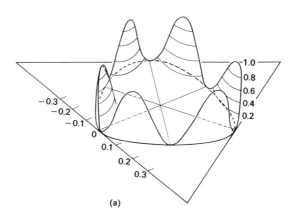

(a)

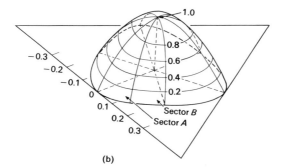

(b)

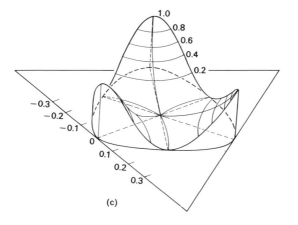

(c)

Fig. 7.7 Dalitz plot densities for the decay into three pions of an ω-meson of spin-parity: (a) 0^- (pseudoscalar), (b) 1^- (vector), (c) 1^+ (axial vector.) The density is represented by the height in the three-dimensional diagrams.

the pion momentum vectors is to build an axial vector from two of them, and take the scalar product of this with the third pion. Thus, one can form terms of the type $(p_1 \wedge p_2) \cdot p_3$. However, since p_1, p_2, and p_3 must be coplanar, in order to conserve momentum, all such terms are zero. Thus, the ω-meson cannot be scalar, nor can any particle decaying to three pions, if parity is conserved in the decay.

iii) $J^P = 1^-$

With this spin-parity, the ω-meson would be a *vector* particle, since it is described by an amplitude or wave function with the spatial properties of a 3-vector. We therefore require A to transform like a vector, but to be even under space inversion, and thus be an axial vector. From the polar vectors p_i, one constructs the requisite form for A as follows:

$$A = (p_1 - p_2) \wedge p_3 + (p_2 - p_3) \wedge p_1 + (p_3 - p_1) \wedge p_2 \qquad (7.9)$$

which, as before, is antisymmetric under pion label interchange. Since $p_1 + p_2 + p_3 = 0$, this is more simply written

$$A = 6p_2 \wedge p_1. \qquad (7.10)$$

Then A must vanish whenever p_1 and p_2 (and therefore p_3 also) are parallel or antiparallel. This is precisely the condition at the boundary of the Dalitz plot; given the numerical values of p_1 and p_2, p_3 must lie in the interval $|p_1 - p_2| < p_3 < |p_1 + p_2|$, the limits obtaining at the boundary, when all three momenta are collinear. It is clear, therefore, that this assignment predicts a Dalitz plot density with a maximum at the center and vanishing at the boundary, as in Fig. 7.7(b). The density has the explicit form

$$A^2 = [1 - (1 + B)(r/r_0)^2 - B(r/r_0)^3 \cos 3\phi], \qquad (7.11)$$

where $B = 2(1 - 3m/M)/(1 + 3m/M)^2$, $r_0 = r_{max} = \frac{1}{3}(M - 3m)$, and m and M are the masses of the π-and ω-mesons. Indeed, the right-hand side of (7.11), when set to zero, is the equation of the boundary. The experimental distribution in Fig. 7.6 is consistent with this prediction.

iv) $J^P = 1^+$

To complete the discussion we consider finally the case when the ω is an *axial vector* meson. Then A has the transformation properties of a polar vector. From the p_i and E_i, such a vector with the required exchange symmetry can be written as

$$A = (p_1 - p_2)E_3 + (p_2 - p_3)E_1 + (p_3 - p_1)E_2. \qquad (7.12)$$

In this case, A vanishes at the center of the Dalitz plot, when $E_1 = E_2 = E_3$. Making use of the fact that $E_1 + E_2 + E_3 = M$ and $p_1 + p_2 + p_3 = 0, (7.12)$

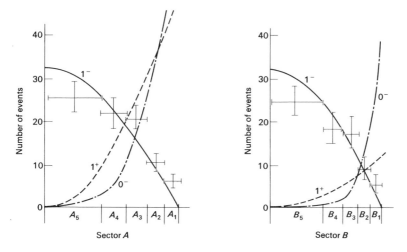

Fig. 7.8 Radial distribution of the Dalitz plot in $\omega \rightarrow 3\pi$ decay. The experimental distribution indicates the spin-parity assignment 1^-. The regions A_1-A_5, B_1-B_5 refer to the sub-divisions in Fig. 7.6.

can also be put in the form

$$A = p_1(M - 3E_2) - p_2(M - 3E_1)$$

which goes to zero along three radial lines parallel to the energy axes, that is, when $p_1 = p_2$ and $E_1 = E_2$. In terms of r, ϕ the density has the form

$$A^2 = r^2[1 - \tfrac{1}{2}(1 - 3m/M) - (r/r_0) \cos 3\phi], \qquad (7.13)$$

and is shown in Fig. 7.7(c). As Fig. 7.6 shows, this is also inconsistent with experiment.

To summarize therefore, the scalar assignment 0^+ is forbidden by parity conservation, and the pseudoscalar 0^- and axial vector 1^+ because they predict zero density at the center of the plot. Only the vector assignment 1^- agrees with experiment. Figure 7.8 summarizes the results on the radial density distribution, integrated over the azimuthal angle ϕ.

7.1.4 Dalitz Plots Involving Three Dissimilar Particles

As an example of a Dalitz plot involving three particles of unequal mass, we take the case of the hyperon resonance $Y_1^*(1385)$–also denoted $\Sigma(1385)$. This state was first observed by Alston *et al.* in 1960 in interactions of 1.15 GeV/c K^--mesons in a liquid hydrogen bubble chamber at the Lawrence Radiation Laboratory, Berkeley. The reaction studied was

$$K^- + p \rightarrow \pi^+ + \pi^- + \Lambda. \qquad (7.14)$$

The kinetic energies of π^+- and π^--mesons, as computed in the overall center-of-mass frame, are plotted along x- and y-axes respectively. If Q is the total available kinetic energy of the three final state particles in this frame, then $T_\Lambda = Q - (T_{\pi^+} + T_{\pi^-})$, and lines of constant T_Λ are thus inclined at 45° to the axes, as shown in Fig. 7.9. Momentum-energy conservation constrains the points representing individual events inside the distorted ellipse shown. If there are no strong correlations in the final state (7.14), the density of points should be uniform, as usual.

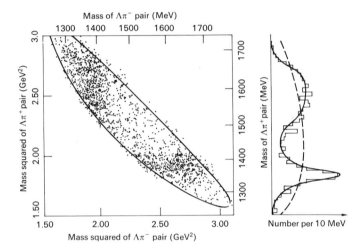

Fig. 7.9 Dalitz plot of the $\Lambda\pi^+\pi^-$ events from reaction (7.14), as measured by Shafer *et al.* (1963), for 1.22 GeV/c incident momentum. The effective $\Lambda\pi^+$ mass spectrum is shown at right. The dotted curve is that expected for a phase-space distribution (ordinate equal to the interval in $M^2_{\Lambda\pi^-}$ between the Dalitz plot boundary), while the full line curve corresponds to a Breit-Wigner resonance expression fitted to the $\Lambda\pi^+$ and $\Lambda\pi^-$ systems.

Figure 7.9 shows clearly the strong departures from uniform density, as evidenced by horizontal and vertical bands on the plot, corresponding to favored values of T_{π^-} and T_{π^+}. This means that the reaction (7.14) is then proceeding as a two-body reaction, the two bodies consisting of one pion (π1) and a resonant state of the Λ with the other pion (π2), with (more or less) unique mass $M_{\Lambda\pi} = 1385$ MeV. Thus, if p is the CMS momentum of each, the total CMS energy will be

$$W = \sqrt{M^2_{\Lambda\pi} + p^2} + \sqrt{m^2_\pi + p^2}, \qquad (7.15)$$

so that, for fixed $M_{\Lambda\pi}$, the momentum p is unique. Obviously, if there is a broad resonance, the quantity T_π will have a spread in values, and thus a band rather than a line on the Dalitz plot. Note that *both* vertical and horizontal bands are found, since either π^+ or π^- can resonate with the Λ-hyperon. Thus

$$Y_1^*(1385) \to \Lambda + \pi^\pm$$

$$I \qquad 1 \qquad 0 \quad 1,$$

and this resonance must have isospin 1, since it decays by strong interaction and isospin is conserved.

In the Dalitz plot it is usual to display $M_{\Lambda\pi^-}^2$ instead of T_{π^+} along the x-axis (and $M_{\Lambda\pi^+}^2$ along the y-axis), so that one can read off the $\Lambda\pi$ invariant mass directly. It is readily shown that

$$M_{\Lambda\pi^-}^2 = W^2 + m_\pi^2 - 2WE_{\pi^+} = a + bT_{\pi^+}, \tag{7.16}$$

so that $M_{\Lambda\pi^-}^2$ is proportional to T_{π^+} for fixed W, the total CMS energy. Figure 7.9 also includes the $\Lambda\pi^+$ mass spectrum. The pure phase-space distribution (7.4) is shown, together with a curve obtained by assuming that either $\Lambda\pi^+$ or $\Lambda\pi^-$ may resonate, the resonance being described by the Breit-Wigner formula (7.31), with a width $\Gamma = 40$ MeV.

Finally, the spin-parity of the $Y_1^*(1385)$ has been determined to be $J^P = \frac{3}{2}^+$; this conclusion is arrived at by analysis of the polarization in the Λ-decay, relative to the production plane of reaction (7.14).

7.2 SCATTERING AMPLITUDE AND THE RESONANCE CONDITION

Consider a beam of particles to be represented by a plane wave traveling in the z-direction and incident on a spinless target particle, or scattering center. Such an incident wave, which we take as of unit amplitude, is represented by

$$\psi_i = e^{ikz},$$

where $k = 1/\lambda$ and $2\pi\lambda$ is the de Broglie wavelength, and where the time dependence $e^{-i\omega t}$ has been omitted for brevity. A plane wave can be represented as a superposition of spherical waves, incoming and outgoing. At a radial distance r from the scattering center, such that $kr \gg 1$, the radial dependence of these spherical waves has the form $e^{\pm ikr}/kr$, so that the flux through a spherical shell is independent of r, as it must be to conserve probability. The angular dependence is determined by the Legendre polynomials $P_l(\cos\theta)$.

The expansion is $(kr \gg 1)$*

$$\psi_i = e^{ikz} = \frac{i}{2kr} \sum_l (2l + 1)[(-1)^l e^{-ikr} - e^{ikr}]P_l(\cos\theta), \tag{7.17}$$

where the first term in square brackets denotes the incoming wave and the

---

* See, for example: F. Mandl, *Quantum Mechanics*, Butterworths, London, 1957, p. 166; L. I. Schiff, *Quantum Mechanics*, McGraw-Hill, New York, 1955, p. 103.

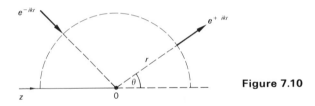

Figure 7.10

second denotes the outgoing wave as in Fig. 7.10 (the sense is clear if one includes the time-dependent term $e^{-i\omega t}$). In all problems with which we deal here, k is typically 10^{13} cm^{-1}, and r, the distance from the scattering center where the particle wave is observed, is many centimeters. Thus, the asymptotic form (7.17) is appropriate.

The scattering center or potential cannot affect the incoming waves, but can in general alter both phase and amplitude of the outgoing wave. The change of phase of the lth partial wave is denoted by $2\delta_l$, and its amplitude by η_l, where $1 > \eta_l > 0$.

The total wave now has the asymptotic form

$$\psi_{\text{total}} = \frac{i}{2kr} \sum_l (2l + 1)[(-1)^l e^{-ikr} - \eta_l e^{2i\delta_l} e^{ikr}] P_l(\cos\theta). \qquad (7.18)$$

Thus, the scattered wave, representing the difference between the outgoing waves with and without the scattering potential, will be

$$\psi_{\text{scatt.}} = \psi_{\text{total}} - \psi_i = \frac{e^{ikr}}{kr} \sum_l (2l + 1) \frac{(\eta_l e^{2i\delta_l} - 1)}{2i} P_l(\cos\theta)$$

$$= \frac{e^{ikr}}{r} F(\theta), \qquad (7.19)$$

where the scattering amplitude

$$F(\theta) = \frac{1}{k} \sum_l (2l + 1) \left(\frac{\eta_l e^{2i\delta_l} - 1}{2i}\right) P_l(\cos\theta). \qquad (7.20)$$

We note that this corresponds to an *elastically* scattered wave, since the wave-number k is taken to be the same before and after scattering. This can be true in the laboratory frame only if the scattering center is infinitely massive. In general, the target (scattering) particle will acquire both momentum and energy; thus the quantities k and λ strictly refer to the properties of the wave in the center-of-momentum frame of the incident and target particles, so that they do not change in an elastic collision. The scattered outgoing flux in solid angle $d\Omega$, through a sphere of radius r, is

$$v_0 \psi_{\text{scatt.}} \psi^*_{\text{scatt.}} r^2 \, d\Omega = v_0 |F(\theta)|^2 \, d\Omega, \qquad (7.21)$$

where v_0 is the velocity of the outgoing particles (relative to the scattering center). But (7.21) is, by definition, the product of the scattering cross section and the incident flux $(= v_i\psi_i\psi_i^* = v_i)$. Since $v_i = v_0$ for elastic scattering,

$$v_0 \, d\sigma = v_0 |F(\theta)|^2 \, d\Omega,$$

or

$$\left(\frac{d\sigma}{d\Omega}\right)_{\text{elastic}} = |F(\theta)|^2. \tag{7.22}$$

The Legendre polynomials P_l obey the orthogonality condition

$$\int P_l P_{l'} \, d\Omega = \frac{4\pi \, \delta_{l,l'}}{(2l + 1)},$$

where

$$\delta_{l,l'} = 1 \qquad \text{for } l = l'$$
$$= 0 \qquad \text{for } l \neq l'.$$

Thus, the total *elastic* scattering cross section, integrated over angle, is, from (7.20) and (7.22),

$$\sigma_{\text{el.}} = 4\pi\lambda^2 \sum_l (2l + 1) \left| \frac{\eta_l e^{2i\delta_l} - 1}{2i} \right|^2. \tag{7.23}$$

When $\eta = 1$, the case for no absorption of the incoming wave, this becomes

$$\sigma_{\text{el.}} = 4\pi\lambda^2 \sum_l (2l + 1) \sin^2 \delta_l. \tag{7.24}$$

Obviously, $\sigma_{\text{el.}}$ is zero when $\delta_l = 0$, corresponding to zero scattering potential. If $\eta < 1$, the *reaction* cross section, σ_r, is then obtained from conservation of probability:

$$\sigma_r = \int (|\psi_{\text{in}}|^2 - |\psi_{\text{out}}|^2) r^2 \, d\Omega,$$

where ψ_{in} is the first term of (7.17) and ψ_{out} is the second term of (7.18). This gives

$$\sigma_r = \pi\lambda^2 \sum_l (2l + 1)(1 - |\eta_l|^2). \tag{7.25}$$

The *total* cross section will be

$$\sigma_T = \sigma_r + \sigma_{\text{el.}} = \pi\lambda^2 \sum_l (2l + 1)2(1 - \eta_l \cos 2\delta_l).$$

Since $P_l(1) = 1$ for all l, (7.20) therefore gives in the *forward direction* $\cos \theta = 1$, $\theta = 0$:

$$\text{Im } F(0) = \frac{1}{2k} \sum_l (2l + 1)(1 - \eta_l \cos 2\delta_l).$$

Comparing the last two equations finally gives us the *optical theorem*

$$\text{Im } F(0) = \frac{k}{4\pi} \sigma_T, \tag{7.26}$$

relating the total cross section to the imaginary part of the forward elastic scattering amplitude.

The relations derived above describe the various cross sections $\sigma_{\text{el.}}$, σ_r, and σ_T in terms of the parameters η and δ. They set bounds on the cross sections imposed by the conservation of probability (often called the unitarity condition). For example, we see from (7.24) that the maximum elastic scattering cross section for the lth partial wave occurs when $\delta_l = \pi/2$, having the value

$$\sigma_{\text{el.}}^{\text{max}} = 4\pi\lambda^2(2l + 1), \tag{7.27}$$

for $\eta_l = 1$, i.e. the case where there is pure scattering without absorption. Similarly, from (7.25) we obtain the maximum absorption or reaction cross section by setting $\eta_l = 0$:

$$\sigma_r^{\text{max}} = \pi\lambda^2(2l + 1).$$

This last equation can be reproduced from a simple classical argument. An orbital angular momentum l corresponds to an "impact parameter" b given by $l\hbar = pb$, or $b = l\lambda$. Particles of angular momentum $l \to l + 1$ therefore impinge on, and are absorbed by, an annular ring of cross sectional area

$$\sigma = \pi(b_{l+1}^2 - b_l^2) = \pi\lambda^2(2l + 1).$$

Note also that, for the case of complete absorption, $\eta = 0$, the elastic cross section is also $\pi\lambda^2(2l + 1)$—see Section 7.7.

The quantity

$$f(l) = \frac{\eta_l e^{2i\delta_l} - 1}{2i} = \frac{i}{2} - \frac{i\eta_l}{2} e^{2i\delta_l} \tag{7.28}$$

of (7.20) is the elastic scattering amplitude (for the lth partial wave). It is a complex quantity, and in Fig. 7.11 we show f plotted as a vector in the complex plane.

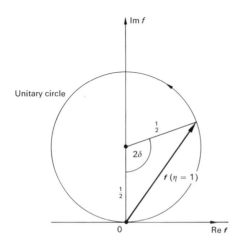

Fig. 7.11 The scattering amplitude f plotted as a vector in the complex plane. Causality requires that, as the energy increases, the vector will trace out the circle in an anticlockwise direction.

For $\eta = 1$, the end of the vector describes a circle of radius $\frac{1}{2}$ and center $i/2$, as the phase shift varies from 0 to $\pi/2$. When $\delta = \pi/2$, f is purely imaginary and has magnitude unity, corresponding to (7.27). If $\eta < 1$, the end of the vector f lies within the circle shown, sometimes referred to as the *unitary circle*. It is so called because, as f is defined, its maximum modulus must be unity if probability is conserved—the intensity in a particular outgoing partial wave cannot exceed that in the corresponding incoming wave.

7.3 THE BREIT-WIGNER RESONANCE FORMULA

Let us now make the connection between the preceding wave-optical discussion and the scattering of two elementary particles—one corresponding to the incident wave and the other to the scattering center. To simplify matters let the two particles be spinless. If the elastic scattering amplitude $f(l)$ passes through a maximum for a particular value of l and for a particular CMS wavelength λ, the two particles are said to *resonate*. The resonant state is then characterized by a unique angular momentum or spin $J = l$, a unique parity and isospin, and a mass corresponding to the total CMS energy of the two particles. A criterion of resonance is that the phase-shift δ_l of the lth partial wave should pass through $\pi/2$. The cross section may also be described in terms of the width Γ or lifetime τ of the resonant state, as follows; dropping the subscript l in (7.28), and with $\eta = 1$, we may rewrite f in the form

$$f = \frac{e^{i\delta}(e^{i\delta} - e^{-i\delta})}{2i}$$

$$= e^{i\delta} \sin \delta = \frac{1}{(\cot \delta - i)}. \tag{7.29}$$

Near resonance $\delta \approx \pi/2$, so that $\cot \delta \approx 0$. If E is the total energy of the two-particle state in the CMS, and E_R is the value of E at resonance ($\delta = \pi/2$), then expanding by a Taylor series

$$\cot \delta(E) = \cot \delta(E_R) + (E - E_R)\left[\frac{d}{dE} \cot \delta(E)\right]_{E=E_R} + \cdots$$

$$\approx -(E - E_R)\frac{2}{\Gamma},$$

where $\cot \delta(E_R) = 0$ and we have defined $2/\Gamma = -[d(\cot \delta(E))/dE]_{E=E_R}$. Neglecting further terms in the series is justified provided $|E - E_R| \approx \Gamma \ll E_R$. Then from (7.29),

$$f(E) = \frac{1}{(\cot \delta - i)} = \frac{\frac{1}{2}\Gamma}{[(E_R - E) - i\Gamma/2]}. \tag{7.30}$$

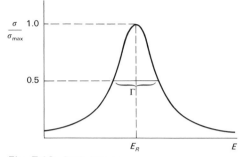

Fig. 7.12 Breit-Wigner resonance curve.

From (7.23) and (7.28), we obtain for the elastic scattering cross section

$$\sigma_{el.}(E) = 4\pi\lambda^2(2l + 1)\frac{\Gamma^2/4}{[(E - E_R)^2 + \Gamma^2/4]}. \tag{7.31}$$

This is known as the *Breit-Wigner formula*. The resonance curve of $\sigma(E)$ is shown in Fig. 7.12. The width Γ is defined so that the elastic cross section $\sigma_{el.}$ falls by a factor 2 from the peak value when $|E - E_R| = \pm\Gamma/2$.

As pointed out in Chapter 1, the width Γ and lifetime τ of the resonant state are connected by the relation $\tau = \hbar/\Gamma$. The energy dependence of the amplitude (7.31) is simply the Fourier transform of an exponential time pulse, corresponding to the radioactive decay of the resonance. This can be seen as follows. If we denote the wave function of the resonant state by ψ, then from (4.59) and with units $\hbar = c = 1$,

$$\psi(t) = \psi(0) \exp\left[-t\left(\frac{\Gamma}{2} + iE_R\right)\right]$$

The Fourier transform of this expression is

$$g(\omega) = \int_0^\infty \psi(t)e^{i\omega t}\, dt,$$

with $\omega = E/\hbar = E$. The amplitude as a function of E is then

$$\chi(E) = \int \psi(t)e^{iEt}\, dt = \psi(0)\int e^{-t[(\Gamma/2)+iE_R-iE]}\, dt$$

$$= \frac{K}{[(E_R - E) - i\Gamma/2]}.$$

where K is some constant.

The final step consists of making the connection between the probability, integrated over time, that the resonant state shall decay with total energy E, and the cross section $\sigma_{el.}(E)$. If the resonance is purely elastic it is clear, from conservation of probability, that the cross section—which measures the probability of forming the resonant state—must be proportional to the probability of decay. Thus we may write $\sigma(l, E) \propto \chi^*\chi$, where $\chi = \chi_l(E)$ and l is the value of the orbital angular momentum involved in forming the resonant state in question. Both of these quantities have a maximum when $E = E_R$ and $\delta_l = \pi/2$, so that

$$(\sigma_{el.})_{max.} = 4\pi\lambda^2(2l + 1) \quad \text{and} \quad (\chi^*\chi)_{max.} = 4K^2/\Gamma^2.$$

Thus, in terms of Γ, measuring the width of a decay process, rather than δ, measuring the phase-shift in a scattering process, we have

$$\sigma_{el.} = 4\pi\lambda^2(2l + 1) \frac{\Gamma^2/4}{[(E - E_R)^2 + \Gamma^2/4]},$$

as in (7.31).

For a spinless projectile hitting a spinless target, $l = J$, the total angular momentum of the resonant state. Therefore, $(2l + 1) \rightarrow (2J + 1)$. For spin 0 particles (e.g. pions, kaons) incident on nucleons (spin $\frac{1}{2}$), the factor $(2J + 1)$ still applies,* except that now only half the target spin states can contribute. The point is that the spin and parity (J^P) of the resonance are fixed, so that only *one* l-value can contribute, and then only if the target particle has the right spin orientation. For example, the $N^*(1238)\pi - p$ resonance has $J^P = \frac{3}{2}^+$ and results from a p-wave $(l = 1)$ pion-nucleon interaction. Thus the $J = l + \frac{1}{2}$ combination forms this state, not $J = l - \frac{1}{2}$ which has $J^P = \frac{1}{2}^+$ (for $l = 1$). A d-wave interaction $(l = 2)$ with $J = l - \frac{1}{2}$ would give the right resonance spin but the wrong parity $(J^P = \frac{3}{2}^-)$. Thus, for a resonance of angular momentum J formed from a nucleon target by a spinless incident particle, (7.31) becomes

$$\sigma_{el.}(E) = \frac{\pi\lambda^2}{2} \frac{(2J + 1)\Gamma^2}{[(E - E_R)^2 + \Gamma^2/4]}. \tag{7.32}$$

In both (7.31) and (7.32) it is assumed that the resonance can only decay *elastically*, i.e. $\pi + n \rightarrow N^* \rightarrow \pi + n$, but not $N^* \rightarrow 2\pi + n$. In general, $\Gamma = \Gamma_{el.} + \Gamma_r$, where $\Gamma_{el.}$ and Γ_r are *partial widths* for decay in the elastic channel and inelastic channel. Then for $\sigma_{el.}(E)$, the Γ^2 numerator in (7.32) should be replaced by $\Gamma_{el.}^2$. The inelastic cross section $\sigma_r(E)$ is the same as (7.32), with the Γ^2 numerator replaced by $\Gamma_{el.}\Gamma_r$.

The scattering amplitude f, in the case of an elastic resonance $(\eta = 1)$, traces out the unitary circle of Fig. 7.11. By applying the concept of causality—the outgoing wave cannot leave the scattering center before the incoming

---

* Intuitively we see that this is just the spin multiplicity factor for a state J. For a proof, see J. M. Blatt and V. F. Weisskopf, *Theoretical Nuclear Physics*, John Wiley, New York, 1952, p. 423.

wave arrives—it may be proved that f traces out a circle in the *anticlockwise* direction (for an attractive potential).

Figure 3.10, p. 103, shows the $\pi^+ p$ and $\pi^- p$ total cross sections as a function of kinetic energy of the incident pion. As remarked in Section 3.11, there is a very obvious $I = \frac{3}{2}$ resonance at $T_\pi = 195$ MeV, corresponding to a pion-proton mass of 1236 MeV. This is further discussed below. It is usually designated $P_{33}(1236)$, meaning that it is a p-wave ($l = 1$) pion-nucleon resonance, of $I = \frac{3}{2}$ and $J = \frac{3}{2}$. One can also distinguish other humps and bumps in σ (total). For example, in the $I = \frac{1}{2}$ channel one can see evidence for the states $D_{13}(1520)$ and $F_{15}(1688)$ as peaks in $\sigma(\pi^- p)$, and $F_{37}(1950)$ as a peak in $\sigma(\pi^+ p)$, i.e. $I = \frac{3}{2}$.

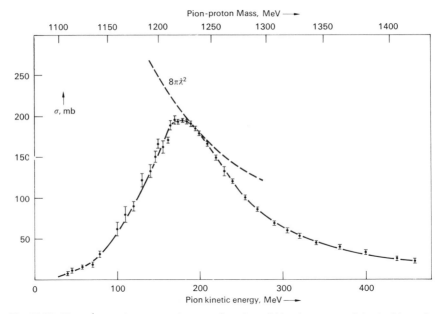

Fig. 7.13 The $\pi^+ p$ total cross section as a function of kinetic energy of the incident pion, or the $\pi^+ - p$ mass, in the region of the 1236 MeV, $I = \frac{3}{2}$, $J^P = \frac{3}{2}^+$ resonance. Not all experimental points have been included. The maximum cross section, $8\pi \lambda^2$, allowed by conservation of probability is shown dotted.

However, it turns out that there are many overlapping pion-nucleon resonances, a recent estimate giving some twenty states of $M_{\pi p} < 2200$ MeV. A mere inspection of the total cross section is therefore misleading. It is necessary to make a sophisticated *phase-shift analysis* of the pion-nucleon elastic scattering data, based on the behavior of the various polynomial coefficients required to fit the angular distributions, as a function of bombarding energy. The best values for the masses, widths, and elasticities of resonant states

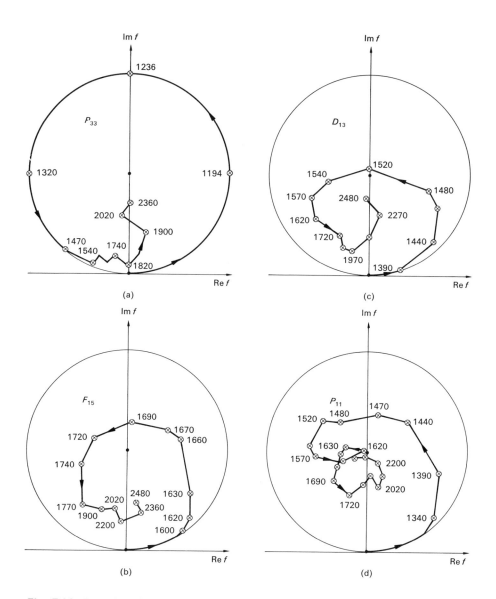

Fig. 7.14 Examples of the behavior of partial wave amplitudes in pion-nucleon scattering, according to a phase-shift analysis at Saclay (Paris) by Ayed *et al.* (1970). The numbers refer to the center-of-mass energy in MeV. (a) The *p*-wave amplitude of $I = \frac{3}{2}$, $J^P = \frac{3}{2}^+$, thus designated P_{33}. The dominant feature is that as the energy increases, the tip of the vector **f** of Fig. 7.11 moves along the unitary circle. The phase-shift δ passes 90° at 1236 MeV, corresponding to the central resonance mass. Reference to (7.30) shows that the width Γ ($\sim$120 MeV) is given by the difference in energy at opposite ends of the diameter parallel to the real axis. Above 1500 MeV, $\eta < 1$, corresponding to the inelastic process $N^* \to N + 2\pi$ as well as $N + \pi$. (b) and (c) The F_{15} and D_{13} amplitudes, the former indicating a resonance of $J^P = \frac{5}{2}^+$, $I = \frac{1}{2}$, of central mass 1690 MeV, and the latter one of $J^P = \frac{3}{2}^-$, $I = \frac{1}{2}$, and mass 1520 MeV. Note that both are strongly inelastic. (d) The P_{11} amplitude ($I = \frac{1}{2}$, $J^P = \frac{1}{2}^+$) demonstrating a resonance with these quantum numbers of central mass 1470 MeV. Note that this is *not* apparent as a bump in cross section in Fig. 3.10 and only phase-shift analysis is able to reveal its existence.

273

are fitted to the data by an iterative procedure. Different analyses give somewhat different solutions. Figure 7.14 shows an example of such an analysis, for the P_{33}, F_{15}, P_{11}, and D_{13} waves.

7.4 A PION-NUCLEON RESONANCE—THE $N^*(1236)$

A discussion of the first pion-nucleon resonance, the $N^*(1236)$, is relatively simple. The amplitude is almost purely elastic because of the low mass, and the effects of higher-lying resonances may be neglected. Application of (7.32) shows that, at the peak, $\sigma_{el.} = 2\pi\lambda^2(2J + 1)$. This limiting value is shown dotted in Fig. 7.13, for $J = \frac{3}{2}$, or $\sigma_{el.} = 8\pi\lambda^2$, proving that the $N^*(1236)$ is a p-wave pion-nucleon resonance of spin-parity $J^P = \frac{3}{2}^+$.

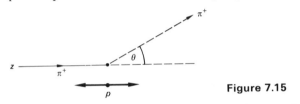

Figure 7.15

In order to confirm this assignment, let us ask what the $\pi^+ - p$ scattering angular distribution will look like near resonance. Take the incident pion direction as the quantization axis (z-axis). The angular momentum wave function of a p-state pion will be $\phi(l, m) = \phi(1, 0)$ since the pion is spinless. For the proton, we will have $\alpha(\frac{1}{2}, \pm\frac{1}{2})$ corresponding to the two possible orientations of spin (Fig. 7.15). The product is the state

$$\psi(j, m) = \psi(\tfrac{3}{2}, \pm\tfrac{1}{2}).$$

When the N^* radiates a pion, the remaining proton may or may not have its spin "flipped." From Table 3.4 of Clebsch-Gordan coefficients, we can write, for the final state wave functions α' and ϕ',

$$\psi(\tfrac{3}{2}, \tfrac{1}{2}) = \sqrt{\tfrac{1}{3}}\phi'(1, 1)\alpha'(\tfrac{1}{2}, -\tfrac{1}{2}) + \sqrt{\tfrac{2}{3}}\phi'(1, 0)\alpha'(\tfrac{1}{2}, \tfrac{1}{2}).$$

Note that, since the scattered pion comes off at some angle θ to the z-direction, it is now possible for its orbital angular momentum to have a finite projection (1 or 0) on the old quantization axis. The ϕ' are simply the spherical harmonics:

$$\phi'(1, 1) = Y_1^1 = -\sqrt{\frac{3}{4\pi}}\sin\theta e^{i\phi}/\sqrt{2},$$

$$\phi'(1, 0) = Y_1^0 = \sqrt{\frac{3}{4\pi}}\cos\theta.$$

The angular distribution of the pions is therefore

$$I(\theta) = \psi\psi^* = \tfrac{1}{3}(Y_1^1)^2 + \tfrac{2}{3}(Y_1^0)^2,$$

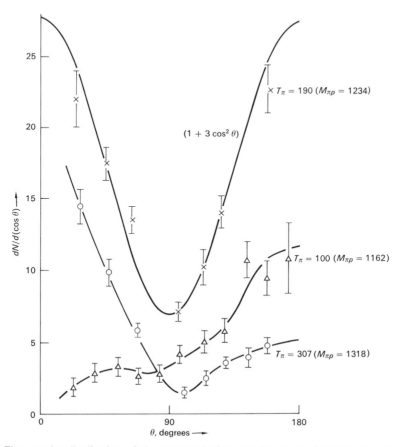

Fig. 7.16 The angular distribution of the scattered pion, relative to the incident pion, in π^+-p elastic scattering, as measured in the center-of-mass frame. In the region of the N^*_{33} resonance of mass 1236 MeV (T_π = 190 MeV), the distribution has the form $(1 + 3\cos^2\theta)$, as in (7.33).

the cross term being zero since Y^1_1 and Y^0_1, as well as $\alpha'(\frac{1}{2}, -\frac{1}{2})$ and $\alpha'(\frac{1}{2}, \frac{1}{2})$, are orthogonal. Thus

$$I(\theta) \propto \sin^2\theta + 4\cos^2\theta = 1 + 3\cos^2\theta. \tag{7.33}$$

The differential cross section is plotted in Fig. 7.16 as a function of CMS angle θ for different values of T_π. At resonance, the dependence is in agreement with (7.33).

7.5 A PION-PION RESONANCE—THE ρ(765)

In the same year (1961) in which the $\omega \rightarrow 3\pi$ resonance was discovered in proton-antiproton annihilation, a two-pion resonance, the ρ-meson, was observed in high energy pion-proton collisions by Erwin *et al.* at Wisconsin,

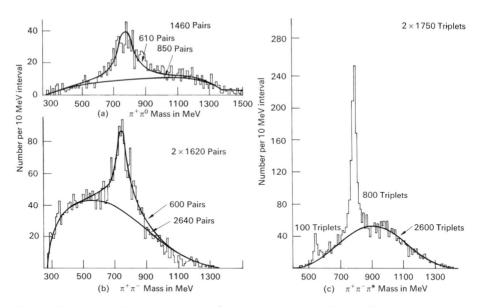

Fig. 7.17 Mass distributions of (a) $\pi^+\pi^0$ pairs from reaction (7.34a), (b) $\pi^+\pi^-$ pairs from reaction (7.34b). The smooth curves indicate the distributions expected from phase space. (c) The $\pi^+\pi^-\pi^0$ invariant mass spectrum from reaction (7.34c). The narrow peak at 785 MeV corresponds to the $\omega \to 3\pi$ resonance, and that at 550 MeV to $\eta \to 3\pi$. (From Alff *et al.*, 1962.)

and by several other groups. Figure 7.17 shows results from a study in a hydrogen bubble chamber of the following reactions, using incident pions of momentum 1.6 to 1.9 GeV/c:

$$\pi^+ + p \to \pi^+ + \pi^0 + p \tag{7.34a}$$

$$\to \pi^+ + \pi^+ + \pi^- + p \tag{7.34b}$$

$$\to \pi^+ + \pi^+ + \pi^- + \pi^0 + p. \tag{7.34c}$$

The mass distributions of $\pi^+\pi^0$ pairs from reaction (7.34a) and $\pi^+\pi^-$ pairs from (7.34b) show a broad peak centered at 765 MeV, with a width about 120 MeV, attributed to the decay $\rho \to 2\pi$. Experiments have failed to detect a similar peak in the $\pi^+\pi^+$ mode.

These results indicate an isospin assignment $I = 1$ for the ρ-meson. Since we have two identical bosons in the final state, $I = 1$ implies that the spatial wave function of the pion pair must be antisymmetric. The simplest possibility is $l = 1$, corresponding to the spin-parity quantum numbers for the ρ-meson: $J^P = 1^-$.

A two-body decay is less informative than a three-body decay process, such as was analyzed in determining the quantum numbers of the ω-meson. To prove the above spin-parity assignment, we have to rely on observations relating to the production process itself. The easiest way to determine the spin is to consider a special class of "peripheral" events, with small momentum transfer to the nucleon. For this purpose, consider the reactions

$$\pi^- + p \rightarrow \pi^- + \pi^0 + p \qquad (7.35a)$$
$$\rightarrow \pi^+ + \pi^- + n. \qquad (7.35b)$$

As discussed in Section 7.8, these events can be described by single pion exchange, as indicated diagrammatically in Fig. 7.18(a). If we now look in the rest frame of the ρ-meson [Fig. 7.18(b)], we are essentially observing the collision of the incident π^- with a virtual π^+ of the meson cloud surrounding the nucleon:

$$\pi^- + \pi^+_{(virtual)} \rightarrow \rho \rightarrow \pi^+ + \pi^-.$$

If we choose the axis of quantization, z, along the incident pion direction, it is clear that in the initial, and hence also the final, state the z-component of angular momentum $m = 0$, since pions have zero spin. Thus, the angular distribution of the scattered pions will be

$$F(\theta) = (Y_l^m)^2 = [P_J^0(\cos \theta)]^2,$$

where J is the ρ-spin and, for $m = 0$, the azimuthal distribution $e^{im\phi}$ is isotropic.

The measured angular distributions for $\pi^-\pi^0$ pairs in (7.35a) is found to have the form

$$F(\theta) = A + B \cos \theta + C \cos^2 \theta. \qquad (7.36)$$

The fact that no terms higher than $\cos^2 \theta$ appear is itself indicative that $J_\rho = 1$. The form (7.36) is interpreted as a coherent superposition of two amplitudes; A_1 due to a $J = 1$ resonance, and A_0 due to an S-wave $\pi\pi$ interaction (of $I = 2$, since $J = 0$) forming a nonresonant background term. Then

$$F(\theta) = |A_0 + A_1 \cos \theta|^2$$
$$= |A_0|^2 + |A_1|^2 \cos^2 \theta + 2 \operatorname{Re} A_0 A_1^* \cos \theta, \qquad (7.37)$$

where we have used the fact that $P_0^0 = 1$, $P_1^0 = \cos \theta$. Now, as indicated in Fig. 7.11, the amplitude A_1 must become purely imaginary at the ρ-peak, so

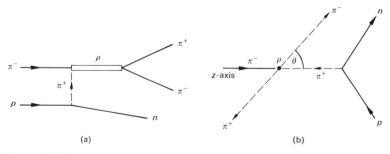

(a)   (b)

Figure 7.18

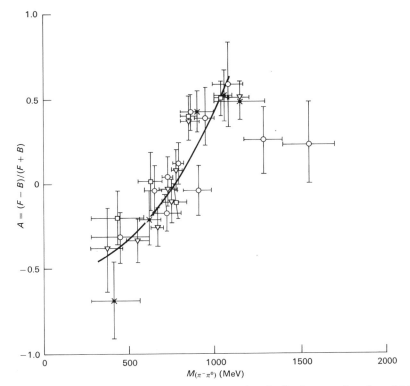

Fig. 7.19 Forward-backward ratio in the pion angular distribution as a function of $M_{\pi^0\pi^-}$ in reaction (7.35a), as measured by various experiments for incident pion momenta from 2.75 to 6 GeV/c. The forward-backward asymmetry goes to zero at the ρ-mass ($M_{\pi^-\pi^0} = 765$ MeV). (After Baton *et al.*, 1965.)

the interference (cos θ) term will change sign as one goes through it. This can be seen in terms of the forward-backward ratio of Fig. 7.19.

The assumption of single-pion exchange, necessary to obtain the above result, is borne out by comparing the rates for reactions (7.35a) and (7.35b). Using the table of Clebsch-Gordan coefficients (Appendix C) for adding two isospins of unity, one finds

$$\phi(\pi^-\pi^0) = \frac{1}{\sqrt{2}}\psi(1,-1) + \frac{1}{\sqrt{2}}\psi(2,-1),$$

$$\phi(\pi^+\pi^-) = \frac{1}{\sqrt{3}}\psi(0,0) + \frac{1}{\sqrt{2}}\psi(1,0) + \frac{1}{\sqrt{6}}\psi(2,0),$$

where $\psi(I, I_3)$ denotes a particular isospin state and its z-component. For

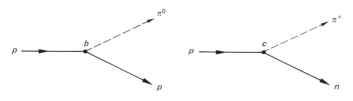

Figure 7.20

$I = 1$, we have

$$\frac{\sigma(\pi^- + \pi^0 \to \rho^- \to \pi^- + \pi^0)}{\sigma(\pi^+ + \pi^- \to \rho^0 \to \pi^+ + \pi^-)} = \frac{\left| \frac{1}{\sqrt{2}} \psi(1,-1) \right|^2}{\frac{1}{\sqrt{2}} \psi(1,0)} = 1.$$

We must remember that one pion in the initial state is virtual. Referring to Section 3.11, we found for the ratio of the couplings $b/c = 1/\sqrt{2}$. Reactions (7.35a) and (7.35b) involve the left- and right-hand diagrams of Fig. 7.20 respectively, so that our final predicted ratio is

$$\frac{\sigma(\pi^- p \to \rho^0 n)}{\sigma(\pi^- p \to \rho^- p)} = \left(\frac{c}{b} \right)^2 = 2.$$

The observed ratio is 1.8 ± 0.2, in agreement with this.

It may also be remarked that the angular distribution of pairs from (7.35b) does *not* follow the form of Fig. 7.19; i.e. the $\cos \theta$ term is not zero at the ρ-peak (see Fig. 7.21). This implies that the $J = 0$ amplitude, A_0, *also* has an imaginary component, corresponding to resonant behavior. In this case, since $I_3 = 0$, this must have the quantum numbers $J = 0, I = 0$ (not $I = 2$, as for the $\pi^- \pi^0$ case, which is a nonresonant amplitude). This possible resonance is referred to as the ε-meson. Its existence has not yet clearly been established. It could be more clearly seen in the $\pi^0 \pi^0$ state, but this is experimentally difficult.

To summarize, the ρ-meson has $I = 1, G = +1, J^P = 1^-$. Its production in pion-nucleon interactions may be pictured in terms of collision of the incident pion with a single virtual pion from the nucleon target. This is discussed further in Section 7.8.

It has been mentioned in Chapter 5 that the ρ-meson may also be produced in electron-positron colliding beam experiments. In such experiments, the width of the ρ-resonance appears to be somewhat smaller ($\Gamma \approx 100$ MeV) than that found for ρ's produced in strong interactions such as (7.34), where $\Gamma \approx 120$ to 140 MeV. The difference, if real, has to be attributed to the effect of final state interactions in strong production, where other hadrons are involved. Put in another way, the Breit-Wigner amplitude is the Fourier transform of an exponential time pulse (integrated from $t = 0$ to $t = \infty$),

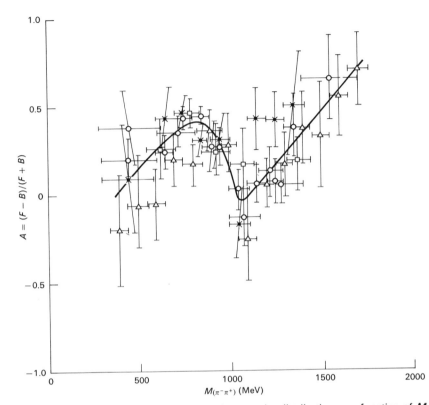

Fig. 7.21 Forward-backward ratio in the pion angular distribution as a function of $M_{\pi^+\pi^-}$ in reaction (7.35b). In comparison with Fig. 7.19, note that the asymmetry does not disappear at the ρ-mass. (After Baton *et al.*, 1965.)

corresponding to a freely decaying state. If, however, such a particle undergoes final state interactions in addition to free decay, one might expect the lifetime to be reduced and the width correspondingly increased.

7.6 EXPERIMENTAL TECHNIQUES IN THE DETECTION OF BOSON AND BARYON RESONANCES

Resonant states can be formed in two ways, in what are called *formation* and *production* experiments respectively. In a formation experiment, the resonant state is formed by combination of the colliding particles, at a CMS energy just equal to the mass of the resonant state. Examples are:

Incident momentum for resonance peak

i) $\pi^+ + p \rightarrow N^*(1236)$ $\qquad\qquad\qquad p_{\pi^+} = 304 \text{ MeV}/c,$
  $\;\llcorner\!\rightarrow \pi^+ + p$

ii) $e^+ + e^- \to \rho^0(765)$ $\quad\quad\quad\quad\quad\quad\quad p_{e^+} = p_{e^-} = \dfrac{m_\rho}{2}$,
$\quad\quad\quad\quad\Large\llcorner\normalsize\,{\to}\, \pi^+ + \pi^-$

iii) $K^- + p \to \Sigma(1520)$ $\quad\quad\quad\quad\quad\quad p_{K^-} = 395 \text{ MeV}/c$.
$\quad\quad\quad\quad\quad\quad\Large\llcorner\normalsize\,{\to}\, \Lambda + \pi^- + \pi^+$

Reaction (i) was used to establish the existence of the first high energy resonant state to be observed. We have already seen that this $I = \frac{3}{2}$ pion-nucleon resonant state is manifested by a strong peak in the $\pi^+ p$ elastic scattering cross section. Although both bubble chamber and counter techniques have been used to detect resonances in formation experiments, they are particularly suited to the counter method and, indeed, most of our present information on the pion-nucleon resonances, their masses, and quantum numbers has come from the use of this technique, particularly in conjunction with polarized proton targets. Reaction (ii) has already been discussed in Section 5.8.

In production experiments, a resonance is formed in the final state, together with at least one other particle. Some examples are:

iv) $\pi^- + p \to p + \rho^-$
$\quad\quad\quad\quad\Large\llcorner\normalsize\,{\to}\, \pi^- + \pi^0$,

v) $\bar{p} + p \to \omega + 2\pi$
$\quad\quad\quad\quad\Large\llcorner\normalsize\,{\to}\, \pi^+ + \pi^- + \pi^0$,

vi) $K^+ + p \to p + K^{*+}(890)$
$\quad\quad\quad\quad\quad\quad\Large\llcorner\normalsize\,{\to}\, K^0 + \pi^+$,

vii) $K^- + p \to K^+ + \pi^0 + \Xi^{*-}(1530)$
$\quad\quad\quad\quad\quad\quad\quad\quad\Large\llcorner\normalsize\,{\to}\, \Xi^0 + \pi^-$.

A few general remarks can be made here. First of all, we note that in formation experiments, with incident beams of pions or kaons, we are confined to baryon resonances of strangeness 0 or -1. Cascade hyperon (Ξ^*) resonances, with $S = -2$, can only be generated by K^- if accompanied by a K^+ or K^0 to conserve strangeness. Boson resonances can only be produced in a formation process either electromagnetically, as in reaction (ii), or conceivably in nucleon-antinucleon processes (for which, however, the CMS energy always exceeds 1.9 GeV). Thus, most of our information on boson resonances comes from the analysis of production reactions, for which there is no unique incident momentum, since at least two bodies are present in the final state. Thus the initial cataloging of most of the boson resonances, and the majority of the baryon

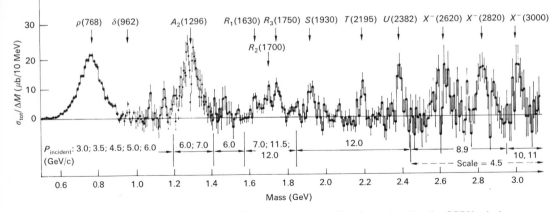

Fig. 7.22 Boson mass spectrum from the reaction $\pi^- + p \rightarrow p + X^-$, determined by the CERN missing-mass spectrometer experiments. Note that the different mass ranges are covered by different incident momenta. Note also that the ρ-meson is a broad resonance, and that the states S, T, U, and X of high mass have relatively narrow widths, ~ 20 MeV. (After Maglic, 1969.)

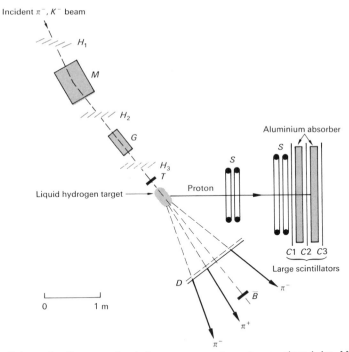

Fig. 7.23 Schematic diagram of missing-mass spectrometer employed by Maglic and co-workers at CERN. The staggered scintillator hodoscopes H_1, H_2, and H_3, and the bending magnet M determine the incident particle momentum, and the gas Cerenkov counter G registers the nature of the particle (pion or kaon). The interaction $\pi^- + p \rightarrow p + X^-$ occurs in the liquid hydrogen target. The angle of emission of the recoiling proton is determined by means of sonic spark chambers S, and its momentum from a measurement of range in aluminum absorbers and by the time-of-flight ($T \rightarrow C1$). The decay products of the boson X^- were registered in the matrix of 72 small counters, D. $\bar{B}$ is a beam anticounter.

282

resonances, has been in the province of the bubble chamber technique. The great advantage of this method, as described in Section 2.7.3, is that complicated final states involving many particles can be analyzed in detail. The drawback of the method is that rare events have to be searched for, picture by picture, usually using only a small fraction of the available beam intensities from present-day accelerators. Once a particular reaction is understood, detailed examination can often be more profitably undertaken by counter/spark chamber methods, which can accept enormously larger incident particle fluxes and trigger just on events of interest. An example of this method is the determination of the charge asymmetry in the η-decay, described in Section 3.15.

We may mention here one powerful counter technique which has been employed to investigate boson resonances in production experiments, the reactions studied involving in the final state just the resonance and a recoil proton:

$$\pi + p \rightarrow p + X.$$

If the incident pion momentum is accurately known, then a precise measure of the momentum vector of the recoil proton gives a value for the mass of the missing boson state X (the decay products of which may not in general be measured). This method is known as the missing-mass spectrometer technique. A typical boson mass spectrum is shown in Fig. 7.22, and the experimental set-up in Fig. 7.23.

7.7 TOTAL AND ELASTIC CROSS SECTIONS; THE BLACK DISC MODEL

Figure 7.24 shows measured total cross sections for a variety of incident particles on proton and neutrons as a function of incident momentum. The technique employed is simply to measure the attenuation of the incident beam through a target of known thickness. The cross sections on neutrons are obtained by taking the difference with deuterium and hydrogen targets and applying a small "shadowing" correction ($\leqslant 10\%$). The main features of these results are that, firstly, the cross sections are finite, and seem to tend to rather constant values at sufficiently high energy. The magnitude of the cross section varies with the type of particle, but is in the region of 20 to 40 mb. If one equates this value to a "geometrical" cross section πR^2, one obtains $R \sim 10^{-13}$ cm $= 1$ F as the "range" of the strong interaction. Secondly, especially in the lower energy range, one observes a large difference in the $\bar{p}p$ and pp cross sections, and this is not unexpected in view of the larger number of isospin channels open for the nucleon-antinucleon process, as well as the higher available energy following annihilation. Similar remarks apply in comparing π^-p with π^+p, and K^-p with K^+p. At very high energies there is a prediction from quantum field theory, known as the Pomerancuk theorem, that the cross sections should become the same for particle and antiparticle,

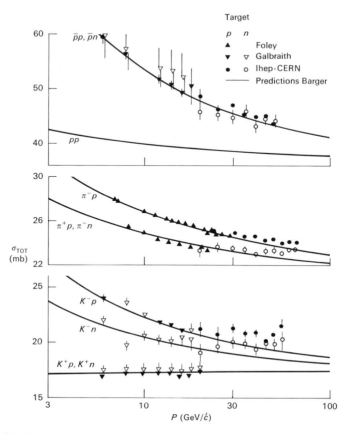

Fig. 7.24 Total cross sections on proton and neutron targets, for incident protons, antiprotons, positive and negative pions, and positive and negative kaons. Note that the momentum scale is logarithmic. The curves are from a Regge-pole fit to the low energy data ($p < 30$ GeV/c). (After Allaby *et al.*, 1969.)

and moreover isospin independent. Thus, cross sections for $\pi^- p$ and $\pi^+ p$ ($\equiv \pi^- n$ from charge independence) should become equal on both counts. Although there is an indication from the data that particle and antiparticle cross sections are coming together, it is clear that the asymptotic region where they should merge (if there is one) is well above 50 GeV.

The simplest possible model of absorption and scattering is that of a totally absorbing black disc of well-defined radius R. Setting $\eta_l = 0$ in (7.23) and (7.25) for this case, one obtains

$$\sigma_{\text{elastic}} = \pi \lambdabar^2 \sum (2l + 1) = \pi R^2,$$

$$\sigma_{\text{inelastic}} = \pi \lambdabar^2 \sum (2l + 1) = \pi R^2,$$

$$\sigma_{\text{total}} = 2\pi \lambdabar^2 \sum (2l + 1) = 2\pi R^2,$$

where λ is the de Broglie wavelength of the colliding particles in their center-of-momentum system. In this model, the angular momentum contributed by the incident wave, l, varies from 0 to $l_{max} = R/\lambda$. For example, for $R = 1$ F, and 20 GeV/c incident momentum, $\lambda = 0.01$ F and $l_{max} = 100$. Then

$$\sum (2l + 1) = (l_{max} + 1)^2 \approx l_{max}^2 = R^2/\lambda^2.$$

As expected, the inelastic or reaction or absorption cross section is simply the geometrical area of the disc, πR^2. We note that $\sigma_{elastic} = \sigma_{inelastic}$ and represents the diffraction or shadow scattering, which is familiar in optics when a plane wave is interrupted by a completely absorbing obstacle. The angular distribution of the elastic scattering is the Fourier transform of the spatial distribution of the obstacle. For the lth partial wave, it is represented by the polynomial $P_l(\cos \theta)$. The sum over Legendre polynomials can be approximated, for small scattering angles, by a Bessel function of first order. Actually, it is more useful to discuss the differential elastic cross section in terms of momentum transfer q, rather than angular deflection θ. Referring to Fig. 7.25, we have

$$q = 2k \sin \frac{\theta}{2}, \tag{7.38}$$

where, with units $\hbar = c = 1$, $k = 1/\lambda = p$, the CMS momentum of each of the colliding particles. Then it may be shown that, in the small angle approximation,

$$\frac{d\sigma_{el.} \text{ (black disc)}}{dq^2} = \pi R^2 \left| \frac{J_1(Rq)}{Rq} \right|^2 \tag{7.39}$$

$$\approx \frac{\pi R^4}{4} \exp \left(-\frac{R^2 q^2}{4} \right), \tag{7.40}$$

where the second expression is accurate, with $R = 1$ F, for values of $q^2 < 0.2$ (GeV/c)^2. For larger values of Rq, the Bessel function $J_1(Rq)$ undergoes

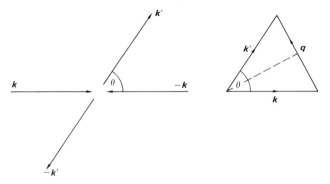

Fig. 7.25 Elastic collision as observed in the CMS frame. $|k| = |k'|$ is the momentum of each particle. The momentum transfer $q = 2k \sin \theta/2$.

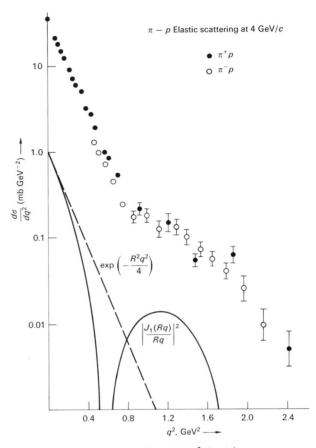

Fig. 7.26 Elastic scattering cross section $d\sigma/dq^2$ for $\pi^\pm$ on protons, at 4 GeV/c incident momentum. The curve shows the distribution (Bessel function) expected for the black disc model, [Eq. (7.39)] with $R = 1$ F. Data from Coffin *et al.* (1966).

maxima and minima characteristic of the diffraction phenomenon. For $R = 1$ F, the first zero is at $q^2 \sim 0.6\,(\text{GeV}/c)^2$. Some experimental data on $\pi^\pm p$ elastic scattering are shown in Fig. 7.26, and compared with the black disc model. The data show no secondary maxima and minima, and this seems to be true for most elastic processes in the high energy region (> 5 GeV/c). However, one can often see structure, in the sense that one observes breaks or changes in the slope of the curves. For $q^2 < 2\,(\text{GeV}/c)^2$, all elastic cross sections seem to be well fitted by the empirical formula

$$\frac{d\sigma}{dq^2}\bigg/\frac{d\sigma(0)}{dq^2} = \exp\left[-(Aq^2 - Bq^4)\right], \qquad (7.41)$$

A and B being both positive [i.e. when $\log(d\sigma/dq^2)$ is plotted against q^2, the curve is concave upward]. Equation (7.40) reproduces the first term of (7.41) which dominates at small q^2, with $R = \sqrt{4A}$. For $\pi^\pm p$ scattering, $A \approx 8$ $(\text{GeV}/c)^{-2}$—corresponding to $R = 1$ F—and is energy independent. However, in pp (and K^+p) scattering, A increases weakly with bombarding energy, corresponding to a shrinkage of the diffraction peak, or an increase in the size of the diffracting obstacle—see Fig. 7.27. For $\bar{p}p$ scattering, A decreases with increase in momentum. Even more significant is the discrepancy between the expected ratio $\sigma_{\text{el.}}/\sigma_{\text{total}} = 0.5$ and that observed. In all cases, $\sigma_{\text{el.}}/\sigma_{\text{total}} < 0.5$ and falls off with increasing energy.

To summarize, therefore, the classical optical model of absorption and scattering by an opaque disc of radius ~ 1 F describes some of the gross features of the observations, but fails to account in detail for the shape of the elastic scattering at large momentum transfer, the value of $\sigma_{\text{el.}}/\sigma_{\text{total}}$, and the dependence on bombarding energy. We may also remark that such a model does not incorporate spin, and thus is unable to account for transverse polarization in the scattering process (typically of order 10%). Somewhat better fits to the data can be obtained from Regge pole theory. Before we describe these results, we shall briefly discuss the notion of particle exchange in simple two-body reactions.

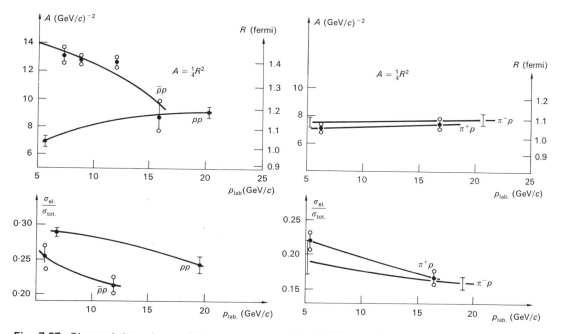

Fig. 7.27 Observed dependence of the parameter A of Eq. (7.41) and of the ratio of elastic to total cross section, for $pp, \bar{p}p, \pi^+p$, and π^-p scattering. (After Svensson, 1967.)

7.8 THE ONE-PARTICLE-EXCHANGE (OPE) MODEL

It will be recalled, from Section 5.3, that we found that the Coulomb (Rutherford) scattering of electrons by protons had the form, at low momentum transfer:

$$\frac{d\sigma}{dq^2} \propto \frac{e^4}{q^4}. \tag{7.42}$$

Here, q^2 is the four-momentum transfer squared: $q^2 = (\Delta p)^2 - (\Delta W)^2$, where Δp is the three-momentum transfer, ΔW the energy transfer. When Δp is small, or when it is measured in the CMS of an elastic collision, ΔW is negligible or zero, and q^2 is then equal to the three-momentum transfer, squared.

We note that in (7.42) the cross section "blows up," and the scattering amplitude $f(q^2) \propto 1/q^2$ is said to have a singularity or *pole* at $q^2 = 0$. We can interpret the electromagnetic interaction as due to exchange of a virtual photon between electron and proton [Fig. 7.28(a)]. $q^2 = 0$ would correspond to a real photon of zero rest mass. Thus, $-q^2$ is the (mass)^2 of the exchanged particle; for a scattering process $q^2 > 0$. Next consider the process of nucleon-nucleon scattering via single-pion exchange, as indicated in Fig. 7.28(b). If we could treat this interaction by nonrelativistic perturbation theory, we would represent the interaction by the Yukawa potential (Section 1.2)

$$V = \frac{ge^{-\mu r}}{r}, \tag{7.43}$$

where μ represents the mass of the Yukawa particle, here identified with the pion. It will be recalled that in the evaluation of the Rutherford scattering (Section 5.3), we took for the Coulomb potential $V = e[\exp(-r/a)]/r$ and eventually set $a \to \infty$. All we have to do is then to replace the coupling e^2 for the diagram of Fig. 7.28(a) by g^2, for the pion-nucleon case, and set $a^{-1} = \mu$, when (5.19) gives us for the matrix element or scattering amplitude:

$$f(q^2) = \text{const.} \frac{g^2}{(\mu^2 + q^2)}. \tag{7.44}$$

Since $q^2 > 0$ for a real scattering process, the denominator is always finite.

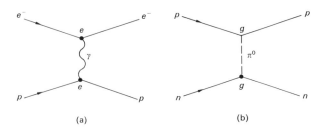

(a)   (b)

Figure 7.28

However, $f(q^2)$ has a singularity or *pole* in the *unphysical* region at $q^2 = -\mu^2$. We note that, were we to substitute a different value for μ—corresponding say to the exchange of a ρ-meson (mass 765 MeV, compared with 140 MeV for the pion)—the physically accessible region for nucleon-nucleon scattering would be still further from the pole. For small q^2, therefore, it is assumed that, if pion exchange is an allowed process, it should dominate the scattering amplitude, so that another way to phrase this result is that the most distant collisions—corresponding to the higher angular momentum states—will be dominated by single pion exchange, because pions are the lightest mesons. These considerations are the basis of the one-particle-exchange model. Although our arguments stem from nonrelativistic potential theory, a relativistic calculation (by perturbation methods) also includes the factor $(\mu^2 + q^2)^{-1}$ from (7.44), and this so-called "pion-propagator" term dominates the cross section, at least at small q^2.

We can illustrate what comes out with two examples. First, we see that, from the measured cross sections, we ought to be able to determine the pion-nucleon coupling constant g^2 in (7.44). The data on $p - p$ elastic scattering at 100 MeV has been analyzed by fitting the angular distributions to a set of arbitrary and smoothly varying phase-shifts for the lower partial waves only, and then relying on the OPE model to supply the scattering amplitude for the remaining terms of high l. The analysis gave a value for $g^2/\hbar c = 15$. This number is also defined alternatively in the form $f^2/\hbar c = (g^2/\hbar c)(m_\pi/2m_p)^2 = 0.08$. The quantity g^2 can also be calculated from low energy pion-nucleon scattering, with the same result.

As a second example, consider the process

$$\pi^- + p \to \rho^- + p,$$

which we discussed in Section 7.5. There we saw that the decay angular distribution of the ρ-meson and decay branching ratios were consistent with single pion exchange. Another consequence follows since the exchanged pion has zero spin. It is that there should be no correlation between the plane defined by the momentum vectors of the ρ-meson and the incident pion on the one hand, and that formed by the initial and final proton momenta on the other. A spinless particle cannot carry information about vectors normal to its line of flight. Experimentally it is found that the angle (called the Treiman-Yang angle) between these planes is isotropic for small momentum transfers, $q^2 < 0.2(\text{GeV}/c)^2$, but not for large momentum transfers. In order to conserve parity at the lower vertex (Fig. 7.18), the pion must be emitted in a p-wave. This circumstance gives additional terms in q^2 in the numerator of (7.44) so that the predicted q^2-dependence is weaker. The actual distribution of momentum transfer for the reaction $\pi^+ + p \to \rho^+ + p$ at 2.75 GeV/c incident momentum is shown in Fig. 7.29, together with the predictions of the OPE

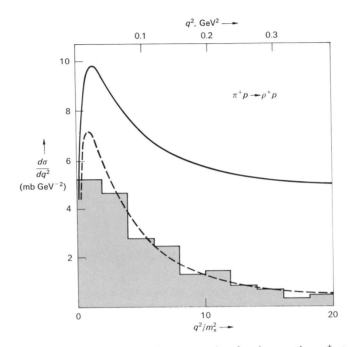

Fig. 7.29 The observed differential cross section for the reaction $\pi^+ + p \rightarrow \rho^+ + p$ at 2.75 GeV/c, compared with the predictions of the OPE model (full curve), and the OPE model with absorption corrections (dotted curve). (After Jackson, 1965.)

model. Although the observed forward peaking ($q^2/m_\pi^2 < 2$) is roughly reproduced, the predicted variation at large q^2 is quite wrong. It seems likely therefore that the OPE model gives a reasonable description of the process if one only considers the longest-range part of the interaction, i.e. the high partial waves of the incident beam, giving the forward peak, and involving small q^2. On the other hand, for the partial waves of small l, corresponding to more "central" collisions, the one-pion exchange picture is incorrect. These waves can react in many other ways and thus do not generally appear in the $\rho^+ p$ final state. As indicated in Fig. 7.29, allowance for such absorption effects improves agreement with experiment, although the calculation of such corrections is difficult to make in an unambiguous manner.

To summarize, in processes where the quantum numbers allow single pion exchange, a first-order perturbation calculation works reasonably well for the high partial waves (small momentum transfers). Often, π-exchange is disallowed; for example, in the charge-exchange scattering $\pi^- p \rightarrow \pi^0 n$, the exchanged particle must have even G-parity and the simplest possibility is a ρ-meson. In such reactions, the single-particle-exchange model gives predictions in serious disagreement with the data.

7.9 THE REGGE POLE MODEL

A basic problem in the OPE model is that the poles in the scattering amplitude, which are assumed to dominate the scene, correspond to exchange of particles carrying fixed angular momentum, J. Thus, in the pion exchange discussed above, $J = 1$ (p-wave). As a consequence, it is found that at high incident energies the amplitude contains a factor E^J so that it always "blows up" if J is large enough. A possible way out of this difficulty was suggested by Regge in 1959, who proposed to treat the angular momentum as a continuous, complex variable—although, clearly, physically observable states must have integral, or half-integral, angular momentum. Although Regge worked in nonrelativistic potential theory, his ideas were quickly applied to high energy particle physics by Chew et al. (1962). These objects with complex angular momentum are referred to as *Regge poles*.

First let us consider the general collision process $a + b \rightarrow c + d$, where a, b, c, and d are hadronic states. These particles have four-momenta denoted by $p_a, p_b, p_c,$ and p_d respectively [Fig. 7.30(a)]. The four-momentum $p_a = (\mathbf{p}_a, iE_a)$, where $\mathbf{p}$ is the three-momentum and E the total energy of the particle.

There are two independent Lorentz invariant quantities one can form from these four-vectors (apart from the particle rest mass $p^2 = -m^2$). These are denoted by the so-called Mandelstam variables

$$s = -(p_a + p_b)^2 = -(p_c + p_d)^2 = E^2, \tag{7.45}$$

$$t = -(p_a - p_c)^2 = -(p_b - p_d)^2 = -q^2. \tag{7.46}$$

In the CMS of the collision, a and b have equal and opposite three-momentum, so that $s = E^2$, the square of the total CMS energy [Fig. 7.30(b)]. t is the square of the four-momentum transfer between a and c, or b and d. For an *elastic collision*, we note that in the CMS no energy is transferred and $|\mathbf{k}| = |\mathbf{k}'|$, so that

$$t = -(\mathbf{k} - \mathbf{k}')^2 = -2k^2(1 - \cos\theta), \tag{7.47}$$

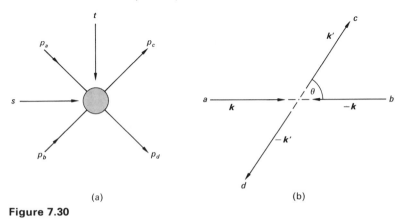

(a)   (b)

Figure 7.30

where θ is the scattering angle in the CMS [see (5.21)]. Note that $t = -q^2$ is negative for a scattering process. We emphasize that s and t are invariants and therefore have the same values in both laboratory and center-of-mass systems. It is also possible to consider the crossed momentum transfer, defined by $u = -(p_a - p_d)^2$. It is left as an exercise to show that

$$s + t + u = m_a^2 + m_b^2 + m_c^2 + m_d^2,$$

so that u is not an independent quantity.

When we are interested in the properties of the intermediate state formed from a and b (for example, a resonance of a and b), it is natural to describe the partial wave scattering amplitude in terms of s-channel quantities, i.e. as $f(l, E)$ where l and E are the angular momentum and energy of the combination a and b. At low energies (< 1 GeV) we know that pion-nucleon elastic scattering is dominated by s-channel resonances. However, at very high energies (above 10 GeV) the cross section varies very smoothly with energy and such a description is not appropriate. Rather, we think of the scattering as dominated by exchange of poles in the momentum transfer or t-channel (just as for electron-proton scattering due to photon exchange). Since t is negative, these exchanged particles are outside the physical region for the reaction $a + b \rightarrow c + d$. But now suppose that we replace b and c by their antiparticles $\bar{b}$ and $\bar{c}$ and reverse their momenta (Fig. 7.31).

Then in (7.45) and (7.46), s and t change sign. t is now the energy variable for the t-channel reaction $a + \bar{c} \rightarrow \bar{b} + d$, and we call it $\bar{s}$. s is negative, it is the momentum transfer variable for the t-channel reaction, and we call it $\bar{t}$. The reaction $a + \bar{c} \rightarrow \bar{b} + d$ is called the *crossed reaction* to $a + b \rightarrow c + d$. The point is that the t-channel exchange poles of the original reaction now become $\bar{s}$-channel resonances, in the physical region of the crossed reaction. The important *principle of crossing symmetry* states that both reactions are

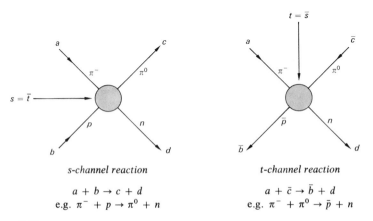

s-channel reaction

$$a + b \rightarrow c + d$$
$$\text{e.g. } \pi^- + p \rightarrow \pi^0 + n$$

t-channel reaction

$$a + \bar{c} \rightarrow \bar{b} + d$$
$$\text{e.g. } \pi^- + \pi^0 \rightarrow \bar{p} + n$$

Figure 7.31

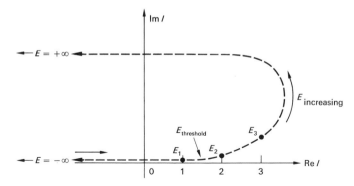

Fig. 7.32 Typical behavior of a Regge trajectory in the complex angular momentum plane.

described by one and the same amplitude; we can write this as

$$\bar{F}(\bar{s}, \bar{t}) = F(t = \bar{s}, s = \bar{t}). \tag{7.48}$$

Obviously, this principle requires that the function F can be continued into the unphysical regions in the s, t plane. For example, $\bar{s} > 0$ for the t-channel corresponds to unphysical values of $t > 0$ or, from (7.47), $\cos \theta > 1$ in the s-channel. Crossing symmetry is of vital importance in seeking to describe high energy processes in the Regge picture.

Let us now turn to the s-channel, and consider the behavior of the amplitude $f(l, E)$ according to the Regge hypothesis. [The index l, for the lth partial wave contributing to the scattering, implies the usual angular dependence $P_l(\cos \theta)$.] The angular momentum is written as a function of energy, $\alpha(E)$, with real and imaginary parts: $l = \text{Re } \alpha(E)$. The trajectory described by α as E varies is called a Regge trajectory. Figure 7.32 shows an example of the form such a trajectory might have in the complex angular momentum plane.

The trajectory starts off along the negative real axis, and, depending on the strength of the scattering potential, may cross over the origin along the positive real axis as E increases. The particles a and b may possess a bound state of total energy E_1 and $l = 1$, say. This corresponds to $\text{Im } \alpha(E) = 0$ and $\text{Re } \alpha(E) = n$, an integer. Such a bound state or pole in $\alpha(E)$ will necessarily occur at $E < E_{\text{threshold}}$, where $E_{\text{threshold}} = m_a + m_b$. Increasing E still further through $E_{\text{threshold}}$, the trajectory now leaves the real axis, $\alpha(E)$ acquiring a positive imaginary part. Whenever the trajectory passes $\text{Re } \alpha(E) = l = n$, an integer, one can have an unbound state of energy E_n and angular momentum n, which is identified as an s-channel resonance. This can be seen as follows. If $\text{Im } \alpha \ll \text{Re } \alpha$, we may write a Taylor expansion for $E \approx E_n$ of the form

$$\alpha(E) = \text{Re } \alpha(E) + i \text{ Im } \alpha(E)$$

$$\approx n + (E - E_n)\left[\frac{d}{dE} \text{Re } \alpha(E)\right]_{E = E_n} + i \text{ Im } \alpha(E).$$

Setting

$$\varepsilon = \frac{d}{dE} \, (\mathrm{Re}\,\alpha(E))_{E\,=\,E_n} \quad \text{and} \quad \Gamma = \frac{2}{\varepsilon} \, \mathrm{Im}\,\alpha(E),$$

gives

$$\alpha(E) = n + \varepsilon[(E - E_n) + i\Gamma/2].$$

Whenever l and E have values corresponding to a resonance, the amplitude $f(l, E)$ must have a singularity or *pole*. We express this by writing

$$f(l, E) = \frac{R(E)}{l - \alpha(E)} \tag{7.49}$$

where $R(E)$ is some function (the residue function). Thus

$$f(l = n, E) = -\frac{R}{\varepsilon} \, \frac{1}{[(E - E_n) + i\Gamma/2]}, \tag{7.50}$$

which is the usual Breit-Wigner formula for the amplitude near a resonance of angular momentum n, and energy E_n. Since Im $\alpha(E)$ and Γ are both positive, we require the slope ε of Re $\alpha(E)$ versus E to be positive. A corollary is therefore that when the trajectory turns over and heads to $E \rightarrow +\infty$, as it eventually must, there will be no further resonant states. Thus, the same Regge trajectory connects together bound states and resonances, and so to speak interpolates between the integral (or half integral) values of Re $\alpha(E)$ which these states possess. The number of bound states or resonances on a particular trajectory will depend on the "strength" of the interaction potential.

What other properties do Regge trajectories have, and what evidence is there for them? If we seek to describe a process in these terms, it is clear that all the quantum numbers except angular momentum ought to be the same for all poles on a given trajectory. Thus, the parity, baryon number, isospin, strangeness, and G-parity are defined by conservation laws applying to the particular reaction, independent of how we choose to describe it. Thus $\pi^+ + p \rightarrow$ anything can only proceed via the $I = \frac{3}{2}$, $B = 1$, $S = 0$ channel. The requirement that Regge poles should also describe exchange forces in the t-channel has the effect that the consecutive poles must be separated by two units of angular momentum ($\Delta l = 2$) and this fixes the *parity* of the trajectory. Consider, for example, the scattering of two pions. By Bose symmetry, states of even (odd) isospin must be associated with even (odd) l, so that, given the isospin channel, all two-pion poles must be separated along the trajectory by $\Delta l = 2$. Physically, we can observe the l- or J-values of resonances as a function of the mass $E = M$. This corresponds to a trajectory in the l, E plane, i.e. the variation of Re $\alpha(E)$ with E as we move along the curve of Fig. 7.32. Such a

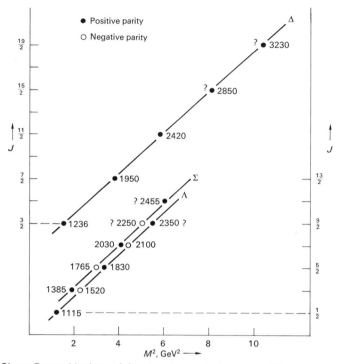

Fig. 7.33 Chew-Frautschi plots of fermion Regge trajectories. The trajectory marked Λ consists of the sequence $I = \frac{3}{2}$, $S = 0$, and $J^P = \frac{3}{2}^+$, $\frac{7}{2}^+$, $\frac{11}{2}^+ \cdots$; that marked Λ of the sequence $I = 0$, $S = -1$, $J^P = \frac{1}{2}^+$, $\frac{3}{2}^-$, $\frac{5}{2}^+ \cdots$; and that marked Σ of the sequence $I = 1$, $S = -1$, $J^P = \frac{3}{2}^+$, $\frac{5}{2}^-$, $\frac{7}{2}^+ \cdots$; resonances for which the spin-parity is not firmly established are indicated by a question mark.

plot is called a Chew-Frautschi diagram. Figure 7.33 gives some examples of fermion trajectories. For example, the so-called Δ-trajectory links together baryon states of $I = \frac{3}{2}$, $B = 1$, $S = 0$, and positive parity. The trajectory in the l, E plane is found to be a straight line if J is plotted against M^2, with slope $dJ/dM^2 = 0.9$ GeV^{-2}. Note the $\Delta J = 2$ separations. The Σ- and Λ-trajectories have a similar slope. Here, even and odd parity states appear to fit on the same trajectory, which is thereby populated by states with $\Delta J = 1$ instead of 2—a phenomenon called exchange degeneracy. The situation for boson states is depicted in Fig. 7.34. Again, observed resonances appear to fit a line of slope 0.9 GeV^{-2}. It should be remarked that the spin-parity of many of the boson states of high mass have not yet been definitely established. It is not clear why the dependence of J on M^2 should be so remarkably linear, since Regge theory makes no prediction on this point. These plots do not, of course, compel the acceptance of the Regge pole model, but they are consistent with it, and in any case quite remarkable in their own right. It may be mentioned that the quadratic dependence of angular momentum on the resonance mass is expected

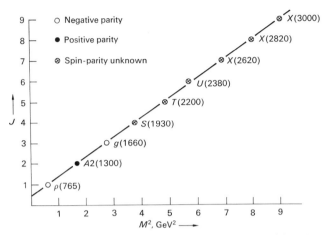

Fig. 7.34 Chew-Frautschi plot of nonstrange meson resonances, of $I = 1$ and spin, parity and G-parity, $J^{PG} = 1^{-+}, 2^{+-}, 3^{-+} \cdots$. The quantum numbers of only the first three states are known at present, the remainder having been plotted at the nearest integer spin value. The masses of the S, T, U, and X bosons are taken from Fig. 7.22.

under certain assumptions in the framework of the quark model of hadrons. High angular momentum states are then interpreted in terms of relative orbital motion of the quark constituents, in a harmonic oscillator binding potential.

7.10 APPLICATION OF REGGE POLES TO HIGH ENERGY REACTIONS

Perhaps the most important application of the Regge pole model is to the study of t-channel exchange processes in high energy reactions, where s is large and t is small. Let us consider the elastic scattering process $\pi^+ + p \to \pi^+ + p$. From (7.20) and (7.28) we may write for the scattering amplitude summed over all contributing partial waves

$$F(\theta) = F(E, \cos \theta) = \frac{1}{k} \sum_l (2l + 1)f(l, E)P_l(\cos \theta), \qquad (7.51)$$

where $E = \sqrt{s}$ is the center-of-mass total energy.

In the Regge pole model, this partial wave expansion may be transformed, by a method due to Watson and Sommerfeld, to a sum over Regge poles, plus a background integral, which we shall ignore hereafter. The new form is

$$F(E, \cos \theta) = F(s, t) = \frac{1}{k} \sum_i \frac{\beta_i(s)}{\sin \pi \alpha_i(s)} [P_{\alpha_i(s)}(-\cos \theta)] + \text{B.I.} \qquad (7.52)$$

This result follows from the properties of certain contour integrals in the complex angular momentum plane, which we shall not discuss here. The quantity

$$\beta_i(s) = -\pi[2\alpha_i(s) + 1]R_i(s), \qquad (7.53)$$

where $R_i(s)$ is defined in (7.49) as the residue at the pole. The denominator term $1/(\sin \pi l)$ has poles at $l = 0, \pm 1, \pm 2 \cdots$, with residue $(-1)^l/\pi$. Furthermore, $P_l(-\cos \theta) = (-1)^l P_l(\cos \theta)$. Using these facts, (7.51) and (7.52) are seen to be entirely equivalent. The angle θ, of course, refers to the scattering angle in the s-channel. For the elastic scattering of particles of mass m and M, (7.47) gives after a little algebra

$$\cos \theta = 1 + \frac{t}{2k^2} = 1 + \frac{2ts}{[-4m^2s + (s - M^2 + m^2)^2]}. \qquad (7.54)$$

As mentioned previously, the description (7.52) in terms of s-channel poles is not particularly useful for studying the processes at high s and small t. We are interested in *exchange* (t-channel) poles, as in the OPE model. Let us instead go to the crossed (t-channel) reaction, i.e. $\pi^+ \pi^- \to p\bar{p}$. Then

$$\bar{F}(\bar{s}, \bar{t}) = F(t, s) = \frac{1}{k} \sum \frac{\beta_i(t)}{\sin \pi \alpha_i(t)} [P_{\alpha_i(t)}(-\cos \theta_t)], \qquad (7.55)$$

where

$$\cos \theta_t = 1 + \frac{2st}{[-4m^2t + (t - M^2 + m^2)^2]} \qquad (7.56)$$

is now in the unphysical region. Thus, when t is small and $s \to \infty$, $\cos \theta_t \propto (-s)$, and it may be shown that $P_{\alpha(t)}(-\cos \theta_t) \propto (s/s_0)^{\alpha(t)}$. If we assume that one Regge pole dominates everything, we will have

$$F(s, t) \propto \frac{1}{k} \beta_1(t) \left(\frac{s}{s_0}\right)^{\alpha_1(t)}, \qquad (7.57)$$

where s_0 is a parameter with dimensions (energy)^2. The cross section then has the form, from (7.22) and (7.47),

$$\frac{d\sigma}{dt} = \frac{d\sigma}{d\Omega} \frac{d\Omega}{dt} = [F(s, t)]^2 \frac{\pi}{k^2}.$$

In the limit $s \to \infty$, $s \approx 4k^2$, since the masses of the particles involved may be neglected [refer to Fig. 7.30(b)]. Then

$$\frac{d\sigma \text{ (elastic)}}{dt} = D_1(t) \left(\frac{s}{s_0}\right)^{[2\alpha_1(t) - 2]}, \qquad (7.58)$$

where D is some function of t only. From the optical theorem (7.26) we know that the total cross section

$$\sigma_T = \frac{4\pi}{k} \text{Im } F(t = 0).$$

Thus, if σ_T is to be nearly constant at high s, one needs $\alpha_1(0) = 1$ in (7.57). This then is one property required of the trajectory of a single exchanged (t-channel) Regge pole if it is to account for elastic scattering at high energy. Of

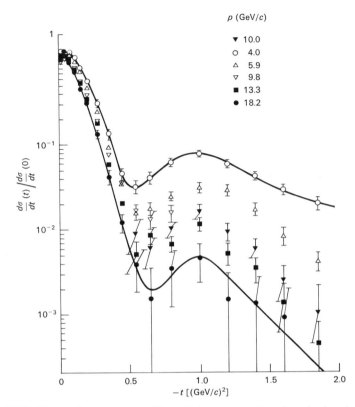

Fig. 7.35 Observations on the differential cross section for elastic charge-exchange, $\pi^- p \to \pi^0 n$, compiled by Sonderegger *et al.* (1966). The shrinkage of the *t*-distribution with increasing incident momentum, *p*, yields $\alpha_\rho(t) = 0.6 + 0.9t$ for the ρ-exchange trajectory.

course, this one trajectory ought to account for all elastic scattering phenomena, and it is not difficult to see that, since no quantum numbers (strangeness, isospin, baryon number, etc.) apart from angular momentum may be exchanged in such processes, this trajectory must have the quantum numbers of the vacuum. It is therefore called the vacuum trajectory. The implication is that all other Regge trajectories must have $\alpha(t = 0) < 1$, so that the vacuum trajectory can dominate elastic scattering in the high energy limit, as in (7.58). Since exchange of a vacuum pole implies, for example, that $\sigma_{\pi^+ p}$ (total) $= \sigma_{\pi^- p}$ (total) as predicted by the Pomerancuk theorem, this trajectory is also called the Pomerancuk trajectory. It is often denoted α_P.

In (7.58) the term containing *s* must dominate both the *s*- and *t*-dependence at large values of *s*. If *b* is the (positive) slope at small *t*, and the trajectory is linear in this small region, then

$$\alpha_P(t) = 1 + bt,$$

so that

$$\frac{d\sigma \text{ (elastic)}}{dt} = \left(\frac{d\sigma}{dt}\right)_{t=0} \exp\left[2b \log (s/s_0)\right]t. \tag{7.59}$$

Comparing with (7.41), and with $q^2 = -t$, we see that this correctly predicts the exponential fall-off of the elastic cross section with momentum transfer, for small negative t. It also predicts that the width of the forward diffraction peak should increase logarithmically with energy (unfortunately, the scale factor s_0 is not given by the Regge theory). The observation of an increase of A with energy in (7.41) for elastic p-p scattering was therefore an initial, albeit momentary, success for the Regge model. Unfortunately pion-proton scattering shows no change in A with s, and p-$\bar{p}$ scattering shows a decrease. Even for a given elastic scattering process, the inadequacy of (7.59) has been emphasized by recent experiments on p-p scattering at the CERN ISR (Table 2.2). The older data extended to $s = 150$ GeV^2, and were consistent with a constant value of $b = 0.6$. The new ISR data extend the range of s by more than one order of magnitude, and yield values of the coefficient of the t-distribution which are significantly less than expected from the extrapolation of (7.59) using the low energy data. Possibly the scale factor s_0 is so large that presently available energies are not high enough to reveal the simple structure. This point is also borne out by Fig. 7.24 which certainly indicates that the asymptotic region of the Pomerancuk theorem is at very high energy. If this is so, one should include other exchange trajectories for particular reactions at present

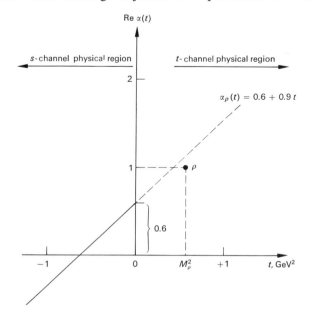

Fig. 7.36 The ρ Regge trajectory.

energies, and the fact that one gets a different s-dependence of the t-distribution is not too surprising.

Let us now look at pion-nucleon charge-exchange scattering, $\pi^- p \rightarrow \pi^0 n$. The exchanged particle must have G-parity $+1$, $I = 1$, spin-parity $1^-, 3^-$, $5^- \cdots$, and it is therefore identified with the ρ-trajectory; this is the only one which can obviously contribute. Formula (7.57) applies, with $\alpha_1 = \alpha_\rho$. The data (Fig. 7.35) appear to show a shrinkage, and from its magnitude one can compute $\alpha_\rho(t)$; empirically

$$\alpha_\rho(t) = 0.6 + 0.9t,$$

for $-t < 1$ $(\text{GeV}/c)^2$. This is shown by the solid line in Fig. 7.36, for $t < 0$ (the s-channel physical region). If we extrapolate linearly into the physical region of the crossed channel $\pi^- \pi^0 \rightarrow \bar{p}n$, we expect to get the physical resonances lying on the ρ-trajectory. The above formula gives

$$\alpha_\rho(t = m_\rho^2 = (0.76 \text{ GeV})^2) \simeq 1.1,$$

which is, within the errors, equal to the spin of the ρ-meson. We also note from Fig. 7.35 that $d\sigma/dt$ has a minimum at $-t = 0.6 \text{ GeV}^2$. This is interpreted as the effect of the spin-flip term in the amplitude. One can show that this is proportional to $\alpha(t)$, and should be zero when $\alpha(t) = 0$. From the above formula, this indeed occurs at $-t = 0.7 \text{ GeV}^2$.

We have only touched here on some of the features of the Regge pole model. Its most obvious and unambiguous predictions are in the ultra high energy region, but even here, there are some discrepancies. The unpleasant features of the OPE model are essentially avoided in the Regge picture by postulating that the exchanged "particle" does not have to carry integral angular momentum.

7.11   OTHER DEVELOPMENTS

We may mention in conclusion two recent developments. One is the concept of *duality*. The foregoing discussion of the Regge model has indicated that the t-dependence of the amplitude at high energy can be described in terms of resonances in the crossed channel. The contributing resonant states in $\bar{s}$ are of predominantly low energy. Thus, the resonant behavior at low energy in the t-channel is determining the high energy behavior in the s-channel. Can one go further than this, and, from the high energy behavior, actually predict the low energy resonant states? Very crudely, the idea of duality is that one can do this, and that, in some average sense, the s-channel resonances are indeed built up from t-channel Regge poles. In certain cases, e.g. pion-nucleon charge exchange, it has been possible, starting off from the properties of the appropriate t-channel (boson) resonances, to compute the s-channel amplitude and to reproduce the positions of the s-channel pion-nucleon resonances in remarkable detail (Dolen, Horn, and Schmid, 1968). This suggests that the duality hypothesis should be built into the amplitude *ab initio*, by making it symmetric in the Mandelstam variables s and t. Veneziano (1968) did just this, writing the amplitude $F(s, t)$ quite arbitrarily as the product of gamma functions. The remarkable properties possessed by the Veneziano amplitude include Regge behavior at high energies, resonances at low energy, duality and crossing symmetry. Furthermore, it is possible to extend the prescription from two-body to many-body amplitudes. On the debit side, in common with the usual Regge model, it predicts states on "daughter" trajectories which are not observed. The qualified success of this empirical approach suggests that it is an important step towards a theory of strong interactions. Equally it is clear that very many such steps will be required before the goal is achieved.

BIBLIOGRAPHY

Blatt, J., and V. Weisskopf, *Theoretical Nuclear Physics*, John Wiley, New York, 1952.

Dalitz, R. H., "Strange-particle resonant states," *Ann. Rev. Nucl. Science* **13,** 339 (1963).

Fraser, W. R., *Elementary Particles*, Prentice-Hall, Englewood Cliffs, New Jersey, 1966.

Galbraith, W., "Hadron-nucleon total cross sections at high energies," *Rep. Prog. Phys. (Lond.)* **69,** 547 (1969).

Källén, G., *Elementary Particle Physics*, Addison-Wesley, Reading, Mass., 1964.

Svensson, B. E. Y., "High energy phenomenology and Regge poles," 1967 CERN School of Physics, Vol. II (CERN 67-24).

Tripp, R. D., "Spin and parity of elementary particles," *Ann. Rev. Nucl. Science* **15,** 325 (1965).

Wetherell, A. M., "High energy elastic scattering," Proc. 1965 CERN School of Physics, Vol. 1 (CERN 65-24).

PROBLEMS

7.1 Using the result of Problem 4.12, include the correct (nonrelativistic) phase-space factors to compute the ratio of decay rates

$$\frac{(K^+ \to \pi^+\pi^0\pi^0)}{(K^+ \to \pi^+\pi^+\pi^-)}.$$

Compare your results with the data of the Table in the Preface.

7.2 Show that, in $K \to 3\pi$ decay, the relativistic factor $E_1 E_2 E_3$ is constant within a range of $\pm 1\%$, over different regions of the Dalitz plot.

7.3 In what process could you observe the creation of the $Y_0^*(1405)$ (or $\Lambda(1405)$) resonance? Discuss any limitations on the energies of the colliding particles. List the allowed decay modes of this resonance and state which are the most probable. List also those decay modes which conserve strangeness but are forbidden by other conservation laws.

7.4 In the decay of a resonance of mass M into three particles of momenta p_1, p_2, and p_3, show that the boundary of the Dalitz plot is given by the condition $|p_1| + |p_2| - |p_3| = 0$. Deduce the equation of the boundary in terms of E_1 and E_2 in the case where the three decay products have the same rest mass m. Show that, if $m \ll M$, the boundary of the Dalitz plot becomes the inscribed triangle of Fig. 7.3.

7.5 Using the condition $p_1 + p_2 - p_3 = 0$ of the previous question, show that the matrix element (7.10) for three-body decay of a particle of $J^P = 1^-$ leads directly to the vanishing of the Dalitz plot density at the boundary.

7.6 In a reaction $A + B \rightarrow C + D + E$, at a fixed bombarding energy, the quantities m_{CD}^2 and m_{DE}^2 are displayed along the x- and y-axes in a Dalitz plot (m_{CD} is the mass of particles C and D, etc.). If θ is the direction of particle E with respect to either C or D in the CD center-of-mass frame, show that, when C and D resonate at a fixed mass, the quantity $m_{DE}^2 = \alpha - \beta \cos \theta$, where α and β are constants.

7.7 In the following reaction in hydrogen,

$$\pi^- + p \rightarrow X^- + p,$$

a boson resonance X is observed with mass 2.4 GeV. The incident pion momentum is 12 GeV/c. Calculate the maximum angle of emission of the recoil proton with respect to the beam direction, and its momentum. Calculate also the angle and momentum of the proton when the four-momentum transfer is a maximum, and compute q_{max}^2.

7.8 In interactions of high energy (>5 GeV) pions with protons, it is observed that a certain boson resonance X of unit charge is produced with a cross section which varies very weakly with pion energy. By comparison with the behavior of p-p elastic scattering, what statement could be made about the nature of the particle (or particles) exchanged in the t-channel? What do you expect for (i) the G-parity of X, assuming it to decay into pions, (ii) the relation between the spin and parity of X, (iii) the angular distribution of X with respect to the beam direction? Give an example of such a state, from the Table in the Preface.

7.9 (a) From Fig. 3.10, p. 103, try to estimate the $\pi^- p$ total cross section due to the $N^*(1688)$ resonance, which is known to have $I = \frac{1}{2}$, $J^P = \frac{5}{2}^+$ (make an eyeball subtraction of background under the peak). Deduce an approximate value for the elasticity x of this resonance, where $x = \Gamma_e/\Gamma$, Γ being the total width and Γ_e the elastic partial width. Compare your result with the decay branching ratios in the Table in the Preface. (b) Is there a more reliable way to determine x?

7.10 The Breit-Wigner formula (7.31) describes a resonance of width $\Gamma \ll E_0$, the peak energy. For a broad resonance, this formula is not exact, since the phase-space available for two-body decay changes appreciably as one goes through the resonance. Show that for an S-wave resonance this effect may be accounted for by replacing Γ by $\Gamma_0 \cdot (pE_0/Ep_0)$, where p is the CMS momentum of either particle, and Γ_0, p_0, and E_0 refer to the resonance peak. What additional factors would you expect for a resonance decaying to two particles with orbital angular momentum l?

Appendixes

Relativistic Kinematics

The relativistic relation between total energy E, momentum p, and rest mass m of a particle is

$$E^2 = m^2 c^4 + p^2 c^2, \tag{A.1}$$

or, taking the velocity of light $c = 1$,

$$E^2 = p^2 + m^2.$$

The particle velocity in these units is

$$v = \beta c = \beta,$$

and the Lorentz factor

$$\gamma = (1 - \beta^2)^{-1/2},$$

so that

$$E = \gamma m,$$
$$p = \gamma \beta m = \sqrt{(\gamma^2 - 1)}m, \tag{A.2}$$
$$\beta = \sqrt{\gamma^2 - 1}/\gamma.$$

E, p can be written as components of a four-vector p_μ, where $\mu = 1 \cdots 4$ and $p_1 = p_x$, $p_2 = p_y$, $p_3 = p_z$, the Cartesian space components. Then $p_4 = iE$, the time component. Thus the square of the "length" of this four-vector is

$$p^2 = \sum_\mu p_\mu^2 = p_1^2 + p_2^2 + p_3^2 + p_4^2 = -m^2 = \text{invariant.} \tag{A.3}$$

Suppose the values E, p refer to properties of a particle measured in the laboratory frame Σ. In a different reference frame Σ', moving with constant velocity β^* along the x-axis, the new values E', p' can be found from the rules for transforming components of four-vectors. In matrix form the result is:

$$
\begin{vmatrix} p'_1 \\ p'_2 \\ p'_3 \\ p'_4 \end{vmatrix} = \begin{vmatrix} \gamma^* & 0 & 0 & i\beta^*\gamma^* \\ 0 & 1 & 0 & 0 \\ 0 & 0 & 1 & 0 \\ -i\beta^*\gamma^* & 0 & 0 & \gamma^* \end{vmatrix} \begin{vmatrix} p_1 \\ p_2 \\ p_3 \\ p_4 \end{vmatrix} \tag{A.4}
$$

or in space/time coordinates:

$$
p'_x = \gamma^*(p_x - \beta^*E),
$$
$$
p'_y = p_y,
$$
$$
p'_z = p_z,
$$
$$
E' = \gamma^*(E - \beta^*p_x), \qquad \text{where} \quad \gamma^* = (1 - \beta^{*2})^{-1/2}.
$$

Transformations of angle are found quite simply from these relations. Suppose a particle travels at angle θ relative to the x-axis in frame Σ, and at θ' in Σ'. Writing the transverse momentum component

$$
p_T = \sqrt{p_y^2 + p_z^2} = p'_T,
$$

then we find

$$
\tan \theta' = \frac{p_T}{p'_x} = \frac{1}{\gamma^*} \frac{p_T/p}{[(p_x/p) - \beta^*E/p]} = \frac{1}{\gamma^*} \frac{\sin \theta}{(\cos \theta - \beta^*/\beta)}, \tag{A.5}
$$

and

$$
\tan \theta = \frac{1}{\gamma^*} \frac{\sin \theta'}{(\cos \theta + \beta^*/\beta')}.
$$

When $\beta' < \beta^*$, i.e. the particle velocity in Σ' is less than the velocity of Σ' in Σ, there is a *maximum angle* θ_{max} at which the particle can be emitted in Σ. Differentiating (A.5), we have

$$
\frac{\partial}{\partial\theta'} (\tan \theta) = 0 \qquad \text{when} \quad \theta = \theta_{max},
$$

yielding

$$
(\cos \theta')_{\theta_{max}} = -\beta'/\beta^*,
$$

and

$$
\gamma^* \tan \theta_{max} = \beta'/(\beta^{*2} - \beta'^2)^{1/2}. \tag{A.6}
$$

This existence of the so-called "rainbow angle" has been employed in the missing-mass spectrometer (Section 7.6). When $\theta \approx \theta_{max}$, the angular dependence of the scattered intensity tends to zero, while the momentum dependence becomes extremely large.

Equations (A.5), with $\beta' = \beta = 1$, give the familiar relativistic aberration formulae in optics, while the Doppler shift is given by the last of equations (A.4).

The four-vector notation is extremely useful in threshold calculations in relativistic kinematics. Suppose we wish to compute the threshold energy for the reaction produced when an incident particle of mass m, total energy E, momentum p, hits a stationary target of mass M, resulting in a final state M^*:

$$m + M \rightarrow M^* \rightarrow m_1 + m_2 + \cdots,$$

where $M^* > (m + M)$. The threshold energy E will correspond to the total energy of m and M in the center of momentum system just being equal to M^*.

The total four-momentum, squared, of m and M, calculated in the laboratory frame, will be

$$P^2 = (p + 0)^2 - (E + M)^2$$
$$= p^2 - E^2 - M^2 - 2ME$$
$$= -(M^2 + m^2 + 2ME).$$

In the CMS, the total three-momentum is $p^* - p^* = 0$, and the total energy is M^*. Thus, we also have that

$$(P^2)_{\text{threshold}} = -(M^*)^2.$$

Hence

$$E_{\text{threshold}} = \frac{M^{*2} - M^2 - m^2}{2M} \tag{A.7}$$

Note that, when $E \gg M, m$, the total CMS energy is

$$E^* \approx \sqrt{2ME} \approx 2p^* \tag{A.8}$$

and thus rises as the square root of the product of the target mass and the energy of the incident particle.

The Dirac Equation

An account of the Dirac equation can be found in most standard texts on quantum mechanics. We append a brief account here for convenience only. We start off with the de Broglie prescription of a plane wave representing a free particle. In units $\hbar = c = 1$ this is

$$\psi = e^{i(\mathbf{k}\cdot\mathbf{r} - \omega t)}, \tag{B.1}$$

with $\mathbf{k}$ and ω the propagation vector and angular frequency respectively, which in our units are numerically equal to the momentum $\mathbf{p}$ and energy E. The wavefunction (B.1) is a solution of the *Klein-Gordon equation*:

$$\frac{\partial^2\psi}{\partial t^2} = \frac{\partial^2\psi}{\partial x^2} + \frac{\partial^2\psi}{\partial y^2} + \frac{\partial^2\psi}{\partial z^2} - m^2\psi$$

$$= (\nabla^2 - m^2)\psi, \tag{B.2}$$

suitable for describing free relativistic spinless particles. If we insert (B.1) in (B.2) we get the familiar relation

$$E^2 = p^2 + m^2. \tag{B.3}$$

In the nonrelativistic case, expanding (B.3) in powers of p/m gives to first order $E = m + p^2/2m$. In this approximation (B.1) becomes, on dividing by a factor e^{-imt},

$$\phi = e^{i(\mathbf{p}\cdot\mathbf{r} - (p^2/2m)t)}, $$

which satisfies the *Schrödinger equation* describing nonrelativistic particles:

$$\frac{\partial\phi}{\partial t} - \frac{i}{2m}\nabla^2\phi = 0. \tag{B.4}$$

Dirac formulated his relativistic wave equation under the assumption that derivatives of *both* time and space coordinates should occur to *first order*. It turns out that such an equation describes particles of spin $\frac{1}{2}$. The simplest form that we could write would be the two *Weyl equations*

$$\frac{\partial\psi}{\partial t} = \pm\left(\sigma_x\frac{\partial\psi}{\partial x} + \sigma_y\frac{\partial\psi}{\partial y} + \sigma_z\frac{\partial\psi}{\partial z}\right) = \pm\boldsymbol{\sigma}\cdot\frac{\partial}{\partial\mathbf{r}}\,\psi, \tag{B.5}$$

where the σ's are constants. The condition that this should satisfy (B.2), so that (B.3) may hold, is found by squaring (B.5) and equating coefficients, when we obtain

$$\sigma_x^2 = \sigma_y^2 = \sigma_z^2 = 1,$$
$$\sigma_x\sigma_y + \sigma_y\sigma_x = 0,\ \text{etc.}, \tag{B.6}$$
$$m = 0.$$

Thus the Weyl equations describe massless particles (in fact, neutrinos). The σ's cannot be numbers, since they do not commute. They can, however, be represented by *matrices*. The 2×2 Pauli spin matrices are a suitable choice (not the only one). They are

$$\sigma_x = \begin{vmatrix} 0 & 1 \\ 1 & 0 \end{vmatrix}, \quad \sigma_y = \begin{vmatrix} 0 & -i \\ i & 0 \end{vmatrix}, \quad \sigma_z = \begin{vmatrix} 1 & 0 \\ 0 & -1 \end{vmatrix}, \tag{B.7}$$

where

$$\sigma_x^2 = \sigma_y^2 = \sigma_z^2 = \begin{vmatrix} 1 & 0 \\ 0 & 1 \end{vmatrix},$$

as required. The wave function ψ must now have two components, since it is operated on by 2×2 matrices. Thus

$$\psi = \begin{vmatrix} \psi_1 \\ \psi_2 \end{vmatrix}.$$

Now we wish to include a mass term, and therefore a further matrix. The simplest set of four anticommuting matrices are 4×4, and ψ is then a four-component spinor. First we write down the space-time coordinates as components of a four-vector (i.e. in covariant notation),

$$x_1 = x, \quad x_2 = y, \quad x_3 = z, \quad x_4 = it.$$

Generally any component is denoted by x_μ, where $\mu = 1\cdots4$, and the relativistically invariant scalar product of two four-vectors is $x_\mu \cdot X_\mu$ where a repeated index implies a summation over μ. Now we write down the *Dirac equation* as

$$\left(\gamma_\mu\frac{\partial}{\partial x_\mu} + m\right)\psi = 0, \tag{B.8}$$

which is just generalizing (B.5) by inserting a mass term. The γ's are 4×4 matrices to be determined. We can do this by requiring (B.8) to satisfy the

Klein-Gordon equation (B.2), which we can rewrite in covariant form

$$\left(\frac{\partial^2}{\partial x_\mu \, \partial x_\mu} - m^2 \right) \psi = 0. \tag{B.9}$$

Multiplying (B.8) by $[\gamma_\nu(\partial/\partial x_\nu) - m]$, we get

$$\left(\gamma_\nu \frac{\partial}{\partial x_\nu} - m \right)\left(\gamma_\mu \frac{\partial}{\partial x_\mu} + m \right)\psi = \left(\gamma_\nu\gamma_\mu \frac{\partial^2}{\partial x_\nu \, \partial x_\mu} - m^2 \right)\psi = 0, \tag{B.10}$$

again a sum over μ, ν being understood. Equation (B.10) can be verified by writing out the individual components μ, $\nu = 1 \cdots 4$ and proving that the cross terms cancel. Comparing (B.10) with (B.9) we therefore require the conditions on the γ's:

$$(\gamma_\nu\gamma_\mu + \gamma_\mu\gamma_\nu) = 2\delta_{\mu\nu} \qquad \text{where} \quad \delta_{\nu\mu} = 1, \quad \nu = \mu$$
$$= 0, \quad \nu \neq \mu. \tag{B.11}$$

The usual representation of the γ-matrices obeying these commutation relations is

$$\gamma_k = \begin{vmatrix} \sigma & -i\sigma_k \\ i\sigma_k & 0 \end{vmatrix} \qquad k = 1, 2, 3, \tag{B.12}$$

$$\gamma_4 = \begin{vmatrix} 1 & 0 \\ 0 & -1 \end{vmatrix},$$

where each element stands for a 2×2 matrix, and the σ_k are defined in (B.7). For example,

$$\gamma_1 = \begin{vmatrix} 0 & 0 & 0 & -i \\ 0 & 0 & -i & 0 \\ 0 & i & 0 & 0 \\ i & 0 & 0 & 0 \end{vmatrix}, \qquad \gamma_4 = \begin{vmatrix} 1 & 0 & 0 & 0 \\ 0 & 1 & 0 & 0 \\ 0 & 0 & -1 & 0 \\ 0 & 0 & 0 & -1 \end{vmatrix},$$

and so on. It is also useful to define the product matrix

$$\gamma_5 = \gamma_1\gamma_2\gamma_3\gamma_4 = \begin{vmatrix} 0 & -1 \\ -1 & 0 \end{vmatrix}, \qquad \text{with} \quad \gamma_5^2 = 1, \quad \gamma_5\gamma_\mu + \gamma_\mu\gamma_5 = 0,$$
$$\mu = 1 \cdots 4. \tag{B.13}$$

In order to make quite clear what is happening, let us write out the Dirac equation in full, for the individual components of

$$\psi = \begin{vmatrix} \psi_1 \\ \psi_2 \\ \psi_3 \\ \psi_4 \end{vmatrix}.$$

We have the four simultaneous equations

$$-i\frac{\partial \psi_4}{\partial x_1} - \frac{\partial \psi_4}{\partial x_2} - i\frac{\partial \psi_3}{\partial x_3} + \frac{\partial \psi_1}{\partial x_4} + m\psi_1 = 0,$$

$$-i\frac{\partial \psi_3}{\partial x_1} + \frac{\partial \psi_3}{\partial x_2} + i\frac{\partial \psi_4}{\partial x_3} + \frac{\partial \psi_2}{\partial x_4} + m\psi_2 = 0,$$

$$+i\frac{\partial \psi_2}{\partial x_1} + \frac{\partial \psi_2}{\partial x_2} + i\frac{\partial \psi_1}{\partial x_3} - \frac{\partial \psi_3}{\partial x_4} + m\psi_3 = 0,$$

$$+i\frac{\partial \psi_1}{\partial x_1} - \frac{\partial \psi_1}{\partial x_2} - i\frac{\partial \psi_2}{\partial x_4} - \frac{\partial \psi_4}{\partial x_4} + m\psi_4 = 0.$$

(B.14)

Writing the plane wave solution as

$$\psi(r, t) = u_j e^{i p_\mu x_\mu} \qquad j = 1 \cdots 4,$$

(B.15)

where the u_j are now four-component spinors, and taking the simplest case $p = p_3$, we get from (B.14) four simultaneous equations:

$$(-E + m)u_1 + pu_3 = 0,$$ (B.16a)
$$(-E + m)u_2 - pu_4 = 0,$$ (B.16b)
$$(E + m)u_3 - pu_1 = 0,$$ (B.16c)
$$(E + m)u_4 + pu_2 = 0.$$ (B.16d)

Next use the fact that $p^2 = E^2 - m^2$, and that $\sum u_j^2 = 1$ for reasons of normalization. Then from (B.16a), $u_3/u_1 = p/(E + m)$, and setting $u_1 = 1$, we therefore obtain $u_3 = p/(E + m)$, $u_2 = u_4 = 0$. Equally, using (B.16b), we could have set $u_1 = u_3 = 0$, with $u_2 = 1$, $u_4 = -p/(E + m)$. So two of the four solutions can be denoted

$$u_{++} = \begin{vmatrix} 1 \\ 0 \\ p/(E + m) \\ 0 \end{vmatrix} \qquad \text{and} \qquad u_{+-} = \begin{vmatrix} 0 \\ 1 \\ 0 \\ -p/(E + m) \end{vmatrix}.$$ (B.17a)

Equations (B.16c,d) give the same solutions, with label interchange $1 \leftrightarrow 4$ and $3 \leftrightarrow 2$, and the sign of E reversed. However, an equally good plane wave (B.15) can be written

$$\psi = u_j e^{-i p_\mu x_\mu},$$

i.e. with momentum $-p$ and energy $-E$. This is clearly allowed since $E = \sqrt{p^2 + m^2}$ does not determine the sign of E. So, the other two of the four solutions are really the *negative energy* solutions:

$$u_{-+} = \begin{vmatrix} -p/(|E| + m) \\ 0 \\ 1 \\ 0 \end{vmatrix} \quad \text{and} \quad u_{--} = \begin{vmatrix} 0 \\ p/(|E| + m) \\ 0 \\ 1 \end{vmatrix}. \quad \text{(B.17b)}$$

Combinations of any of these solutions are also permitted. To see what this all means, let us go to the particle rest frame, where $p = 0$. Each pair of u's can then be written as two-component spinors; for example,

$$u_{++} = \begin{vmatrix} 1 \\ 0 \end{vmatrix}, \quad u_{+-} = \begin{vmatrix} 0 \\ 1 \end{vmatrix},$$

identical with the two-component Pauli spinors of the nonrelativistic theory. These are interpreted as positive energy electron states with two possible *spin* directions, up and down—thus to electrons with half-integral spin. u_{-+} and u_{--} are the two negative energy spin states. Dirac supposed that there was a completely filled sea of negative energy states; a "hole" in this sea of electrons was interpreted as a *positron*, which would be produced whenever a negative energy electron received an energy $\geqslant 2mc^2$ and was kicked into a positive energy state ($e^{\pm}$ pair creation, see Fig. B.1).

A word should be said here about the complex conjugate Dirac wave functions. Normally we define particle density by

$$\rho = \psi^*\psi,$$

where the asterisk represents complex conjugation.

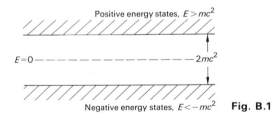

Positive energy states, $E > mc^2$

$E = 0$ — — — — — — — — — — — — — $2mc^2$

Negative energy states, $E < -mc^2$ **Fig. B.1**

If ψ is a column matrix (four-spinor), then by the rules for multiplying any matrices A and B, the number of columns of A and rows of B must be equal. We must therefore represent ψ^* by a row matrix. Thus

$$\psi = \begin{vmatrix} u_1 \\ u_2 \\ u_3 \\ u_4 \end{vmatrix} \quad \text{and} \quad \psi^* = |u_1^* \quad u_2^* \quad u_3^* \quad u_4^*|.$$

Then

$$\rho = \psi^*\psi = \sum_1^4 u_i^* u_i.$$

It is useful to define the spinor

$$\bar{\psi} = \psi^* \gamma_4,$$

so that, from (B.12),

$$\bar{\psi} \gamma_4 = \psi^*$$

and

$$\rho = \bar{\psi} \gamma_4 \psi. \tag{B.18}$$

From the foregoing relations, it is easy to show that $\bar{\psi}$ satisfies the Dirac equation

$$\frac{\partial}{\partial x_\mu} \bar{\psi} \gamma_\mu - m\bar{\psi} = 0.$$

PARITY OF PARTICLE AND ANTIPARTICLE

Suppose $\psi(\mathbf{r}, t)$ satisfies the Dirac equation (B.8). Spatial inversion of this state, simply by replacing $\mathbf{r}$ by $-\mathbf{r}$, does not satisfy the Dirac equation, since it is first order in the space coordinates. $\psi(-\mathbf{r}, t) = \psi(-x_k, x_4)$ satisfies

$$\left(\gamma_4 \frac{\partial}{\partial x_4} - \gamma_k \frac{\partial}{\partial x_k} + m \right) \psi(-\mathbf{r}, t) = 0; \qquad k = 1, 2, 3.$$

On multiplying through from the left by γ_4, and using the rule $\gamma_4 \gamma_k - \gamma_k \gamma_4 = 0$, we get

$$\left(\gamma_\mu \frac{\partial}{\partial x_\mu} + m \right) \gamma_4 \psi(-\mathbf{r}, t) = 0.$$

Thus $\gamma_4 \psi(-\mathbf{r}, t)$ is the inversion of the original state $\psi(\mathbf{r}, t)$, and γ_4 is the parity operator for the Dirac wave function. Now consider a spin-up positive energy state (in the particle rest frame, $p = 0$), Eq. (B.17a):

$$u_{++} = \begin{vmatrix} 1 \\ 0 \\ 0 \\ 0 \end{vmatrix}; \qquad \gamma_4 u_{++} = \begin{vmatrix} 1 \\ 0 \\ 0 \\ 0 \end{vmatrix}.$$

For a spin-up negative energy state (B.17b):

$$u_{-+} = \begin{vmatrix} 0 \\ 0 \\ 1 \\ 0 \end{vmatrix}; \qquad \gamma_4 u_{-+} = - \begin{vmatrix} 0 \\ 0 \\ 1 \\ 0 \end{vmatrix}.$$

Thus, if the positive energy state is assigned an even intrinsic parity, the negative energy state, or, equivalently, the antiparticle, has odd intrinsic parity.

APPLICATION OF DIRAC THEORY TO β-DECAY

The phenomenon of β-decay is described in terms of the interaction of four particles at a point. For example, neutron decay is written $n \rightarrow p + e^- + v^-$.

If we were dealing with scalar particles, we would write for the matrix element for $A + B \rightarrow C + D$

$$M = \text{const. } \psi_A^*(r)\psi_D^*(r)\psi_A(r)\psi_B(r).$$

For spin $\frac{1}{2}$ particles, we deal with four-component wave functions (spinors) and the interaction would be written

$$M = \text{const.} \sum_{ijkl} g_{ijkl}\psi_{C_i}^*\psi_{D_j}^*\psi_{A_k}\psi_{B_l}, \qquad (B.19)$$

i, j, k, and l being the spinor indices. In principle, the interaction can therefore involve $4^4 = 256$ arbitrary constants g_{ijkl}. This number can be reduced drastically by insisting that the laws of β-decay should be Lorentz invariant, which implies that the possible interaction forms should have well-defined properties, i.e. be covariant under Lorentz transformations. It is then found that, using the Dirac matrices γ and the spinors $\bar{\psi}$ and ψ, one can construct just five basic combinations which are named after their Lorentz transformation properties. They are:

		No. of components
$\bar{\psi}\psi (= \psi^*\gamma_4\psi)$ scalar S		1
$\bar{\psi}\gamma_\mu\psi$	four-vector V	4
$i\bar{\psi}\gamma_\mu\gamma_\nu\psi$	six-vector or tensor T	6
$i\bar{\psi}\gamma_5\gamma_\mu\psi$	axial four-vector A	4
$\bar{\psi}\gamma_5\psi$	pseudoscalar P	1.

By way of example, consider the inversion $\psi(-r, t)$ of the state $\psi(r, t)$. As mentioned above, $\psi(-r, t)$ does not satisfy the Dirac equation, but $\gamma_4\psi(-r, t)$, or $\bar{\psi}(-r, t)\gamma_4$, does. Thus under inversion

$$\psi\bar{\psi} \rightarrow \bar{\psi}\gamma_4^2\psi = \bar{\psi}\psi,$$

and is therefore a scalar, while

$$\bar{\psi}\gamma_5\psi \rightarrow \bar{\psi}\gamma_4\gamma_5\gamma_4\psi = -\bar{\psi}\gamma_5\psi$$

changes sign, and is therefore a pseudoscalar. It can be shown that both $\bar{\psi}\gamma_\mu\psi$ and $i\bar{\psi}\gamma_5\gamma_\mu\psi$ behave under rotations like four-vectors; however, the first changes sign under inversions, while the second does not, and is therefore an axial vector.*

Returning to the case of neutron decay, we may write it in the equivalent form

$$v + n \rightarrow e^- + p$$

---

* Rotations are called "proper" Lorentz transformations; inversions are called "improper" transformations.

in which a neutrino and neutron are transformed into electron and proton. We denote the wave functions by

ψ_ν to represent destruction of a neutrino (or creation of antineutrino),
$\bar{\psi}_e$ creation of an electron (or destruction of positron),
ψ_n destruction of a neutron (or creation of antineutron),
$\bar{\psi}_p$ creation of proton (or destruction of antiproton).

We see that this nomenclature is consistent with the physical process of β-decay as in the diagram. The matrix element (interacton energy density) can then be written in the bilinear form

$$M = G(\bar{\psi}_p O_i \psi_n) \times (\bar{\psi}_e O_i' \psi_\nu), \qquad (B.20)$$

nucleon bracket   lepton bracket

where O_i stands for one of the five forms written down above. In general, one expects a linear combination of the O_i ($i = 1 \cdots 5$) for the matrix elements. The above form was first written down by Fermi 35 years ago, with $O_i = \gamma_\mu$ (pure vector interaction). G is a suitable constant specifying the strength of the interaction. If the weak interaction is parity-conserving, we require that M be a scalar quantity under space inversion. Since, experimentally, parity is not conserved in β-decay, M should contain *both* scalar and pseudoscalar parts. We can achieve this by including a factor γ_5 in one of the brackets—conventionally the lepton bracket. Thus the interaction has the general form

$$(\bar{\psi}_p O_i \psi_n)[\bar{\psi}_e O_i(C_i + \gamma_5 C_i')\psi_n], \qquad (B.21)$$

where C and C' are constants, and O_i represents the S, V, T, A, and P interactions. We can rewrite the lepton bracket

$$\bar{\psi}_e O_i(C_i + \gamma_5 C_i')\psi_2 = (C_i + C_i')\bar{\psi}_e O_i(1 + \gamma_5)\psi_\nu \\ + (C_i - C_i')\bar{\psi}_e O_i(1 - \gamma_5)\psi_\nu. \qquad (B.22)$$

To find out the meaning of the operator $(1 \pm \gamma_5)$ on ψ_ν, we note that the Dirac equation (B.8) for a massless neutrino has the form

$$\gamma_\mu \frac{\partial \psi}{\partial x_\mu} = 0, \qquad \text{or } \gamma_4 \frac{\partial \psi}{\partial x_4} = -\gamma_k \frac{\partial \psi_k}{\partial x_k}, \qquad k = 1, 2, 3.$$

We set

$$\gamma_k = i\gamma_4 \gamma_5 \hat{\sigma}_k,$$

where $\hat{\sigma}_k$ are a set of 4×4 matrices formed from the Pauli spinors (B.7):

$$\hat{\sigma}_k = \begin{vmatrix} \sigma_k & 0 \\ 0 & \sigma_k \end{vmatrix}, \qquad (B.23)$$

whence

$$\frac{\partial \psi}{\partial x_4} = -i\gamma_5 \hat{\sigma}_k \frac{\partial \psi}{\partial x_k} = -i\hat{\sigma}_k \gamma_5 \frac{\partial \psi}{\partial x_k},$$

and

$$\frac{\partial \gamma_5 \psi}{\partial x_4} = -i\hat{\sigma}_k \frac{\partial \psi}{\partial x_k},$$

using the fact that $\gamma_5^2 = 1$ and $\gamma_5 \hat{\sigma}_k = \hat{\sigma}_k \gamma_5$.

Adding and subtracting the last two equations, setting $x_4 = ict$, and multiplying through by $i\hbar$ to made things more recognizable, we obtain

$$i\hbar \frac{\partial}{\partial t} [(1 + \gamma_5)\psi_v] = \hat{\sigma}_k cih \frac{\partial}{\partial x_k} [(1 + \gamma_5)\psi_v], \qquad \text{(B.24a)}$$

$$i\hbar \frac{\partial}{\partial t} [(1 - \gamma_5)\psi_v] = -\hat{\sigma}_k cih \frac{\partial}{\partial x_k} [(1 - \gamma_5)\psi_v]. \qquad \text{(B.24b)}$$

Thus, $(1 \pm \gamma_5)\psi_v$ are plane wave eigenstates of the energy operator, $i\hbar \, \partial/\partial t$, with eigenvalues

$$E = -\boldsymbol{\sigma} \cdot \boldsymbol{p}c \qquad \text{for } \phi = (1 + \gamma_5)\psi, \qquad \text{(B.25a)}$$

$$E = +\boldsymbol{\sigma} \cdot \boldsymbol{p}c \qquad \text{for } \chi = (1 - \gamma_5)\psi, \qquad \text{(B.25b)}$$

where $\boldsymbol{\sigma}$ is the spin vector. Since $E = pc$ for neutrinos, it follows that the state $\phi = (1 + \gamma_5)\psi_v$ represents a particle of helicity

$$H = \frac{\boldsymbol{\sigma} \cdot \boldsymbol{p}}{p} = -1,$$

and $\chi = (1 - \gamma_5)\psi_v$ represents one of helicity $+1$. ψ_v in (B.20) defines the neutrino state, since it transforms to a negative electron. Experimentally it is found to have *negative* helicity. Thus, we need the term ϕ, which is associated with left-handed neutrinos (and right-handed antineutrinos), and have to discard the term χ. So we see that this amounts to keeping just the first of the two Weyl equations—in fact (B.24) *are* the Weyl equations [compare (B.5)]. The Dirac equation, describing particles with mass by four-component wave functions, breaks down for massless particles into two decoupled equations, each with two-component wave functions as solutions.

Thus we require $C_i = C_i'$ in (B.21), which thereby becomes

$$\bar{\psi}_e O_i (1 + \gamma_5)\psi_v = \bar{\psi}_e (1 + \gamma_5) O_i \psi_v \qquad \text{for } i = S, T, P$$

$$= \bar{\psi}_e (1 - \gamma_5) O_i \psi_v \qquad \text{for } i = V, A,$$

as may be established from the commutation rules. By extending the argument used for massless particles, it may be shown that the term $\bar{\psi}_e(1 - \gamma_5)$ corresponds to the generation of left-handed electrons with intensity $(1 - \boldsymbol{\sigma} \cdot \boldsymbol{p}/E)$, i.e. a helicity $-\boldsymbol{\sigma} \cdot \boldsymbol{p}/E$, and $\bar{\psi}_e(1 + \gamma_5)$ to right-handed electrons (and conversely for positrons). Experimentally, it is found that electrons have negative helicity, $-\boldsymbol{\sigma} \cdot \boldsymbol{p}/E$. Thus, the experiments show that the β-decay interactions are V, A

and not S, T, or P. The matrix element may now be written

$$M = G[C_V(\bar{\psi}_p\gamma_\mu\psi_n)(\bar{\psi}_e\gamma_\mu(1 + \gamma_5)\psi_\nu) + i^2 C_A(\bar{\psi}_p\gamma_\mu\gamma_5\psi_n)(\bar{\psi}_e\gamma_\mu\gamma_5(1 + \gamma_5)\psi_\nu)]$$

$$= G[(\bar{\psi}_p\gamma_\mu(C_V - \gamma_5 C_A)\psi_n)(\bar{\psi}_e\gamma_\mu(1 + \gamma_5)\psi_\nu)],$$

using $\gamma_5^2 = 1$. We see that, if $C_V = -C_A$, i.e. the axial-vector and vector couplings are equal in magnitude but opposite in sign, we obtain the famous *V-A interaction*. The nucleon bracket can then also be written in the form $\bar{\psi}_p(1 - \gamma_5)\gamma_\mu(1 + \gamma_5)\psi_n$, since

$$(1 + \gamma_5) = \tfrac{1}{2}(1 + \gamma_5)^2$$

and thus

$$\gamma_\mu(1 + \gamma_5) = \tfrac{1}{2}(1 - \gamma_5)\gamma_\mu(1 + \gamma_5).$$

In this theory, nucleons as well as leptons are left-handed, antinucleons right-handed. Since, experimentally, in nucleon decay, $C_A \approx -1.25 C_V$, the degree of longitudinal polarization is in fact only $\sim 0.9\, v/c$ instead of v/c. If we write the combinations $\bar{\psi}(1 - \gamma_5) = \phi$, $(1 + \gamma_5)\psi = \phi$, the exact *V-A* hypothesis of Feynman and Gell-Mann gives

$$M = \text{const.}\, (\bar{\phi}_p\gamma_\mu\phi_n)(\bar{\phi}_e\gamma_\mu\phi_\nu),$$

which is to be contrasted with the original Fermi theory, some 23 years earlier, in which the interaction was pure vector:

$$M = \text{const.}\, (\bar{\psi}_p\gamma_\mu\psi_n)(\bar{\psi}_e\gamma_\mu\psi_\nu).$$

Appendix C

Clebsch-Gordan Coefficients. The Addition of Angular Momenta or Isotopic Spins

Suppose we have two particles of angular momenta j_1 and j_2 with z-components m_1 and m_2. The total z-component is

$$m = m_1 + m_2.$$

The total angular momentum is

$$\mathbf{j} = \mathbf{j}_1 + \mathbf{j}_2,$$

and may therefore lie anywhere inside the limits

$$|j_1 - j_2| \leqslant j \leqslant |j_1 + j_2|.$$

We wish to find the weights of the various allowed j-values contributing to the two-particle state, that is,

$$\phi_1(j_1 m_1)\phi_2(j_2 m_2) = \sum_j C_j \psi(j, m), \qquad \text{with} \qquad m = m_1 + m_2. \quad \text{(C.1)}$$

The C_j are called Clebsch-Gordan coefficients (or Wigner, or vector addition, coefficients). Alternatively, we may want to express $\psi(j, m)$ as a sum of terms of different j_1 and j_2 combinations. We can do this by the use of angular momentum (or isospin)) *shift operators* (also known as "raising" and "lowering" operators).

First let us recall the definition of the x-, y-, and z-component angular momentum operators, in terms of the differential Cartesian operators. Thus

$$J_x = -\frac{ih}{2\pi}\left(y\frac{\partial}{\partial z} - z\frac{\partial}{\partial y}\right),$$

$$J_y = -\frac{ih}{2\pi}\left(z\frac{\partial}{\partial x} - x\frac{\partial}{\partial z}\right), \qquad \text{(C.2)}$$

$$J_z = -\frac{ih}{2\pi}\left(x\frac{\partial}{\partial y} - y\frac{\partial}{\partial x}\right). \qquad \text{(C.2)}$$

318

It is readily verified that the operators

$$J_x, J_y, J_z, \quad \text{and} \quad J^2 = J_x^2 + J_y^2 + J_z^2 \tag{C.3}$$

obey the commutation rules

$$J^2 J_x - J_x J^2 = 0, \quad \text{etc.,}$$

and

$$J_x J_y - J_y J_x = i J_z,$$
$$J_y J_z - J_z J_y = i J_x,$$
$$J_z J_x - J_x J_z = i J_y,$$

where we have used units $\hbar = c = 1$ for brevity. The eigenvalues of the operators J^2 and J_z are given in Eq. (C.5) below.

The *shift operators* are defined as

$$\left. \begin{array}{l} J_+ = J_x + i J_y \\[2mm] J_- = J_x - i J_y \end{array} \right\} \quad \text{whence} \quad \begin{cases} J_z J_+ - J_+ J_z = J_+ \\[2mm] J_z J_- - J_- J_z = -J_- \end{cases} \tag{C.4}$$

Thus:

$$J_z(J_-\phi) = J_z J_- \phi = J_-(J_z - 1)\phi = (m - 1)J_-\phi.$$

Similarly:

$$J_z(J_+\phi) = (m + 1)(J_+\phi).$$

The last equation shows that the wave function $(J_+\phi)$ is an eigenstate of J_z with eigenvalue $(m + 1)$. We can therefore write it as

$$J_+\phi(j, m) = C_+\phi(j, m + 1),$$

where C_+ is an unknown (and generally complex) constant. If we multiply both sides of this equation by $\phi^*(j, m + 1)$, and integrate over volume, we get

$$\int \phi^*(j, m + 1)J_+\phi(j, m) \, dV = C_+ \int \phi^*(j, m + 1)\phi(j, m + 1) \, dV$$

where * indicates complex conjugation.

We choose the normalization of ϕ so that the last integral is unity, and all allowed m-values have unit weight.

So

$$C_+ = \int \phi^*(j, m + 1)J_+\phi(j, m) \, dV,$$

Similarly

$$C_- = \int \phi^*(j, m)J_-\phi(j, m + 1) \, dV$$

$$= \int \phi^*(j, m)J_+^*\phi(j, m + 1) \, dV$$

$$= C_+^*,$$

from Eq. (C.4).

If we neglect arbitrary and unobservable phases, we must have

$$C_+ = C_- = C \quad \text{(a real number)}.$$

Also, from Eq. (C.4),

$$J_+J_- = J_x^2 + J_y^2 - i(J_xJ_y - J_yJ_x) = J_x^2 + J_y^2 - J_z = J^2 - J_z^2 - J_z.$$

Then

$$J_+J_-\phi(j, m) = [j(j + 1) - m^2 - m]\phi(j, m) = C^2\phi(j, m).$$

So,

$$C = \sqrt{j(j + 1) - m(m + 1)}$$

is the coefficient connecting states $(j, m) \leftrightarrow (j, m + 1)$.

To summarize, the angular momentum operators have the following properties:

$$J_z\phi(j, m) = m\phi(j, m), \tag{C.5a}$$

$$J^2\phi(j, m) = j(j + 1)\phi(j, m), \tag{C.5b}$$

$$J_+\phi(j, m) = \sqrt{j(j + 1) - m(m + 1)}\,\phi(j, m + 1), \tag{C.5c}$$

$$J_-\phi(j, m) = \sqrt{j(j + 1) - m(m - 1)}\,\phi(j, m - 1). \tag{C.5d}$$

Example (1)

As an example, we consider two particles of j_1, m_1, and j_2, m_2, forming the combined state $\psi(j, m)$, and we take the case where $j_1 = 1, j_2 = \frac{1}{2}$, and $j = \frac{3}{2}$ or $\frac{1}{2}$.

Obviously the states of $m = \pm\frac{3}{2}$ can only be formed in one way:

$$\psi(\tfrac{3}{2}, \tfrac{3}{2}) = \phi(1, 1)\phi(\tfrac{1}{2}, \tfrac{1}{2}), \tag{C.6}$$

$$\psi(\tfrac{3}{2}, -\tfrac{3}{2}) = \phi(1, -1)\phi(\tfrac{1}{2}, -\tfrac{1}{2}). \tag{C.7}$$

Now we use the operators $J_\pm$ to form the relations:

$$J_-\phi(\tfrac{1}{2}, \tfrac{1}{2}) = \phi(\tfrac{1}{2}, -\tfrac{1}{2}); \qquad J_-\phi/\tfrac{1}{2}, -\tfrac{1}{2}) = 0$$

$$J_-\phi(1, 1) = \sqrt{2}\,\phi(1, 0); \qquad J_-\phi(1, 0) = \sqrt{2}\,\phi(1, -1); \qquad J_-\phi(1, -1) = 0$$

using (C.5c and d).

Now operate on (C.6) with J_- on both sides:

$$J_-\psi(\tfrac{3}{2}, \tfrac{3}{2}) = \sqrt{3}\,\psi(\tfrac{3}{2}, \tfrac{1}{2}) = J_-\phi(1, 1)\phi(\tfrac{1}{2}, \tfrac{1}{2})$$

$$= \sqrt{2}\,\phi(1, 0)\phi(\tfrac{1}{2}, \tfrac{1}{2}) + \phi(1, 1)\phi(\tfrac{1}{2}, -\tfrac{1}{2}).$$

So

$$\psi(\tfrac{3}{2}, \tfrac{1}{2}) = \sqrt{\tfrac{2}{3}}\,\phi(1, 0)\phi(\tfrac{1}{2}, \tfrac{1}{2}) + \sqrt{\tfrac{1}{3}}\,\phi(1, 1)\phi(\tfrac{1}{2}, -\tfrac{1}{2}). \tag{C.8}$$

Similarly, for (C.7):

$$\psi(\tfrac{3}{2}, -\tfrac{1}{2}) = \sqrt{\tfrac{2}{3}}\,\phi(1, 0)\phi(\tfrac{1}{2}, -\tfrac{1}{2}) + \sqrt{\tfrac{1}{3}}\,\phi(1, -1)\phi(\tfrac{1}{2}, \tfrac{1}{2}). \tag{C.9}$$

The $j = \frac{1}{2}$ state can be expressed as a linear sum:

$$\psi(\tfrac{1}{2}, \tfrac{1}{2}) = a\phi(1, 1)\phi(\tfrac{1}{2}, -\tfrac{1}{2}) + b\phi(1, 0)\phi(\tfrac{1}{2}, \tfrac{1}{2})$$

with $a^2 + b^2 = 1$.
   Then

$$J_+\psi(\tfrac{1}{2}, \tfrac{1}{2}) = 0 = a\phi(1, 1)\phi(\tfrac{1}{2}, \tfrac{1}{2}) + b\sqrt{2}\,\phi(1, 1)\phi(\tfrac{1}{2}, \tfrac{1}{2}).$$

Thus, $a = \sqrt{\dfrac{2}{3}}$, $b = -\dfrac{1}{\sqrt{3}}$, and so

$$\psi(\tfrac{1}{2}, \tfrac{1}{2}) = \sqrt{\tfrac{2}{3}}\,\phi(1, 1)\phi(\tfrac{1}{2}, -\tfrac{1}{2}) - \sqrt{\tfrac{1}{3}}\,\phi(1, 0)\phi(\tfrac{1}{2}, \tfrac{1}{2}). \tag{C.10}$$

Similarly,

$$\psi(\tfrac{1}{2}, -\tfrac{1}{2}) = \sqrt{\tfrac{1}{3}}\,\phi(1, 0)\phi(\tfrac{1}{2}, -\tfrac{1}{2}) - \sqrt{\tfrac{2}{3}}\,\phi(1, -1)\phi(\tfrac{1}{2}, \tfrac{1}{2}). \tag{C.11}$$

Expressions (C.6) to (C.11) give the coefficients appearing in Table III (p. 330), for the addition of $J = 1$ and $J = \frac{1}{2}$ states.

Example (2)

As a second example, suppose we wish to know the isospin states corresponding to combination of a Σ^0-hyperon and a π^0-meson. In other words we have to find which states of total isospin can be made by adding two states of $I = 1$ and $I_3 = 0$. We write for the amplitude

$$\phi(\Sigma^0)\chi(\pi^0) = A\psi(2, 0) + B\psi(1, 0) + C\psi(0, 0). \tag{C.12}$$

A, B, and C are to be determined, with $A^2 + B^2 + C^2 = 1$. $\psi(I, I_3)$ is the total isospin state, ϕ and ψ the Σ^0 and π^0 isospin wavefunctions. We use the raising operator

$$I^+\psi(I, I_3) = \sqrt{I(I + 1) - I_3(I_3 + 1)}\,\psi(I, I_3 + 1)$$

Applying to the left-hand side of (C.12) gives

$$I^+[\phi(1, 0)\chi(1, 0)] = [I^+\phi(1, 0)]\chi(1, 0) + [I^+\chi(1, 0)]\phi(1, 0)$$
$$= \sqrt{2}\,[\phi(1, 1)\chi(1, 0) + \chi(1, 1)\phi(1, 0)]. \tag{C.13}$$

Applying to the right-hand side gives

$$I^+[A\psi(2, 0) + B\psi(1, 0) + C\psi(0, 0)] = A\sqrt{6}\,\psi(2, 1) + B\sqrt{2}\,\psi(1, 1) + 0. \tag{C.14}$$

On multiplying each side by its complex conjugate state, and remembering that $\phi(1, 1)\phi^*(1, 1) = 1$ and that $\phi(1, 0)\phi^*(1, 1) = 0$ (orthonormality), we get

$$6A^2 + 2B^2 = 2(\sqrt{2})^2 = 4. \tag{C.15}$$

Applying the operator I^+ once more to (C.13) and (C.14), we obtain

$$A\sqrt{6}\sqrt{4}\psi(2, 2) + 0 = 4\phi(1, 1)\chi(1, 1)$$

or

$$A^2 = \tfrac{2}{3}$$

From (C.15) and (C.12) one then finds $B = 0$ and $C^2 = \tfrac{1}{3}$. Thus

$$\phi(\Sigma^0)\chi(\pi^0) = \sqrt{\tfrac{2}{3}}\,\psi(2, 0) \pm \sqrt{\tfrac{1}{3}}\,\psi(0, 0)$$

which is the decomposition required. The fact that the $I = 1$ state does not contribute can also be understood from a simple vector model.

Lorentz-Invariant Phase Space

The transition rate formula (3.16) is

$$W = \frac{2\pi}{\hbar} |M_{if}|^2 \rho_f, \qquad (D.1)$$

where W is the probability of a transition, per second, between an initial state i and a particular final state f. M_{if} is the matrix element for the transition and the factor ρ_f denotes the number of states available in momentum space, per unit interval of the total energy of the state f, for the particle or particles produced in the final state. The proof of (D.1) is given in most standard texts on quantum mechanics, in the framework of perturbation theory. In this case, the matrix element has the form $M = \int \psi_f^* H_I \psi_i \, dV$ where H_I is the operator corresponding to the interaction energy, such that $H_I \ll H_0$, and H_0 is the "free" Hamiltonian. H_0 has as eigenvalues the energy of the system i before the interaction is turned on. It is common to represent both i and f by free-particle plane wave functions (Born approximation), particularly in collision problems solved by perturbation methods (for example, electron scattering by nuclei, discussed in Chapter 5).

In general, and certainly in strong interactions, the form of M is not known explicitly—in fact (D.1) is effectively a definition of M. In any case, both M and ρ_f refer to some arbitrary normalization volume V (somewhere inside which the interaction occurs), and the wave functions of the individual particles must be normalized to this volume. Thus, if a particle is represented by a plane wave, with $\hbar = c = 1$,

$$\psi = V^{-1/2} \exp\left[i(\mathbf{p} \cdot \mathbf{r} - Et)\right], \qquad (D.2)$$

so that $\int \psi^* \psi \, dV = 1$. In an actual problem, one usually sets $V = 1$ and forgets about it. V must drop out in the final result, since the physical quantities like cross sections or decay rates are quoted per particle and must be independent of V. We include it here because the normalization (D.2) is not correct if we want the matrix element, and ρ_f, to be in a relativistically invariant form, which is useful in many cases, particularly when one can only make general statements about the properties of M, as in S-matrix theory.

First let one of the particles in the interaction be at rest relative to a box of volume V. Then the particle density is $1/V$ in this frame. Now suppose the particle plus box move at velocity v relative to the chosen reference frame, Σ, in which W is to be calculated. If the particle mass is m, its energy in Σ is E, where $E/m = (1 - v^2/c^2)^{-1/2}$. As a consequence of the Lorentz transformation, a length measured in the direction of v in the particle rest-frame is reduced by a factor m/E when measured in the frame Σ, so that the volume of the box is contracted to a value $V' = Vm/E$. If the wave function is integrated over a volume V in Σ, then $\int \psi^*\psi \, dV = E/m$; i.e., the particle density is increased by a factor E/m. Therefore, to ensure that the particle density is correctly normalized in the same way in all frames, we should incorporate a factor $\sqrt{m/E}$ with the wave function ψ, for each particle, where E is the energy in the chosen frame Σ. Several conventions are in use; $\sqrt{m/E}$, $1/\sqrt{2E}$, etc. We shall take $1/\sqrt{2E}$. Setting the phase-space factor

$$\frac{p^2 \, dp \, d\Omega V}{h^3} = \frac{d^3p}{(2\pi)^3} \, ,$$

i.e. with $V = 1$, and units $\hbar = c = 1$, the transition rate becomes

$$W = \frac{2\pi \, |M|^2}{\prod_{\text{initial}} 2E_j} \rho_f, \tag{D.3}$$

with

$$\rho_f = \frac{d}{dE_{\text{total}}} \frac{\int d^3p_1 \, d^3p_2 \cdots d^3p_{n-1}}{(2\pi)^{3(n-1)} \prod_{\text{final}} 2E_k}. \tag{D.4}$$

The products of the E's have been factored into initial and final states. There are n particles in the final state, hence $(n - 1)$ independent three-momenta in phase-space. ρ_f in (D.4) is Lorentz-invariant. This is clearly so for the decay of a single particle ($j = 1$), since the decay rate per unit volume is proportional to γ^{-1} or E^{-1}, so that ρ_f must be invariant. To prove the general case, we first write

$$dN = \int d^3p_1 \, d^3p_2 \cdots d^3p_{n-1} = \int d^3p_1 \, d^3p_2 \cdots d^3p_n \, \delta(p_i - p_f),$$

where δ stands for the Dirac δ-function,* conserving the total momentum $(\mathbf{p}_i = \mathbf{p}^f)$ so that $\int d^3p_n\, \delta(\mathbf{p}_i - \mathbf{p}_f) = 1$. Also, energy conservation implies

$$\int \frac{dN}{dE}\, \delta(E_i - E_f)\, dE = \frac{dN}{dE}\bigg|_{E_i = E_f}.$$

If we write an energy-momentum four-vector

$$p = (p_1, p_2, p_3, p_4) = (\mathbf{p}, p_4) = (\mathbf{p}, iE),$$

then the density of final states per unit total energy is

$$\frac{dN}{dE_{\text{total}}} = \int \frac{\prod_1^n d^3p_k\, \delta(\mathbf{p}_i - \mathbf{p}_f)}{dE} = \int \prod_1^n d^3p_k\, \delta^{(4)}(p_i - p_f),$$

where $\delta^{(4)}$ indicates a δ-function conserving four-momentum;

$$\delta^{(4)}(p_i - p_f) = \delta(p_4^2 - \mathbf{p}^2 - m^2).$$

Now, for any particle of mass m,

$$\int \delta(p_4^2 - \mathbf{p}^2 - m^2)\, dp_4 = \int \frac{1}{2p_4}\, \delta(p_4^2 - \mathbf{p}^2 - m^2)\, d(p_4^2)$$

$$= \left| \frac{1}{2p_4} \right|_{p_4^2 = p^2 + m^2} = \frac{1}{2E}.$$

Hence

$$\rho_f = \frac{dN}{dE} = \frac{\int \prod_n d^4p_k\, \delta(p_k^2 + m_k^2)\, \delta^{(4)}(p_i - p_f)}{\prod_{(n-1)} (2\pi)^3}. \qquad \text{(D.5)}$$

The last term in the numerator takes care of energy-momentum conservation; the second term ensures that the integral over the element of energy and three-momentum, d^4p_k for the kth particle, is such that $p_k^2 + m_k^2 = 0$, i.e. $p_k^2 = \mathbf{p}_k^2 - E_k^2 = -m_k^2$. The above formula, incorporating four-vectors only, is "manifestly covariant."

---

* Defined as: $\int_b^c dx\, \delta(x - a) = 1$ for a in the interval $b \to c$; $= 0$ for a outside the interval $b \to c$; $\delta(x - a) = 0$ for $x \neq a$.

G-parity of the Pion

In Chapter 3 the G operation was defined as

$$G = C \exp(i\pi I_2),$$

that is, a rotation through $180°$ about the y-axis in isospin space, followed by charge conjugation. For the neutral pion, $G = -1$. For charged pions, let us first find the result of the isospin rotation. From the definition of the raising and lowering operators (Appendix C)

$$I^{\pm}\psi(I, I_3) = \sqrt{I(I + 1) - I_3(I_3 \pm 1)} \cdot \psi(I, I_3 \pm 1)$$

where

$$I^{\pm} = I_1 \pm iI_2, \qquad I_2 = -\tfrac{1}{2}i(I^+ - I^-), \qquad I_1 = \tfrac{1}{2}(I^+ + I^-).$$

From these expressions applied to pion states, one finds

$$I_2 |\pi^0\rangle = -\frac{i}{\sqrt{2}}(|\pi^+\rangle - |\pi^-\rangle) = \alpha$$

say,

$$I_2 |\pi^{\pm}\rangle = \pm\frac{i}{\sqrt{2}} |\pi^0\rangle = \pm\beta.$$

Applying the operator I_2 repeatedly gives

$$I_2 |\pi^0\rangle = (I_2)^3 |\pi^0\rangle = \cdots = (I_2)^{2n+1} |\pi^0\rangle = \alpha.$$
$$(I_2)^2 |\pi^0\rangle = \cdots = (I_2)^{2n} |\pi^0\rangle = |\pi^0\rangle.$$
$$I_2 |\pi^{\pm}\rangle = (I_2)^3 |\pi^{\pm}\rangle = \cdots = (I_2)^{2n+1} |\pi^{\pm}\rangle = \pm\beta.$$
$$(I_2)^2 |\pi^{\pm}\rangle = \cdots = (I_2)^{2n} |\pi^{\pm}\rangle = \pm\frac{i}{\sqrt{2}}\alpha = \pm\tfrac{1}{2}(|\pi^+\rangle - |\pi^-\rangle)$$

Then, expanding the operator $R = \exp(i\pi I_2)$ by a power series,

$$\exp(i\pi I_2) \cdot |\pi^+\rangle = \left(1 + i\pi I_2 - \frac{\pi^2}{2!}(I_2)^2 + \cdots\right)|\pi^+\rangle$$

$$= \frac{(|\pi^+\rangle + |\pi^-\rangle)}{2} + \frac{(|\pi^+\rangle - |\pi^-\rangle)}{2}\cos\pi - \frac{|\pi^0\rangle}{\sqrt{2}}\sin\pi.$$

$$= + |\pi^-\rangle. \qquad\qquad\qquad\qquad\qquad\text{(E.1)}$$

Similarly one finds that

$$\exp(i\pi I_2)|\pi^-\rangle = +|\pi^+\rangle \qquad \text{and} \quad \exp(i\pi I_2)|\pi^0\rangle = -|\pi^0\rangle.$$

Since

$$C|\pi^0\rangle = +|\pi^0\rangle,$$

it follows that if we define

$$C|\pi^\pm\rangle = -|\pi^\mp\rangle,$$

then all pion states have the same G-parity:

$$G|\pi^{+,-,0}\rangle = -|\pi^{+,-,0}\rangle.$$

Some texts define the rotation R about the x-axis. Using similar methods to those above, it is then found that

$$\exp(i\pi I_1)|\pi^{+,-,0}\rangle = -|\pi^{+,-,0}\rangle$$

with the same phase for all charge states. In this case, one must define

$$C|\pi^\pm\rangle = +|\pi^\mp\rangle$$

if one requires the same G-parity for all pion states.

A much quicker and neater method of obtaining the result (E.1) is to make use of the fact that the isospin functions $\psi(I, I_3)$ have the same properties under rotations in "isospin space" as the spherical harmonics $Y_l^m(\theta, \phi)$ have in real space. Setting $I = l$ and $I_3 = m$, and referring to Fig. 3.1 and the spherical harmonic tables on p. 329, we note that the rotation R corresponds to the substitution $\theta \to \pi - \theta$, $\phi \to \pi - \phi$. Since

$$Y_1^1(\theta, \phi) = -\sqrt{\frac{3}{8\pi}}\sin\theta e^{i\phi},$$

then

$$Y_1^1(\theta, \phi) \xrightarrow{R} +\sqrt{\frac{3}{8\pi}}\sin\theta e^{-i\phi} = Y_1^{-1}(\theta, \phi)$$

Thus

$$\psi(1, 1) \xrightarrow{R} +\psi(1, -1)$$

TABLE I TABLE OF ATOMIC CONSTANTS

Avogadro's number (N_0)	$6.02217(\pm4) \times 10^{23}$ mole^{-1}
Velocity of light *in vacuo*	$c = 2.99793(\pm1) \times 10^{10}$ cm sec^{-1}
Charge on electron	$e = 4.80325(\pm2) \times 10^{-10}$ esu

Planck's constant reduced $\dfrac{h}{2\pi} = \hbar$ $= 6.58218(\pm2) \times 10^{-22}$ MeV sec

$= 1.054592(\pm8) \times 10^{-27}$ erg sec

Fine structure constant $\alpha = e^2/\hbar c$ $= 1/137.0360(\pm2)$

Mass of electron $\quad m_e c^2 = 9.10956(\pm5) \times 10^{-28}$ g

$= 0.511004(\pm2)$ MeV

Mass of proton $\quad m_p c^2 = 938.259(\pm5)$ MeV $= 1836.11(\pm1)m_e$

1 MeV $= 1.602192(\pm7) \times 10^{-6}$ erg

Classical radius of electron $\quad r_e = e^2/m_e c^2 = 2.81794(\pm1) \times 10^{-13}$ cm

Compton wavelength of electron $\quad \hbar/m_e c = r_e/\alpha = 3.86159(\pm1) \times 10^{-11}$ cm

First Bohr radius for infinitely
heavy nucleus $\quad a_\infty = \hbar^2/e^2 m_e = r_e/\alpha^2 = 0.5291772(\pm8) \times 10^{-8}$ cm

Thomson cross section $\quad \frac{8}{3}\pi r_e^2 = 665.245(\pm6)$ mb $(1 \text{ mb} = 10^{-27}$ cm$^2)$

Bohr magneton $\quad \mu_B = e\hbar/2m_e c = 0.578838(\pm2) \times 10^{-14}$ MeV G^{-1}

Nuclear magneton $\quad \mu_n = e\hbar/2m_p c = 3.15253(\pm2) \times 10^{-18}$ MeV G^{-1}

Unit of precession frequency
$$\begin{cases} \omega = \dfrac{\mu_B}{\hbar} = \dfrac{e}{2m_e c} = 8.79401(\pm3) \times 10^6 \text{ rad sec}^{-1}\text{G}^{-1} \\[2em] \omega = \dfrac{\mu_n}{\hbar} = \dfrac{e}{2m_p c} = 4.78948(\pm3) \times 10^3 \text{ rad sec}^{-1}\text{G}^{-1} \end{cases}$$

Radius of curvature for momentum p: $pc = 0.03H\rho$ $\quad \begin{cases} pc \text{ in GeV} \\ H \text{ in kG} \\ \rho \text{ in m} \end{cases}$

1 Fermi $(= 1 \text{ F}) = 10^{-13}$ cm.

1 Barn $= 1$ b $= 10^3$ mb $= 10^6$ μb $= 10^{-24}$ cm^2.

Useful approximations: $\quad hc \approx 200$ MeV F

Classical radius of electron $e^2/m_e c^2 \approx 2.8$ F $\approx 2 \times \dfrac{\hbar}{m_\pi c}$

$\approx 2 \times$ range of nuclear forces

TABLE II SPHERICAL HARMONICS

$$Y_l^m(\theta, \phi) = \sqrt{\frac{(2l + 1)(l - m)!}{4\pi(l + m)!}} \, P_l^m(\cos \theta)e^{im\phi}$$

$$P_l^m(\cos \theta) = (-1)^m \sin^m \theta \left[\left(\frac{d}{d(\cos \theta)} \right)^m P_l(\cos \theta) \right]; \quad (m \leqslant l)$$

$$P_l(\cos \theta) = \frac{1}{2^l l!} \left[\left(\frac{d}{d(\cos \theta)} \right)^l (-\sin^2 \theta)^l \right]$$

$$Y_l^{-m}(\theta, \phi) = (-1)^m [Y_l^m(\theta, \phi)]^*$$

$$P_l^{-m}(\cos \theta) = (-1)^m \frac{(l - m)!}{(l + m)!} P_l^m(\cos \theta)$$

$l = 0$ $\qquad Y_0^0 = \dfrac{1}{\sqrt{4\pi}}$

$l = 1$ $\qquad Y_1^0 = \sqrt{\dfrac{3}{4\pi}} \cos \theta$ $\qquad\qquad Y_1^1 = -\sqrt{\dfrac{3}{8\pi}} \sin \theta e^{i\phi}$

$l = 2$ $\qquad Y_2^0 = \sqrt{\dfrac{5}{16\pi}} (3 \cos^2 \theta - 1)$ $\qquad Y_2^1 = -\sqrt{\dfrac{15}{8\pi}} \sin \theta \cos \theta e^{i\phi}:$

$\qquad\qquad Y_2^2 = \sqrt{\dfrac{15}{32\pi}} \sin^2 \theta e^{2i\phi}.$

$l = 3$ $\qquad Y_3^0 = \sqrt{\dfrac{7}{16\pi}} (5 \cos^3 \theta - 3 \cos \theta)$ $\qquad Y_3^1 = -\sqrt{\dfrac{21}{64\pi}} \sin \theta(5 \cos^2 \theta - 1)e^{i\phi}$

$\qquad\qquad Y_3^2 = \sqrt{\dfrac{105}{32\pi}} \sin^2 \theta \cos \theta e^{2i\phi}$ $\qquad Y_3^3 = -\sqrt{\dfrac{35}{64\pi}} \sin^3 \theta e^{3i\phi}.$

TABLE III CLEBSCH-GORDAN COEFFICIENTS

As an example of the use of this table, take the case of combining two angular momenta $j_1 = 1$, $m_1 = 1$ and $j_2 = 1$, $m_2 = -1$. We look up the entry for combining 1×1, and the fourth line gives for the coefficients in Eq. (C1) of Appendix C:

$$\phi_1(1, 1)\phi_2(1, -1) = \sqrt{\tfrac{1}{6}}\,\psi(2, 0) + \sqrt{\tfrac{1}{2}}\,\psi(1, 0) + \sqrt{\tfrac{1}{3}}\,\psi(0, 0).$$

This tells us how two particles of angular momentum (or isospin) unity combine to form states of angular momentum $j = 0$, 1, or 2. Alternatively, a state of particular (j, m) can be decomposed into constituents. Thus $j = 2$, $m = 0$ can be decomposed into products of states of $j = 1$, with of course $m_1 + m_2 = m = 0$. The fourth column of the 1×1 table gives

$$\psi(2, 0) = \sqrt{\tfrac{1}{6}}\,\phi_1(1, 1)\phi_2(1, -1) + \sqrt{\tfrac{2}{3}}\,\phi_1(1, 0)\phi_2(1, 0) + \sqrt{\tfrac{1}{6}}\,\phi_1(1, -1)\phi_2(1, 1).$$

The sign convention in the table follows that of Condon and Shortley (1951).

$\tfrac{1}{2} \times \tfrac{1}{2}$

m_1 $\backslash$ m_2	J M	1 $+1$	1 0	0 0	1 -1
$+\tfrac{1}{2}$ $+\tfrac{1}{2}$		1			
$+\tfrac{1}{2}$ $-\tfrac{1}{2}$			$\sqrt{\tfrac{1}{2}}$	$\sqrt{\tfrac{1}{2}}$	
$-\tfrac{1}{2}$ $+\tfrac{1}{2}$			$\sqrt{\tfrac{1}{2}}$	$-\sqrt{\tfrac{1}{2}}$	
$-\tfrac{1}{2}$ $-\tfrac{1}{2}$					1

$1 \times \tfrac{1}{2}$

m_1 $\backslash$ m_2	J M	$\tfrac{3}{2}$ $+\tfrac{3}{2}$	$\tfrac{3}{2}$ $+\tfrac{1}{2}$	$\tfrac{1}{2}$ $+\tfrac{1}{2}$	$\tfrac{3}{2}$ $-\tfrac{1}{2}$	$\tfrac{1}{2}$ $-\tfrac{1}{2}$	$\tfrac{3}{2}$ $-\tfrac{3}{2}$
$+1$ $+\tfrac{1}{2}$		1					
$+1$ $-\tfrac{1}{2}$ / 0 $+\tfrac{1}{2}$			$\sqrt{\tfrac{1}{3}}$ $\sqrt{\tfrac{2}{3}}$	$\sqrt{\tfrac{2}{3}}$ $-\sqrt{\tfrac{1}{3}}$			
0 $-\tfrac{1}{2}$ / -1 $+\tfrac{1}{2}$					$\sqrt{\tfrac{2}{3}}$ $\sqrt{\tfrac{1}{3}}$	$\sqrt{\tfrac{1}{3}}$ $-\sqrt{\tfrac{2}{3}}$	
-1 $-\tfrac{1}{2}$							1

CLEBSCH-GORDAN COEFFICIENTS (CONT'D)

$\frac{3}{2} \times \frac{1}{2}$

m_1	m_2	J → M →	2 +2	2 +1	1 +1	2 0	1 0	2 −1	1 −1	2 −2
$+\frac{3}{2}$	$+\frac{1}{2}$		1							
$+\frac{3}{2}$	$-\frac{1}{2}$			$\sqrt{\frac{1}{4}}$	$\sqrt{\frac{3}{4}}$					
$+\frac{1}{2}$	$+\frac{1}{2}$			$\sqrt{\frac{3}{4}}$	$-\sqrt{\frac{1}{4}}$					
$+\frac{1}{2}$	$-\frac{1}{2}$					$\sqrt{\frac{1}{2}}$	$\sqrt{\frac{1}{2}}$			
$-\frac{1}{2}$	$+\frac{1}{2}$					$\sqrt{\frac{1}{2}}$	$-\sqrt{\frac{1}{2}}$			
$-\frac{1}{2}$	$-\frac{1}{2}$							$\sqrt{\frac{3}{4}}$	$\sqrt{\frac{1}{4}}$	
$-\frac{3}{2}$	$+\frac{1}{2}$							$\sqrt{\frac{1}{4}}$	$-\sqrt{\frac{3}{4}}$	
$-\frac{3}{2}$	$-\frac{1}{2}$									1

$2 \times \frac{1}{2}$

m_1	m_2	J → M →	$\frac{5}{2}$ $+\frac{5}{2}$	$\frac{5}{2}$ $+\frac{3}{2}$	$\frac{3}{2}$ $+\frac{3}{2}$	$\frac{5}{2}$ $+\frac{1}{2}$	$\frac{3}{2}$ $+\frac{1}{2}$	$\frac{5}{2}$ $-\frac{1}{2}$	$\frac{3}{2}$ $-\frac{1}{2}$	$\frac{5}{2}$ $-\frac{3}{2}$	$\frac{3}{2}$ $-\frac{3}{2}$	$\frac{5}{2}$ $-\frac{5}{2}$
$+2$	$\frac{1}{2}$		1									
$+2$	$-\frac{1}{2}$			$\sqrt{\frac{1}{5}}$	$\sqrt{\frac{4}{5}}$							
$+1$	$+\frac{1}{2}$			$\sqrt{\frac{4}{5}}$	$-\sqrt{\frac{1}{5}}$							
$+1$	$-\frac{1}{2}$					$\sqrt{\frac{2}{5}}$	$\sqrt{\frac{3}{5}}$					
0	$+\frac{1}{2}$					$\sqrt{\frac{3}{5}}$	$-\sqrt{\frac{2}{5}}$					
0	$-\frac{1}{2}$							$\sqrt{\frac{3}{5}}$	$\sqrt{\frac{2}{5}}$			
-1	$+\frac{1}{2}$							$\sqrt{\frac{2}{5}}$	$-\sqrt{\frac{3}{5}}$			
-1	$-\frac{1}{2}$									$\sqrt{\frac{4}{5}}$	$\sqrt{\frac{1}{5}}$	
-2	$+\frac{1}{2}$									$\sqrt{\frac{1}{5}}$	$-\sqrt{\frac{4}{5}}$	
-2	$-\frac{1}{2}$											1

CLEBSCH-GORDAN COEFFICIENTS (CONT'D)

1×1

m_1	m_2	J → M →	2, +2	2, +1	1, +1	2, 0	1, 0	0, 0	2, −1	1, −1	2, −2
+1	+1		1								
+1	0			$\sqrt{\frac{1}{2}}$	$\sqrt{\frac{1}{2}}$						
0	+1			$\sqrt{\frac{1}{2}}$	$-\sqrt{\frac{1}{2}}$						
+1	−1					$\sqrt{\frac{1}{6}}$	$\sqrt{\frac{1}{2}}$	$\sqrt{\frac{1}{3}}$			
0	0					$\sqrt{\frac{2}{3}}$	0	$-\sqrt{\frac{1}{3}}$			
−1	+1					$\sqrt{\frac{1}{6}}$	$-\sqrt{\frac{1}{2}}$	$\sqrt{\frac{1}{3}}$			
0	−1								$\sqrt{\frac{1}{2}}$	$\sqrt{\frac{1}{2}}$	
−1	0								$\sqrt{\frac{1}{2}}$	$-\sqrt{\frac{1}{2}}$	
−1	−1										1

$\frac{3}{2} \times 1$

m_1	m_2	J → M →	$\frac{5}{2}$, $+\frac{5}{2}$	$\frac{5}{2}$, $+\frac{3}{2}$	$\frac{3}{2}$, $+\frac{3}{2}$	$\frac{5}{2}$, $+\frac{1}{2}$	$\frac{3}{2}$, $+\frac{1}{2}$	$\frac{1}{2}$, $+\frac{1}{2}$	$\frac{5}{2}$, $-\frac{1}{2}$	$\frac{3}{2}$, $-\frac{1}{2}$	$\frac{1}{2}$, $-\frac{1}{2}$	$\frac{5}{2}$, $-\frac{3}{2}$	$\frac{3}{2}$, $-\frac{3}{2}$	$\frac{5}{2}$, $-\frac{5}{2}$
$+\frac{3}{2}$	+1		1											
$+\frac{3}{2}$	0			$\sqrt{\frac{2}{5}}$	$\sqrt{\frac{3}{5}}$									
$+\frac{1}{2}$	+1			$\sqrt{\frac{3}{5}}$	$-\sqrt{\frac{2}{5}}$									
$+\frac{3}{2}$	−1					$\sqrt{\frac{1}{10}}$	$\sqrt{\frac{2}{5}}$	$\sqrt{\frac{1}{2}}$						
$+\frac{1}{2}$	0					$\sqrt{\frac{3}{5}}$	$\sqrt{\frac{1}{15}}$	$-\sqrt{\frac{1}{3}}$						
$-\frac{1}{2}$	+1					$\sqrt{\frac{3}{10}}$	$-\sqrt{\frac{8}{15}}$	$\sqrt{\frac{1}{6}}$						
$+\frac{1}{2}$	−1								$\sqrt{\frac{3}{10}}$	$\sqrt{\frac{8}{15}}$	$\sqrt{\frac{1}{6}}$			
$-\frac{1}{2}$	0								$\sqrt{\frac{3}{5}}$	$-\sqrt{\frac{1}{15}}$	$-\sqrt{\frac{1}{3}}$			
$-\frac{3}{2}$	+1								$\sqrt{\frac{1}{10}}$	$-\sqrt{\frac{2}{5}}$	$\sqrt{\frac{1}{2}}$			
$-\frac{1}{2}$	−1											$\sqrt{\frac{3}{5}}$	$\sqrt{\frac{2}{5}}$	
$-\frac{3}{2}$	+0											$\sqrt{\frac{2}{5}}$	$-\sqrt{\frac{3}{5}}$	
$-\frac{3}{2}$	−1													1

2×1

m_1	m_2	J = 3, M = +3	3, +2	2, +2	3, +1	2, +1	1, +1	3, 0	2, 0	1, 0	3, −1	2, −1	1, −1	3, −2	2, −2	3, −3
+2	+1	1														
+2	0		$\sqrt{\frac{1}{3}}$	$\sqrt{\frac{2}{3}}$												
+1	+1		$\sqrt{\frac{2}{3}}$	$-\sqrt{\frac{1}{3}}$												
+2	−1				$\sqrt{\frac{1}{15}}$	$\sqrt{\frac{1}{3}}$	$\sqrt{\frac{3}{5}}$									
+1	0				$\sqrt{\frac{8}{15}}$	$\sqrt{\frac{1}{6}}$	$-\sqrt{\frac{3}{10}}$									
0	+1				$\sqrt{\frac{6}{15}}$	$-\sqrt{\frac{1}{2}}$	$\sqrt{\frac{1}{10}}$									
+1	−1							$\sqrt{\frac{1}{5}}$	$\sqrt{\frac{1}{2}}$	$\sqrt{\frac{3}{10}}$						
0	0							$\sqrt{\frac{3}{5}}$	0	$-\sqrt{\frac{2}{5}}$						
−1	+1							$\sqrt{\frac{1}{5}}$	$-\sqrt{\frac{1}{2}}$	$\sqrt{\frac{3}{10}}$						
0	−1										$\sqrt{\frac{6}{15}}$	$\sqrt{\frac{1}{2}}$	$\sqrt{\frac{1}{10}}$			
−1	0										$\sqrt{\frac{8}{15}}$	$-\sqrt{\frac{1}{6}}$	$-\sqrt{\frac{3}{10}}$			
−2	+1										$\sqrt{\frac{1}{15}}$	$-\sqrt{\frac{1}{3}}$	$\sqrt{\frac{3}{5}}$			
−1	−1													$\sqrt{\frac{2}{3}}$	$\sqrt{\frac{1}{3}}$	
−2	0													$\sqrt{\frac{1}{3}}$	$-\sqrt{\frac{2}{3}}$	
−2	−1															1

Answers to Problems

CHAPTER 1

1.1 $\Lambda = 98$ cm, $G \sim 10^{-7}$

1.2 $\Lambda = 213$ cm

1.3 $g/G \sim 10^{-6}$

1.4 $E_\gamma = \frac{1}{2}E_\pi(1 \pm \beta)$

1.5   (i)   Allowed

(ii) Forbidden for free protons, allowed for nuclei when p, n binding energy difference is sufficient

(iii) Forbidden (muon number conservation)

(iv) Forbidden (strangeness conservation)

1.6 $|\Delta e/e| > (KM^2/e^2)^{1/2} = 10^{-18}$

CHAPTER 2

2.1   (i) 0.2   (ii) 0.07   (iii) 0.02

2.2 $y_{\text{rms}} = \dfrac{21}{p\beta\sqrt{2}} \times \dfrac{S^{3/2}}{\sqrt{3}}$ radiation lengths

(i) 4.9 cm   (ii) 1.65 cm

2.3   1.6 km; 4.95 m

2.6 $E = 7\,mc^2$

2.7   (i) $1 - \dfrac{p_f}{M}$ (ii) $1 + \dfrac{p_f}{M}$ (iii) 1

2.9   106 MeV; 6×10^{-3}

CHAPTER 3

3.2   As for $\pi^- p \rightarrow n\pi^0, \pi^- p \rightarrow p\pi^-, \pi^+ p \rightarrow \pi^+ p$

3.3   1 : 2

3.4 $2:1$

3.5 (a) No. $I = 0$ or 2 only.
  (b) Yes
  (c) No. $I \geqslant 2$, since $I_3 = 2$.
  (d) No. $I = 0$ or 2 only.
  (e) Yes

3.6 (a) $I = 0, 1, 2,$ or 3
  (b) $I = 1$ or 3

3.7 $I = 0$ or 1. $\sigma_{(i)}/\sigma_{(ii)} = 1$ if $I = 0$ only
  $\phantom{I = 0 \text{ or } 1. \sigma_{(i)}/\sigma_{(ii)}} = 0$ if $I = 1$ only.

3.8 (a) $+1$; $I = 0$; $^3P_0, {}^3P_2, {}^3F_0, {}^3F_2 \cdots$
  $ I = 1$; $^3S_1$; $^3D_1, {}^3D_3 \cdots$.
  (b) $+1$; $I = 0$; $^3P_0, {}^3P_2, {}^3F_0, {}^3F_2 \cdots$.
  (c) -1; $I = 1$; $^1S_0, {}^1D_2 \cdots {}^3P_1, {}^3P_2 \cdots$.
  $p\bar{p} \leftrightarrow 2\pi^0$ implies annihilation from S-states only.

3.9 Ratio $= 1$

3.10 ρ-ω interference occurs in $\pi^+\pi^-$ mode, with amplitude of order α, typical of G-violating electromagnetic interactions. A narrow dip (or peak) will therefore occur in the $\pi^+\pi^-$ mass spectrum in the ω-region.

3.12 (a) J even   (b) none (parity is not conserved in weak decay)

CHAPTER 4

4.4 $\sim 10^{-4}$ (experiment 0.7×10^{-4})

4.5 (i) $2:1$ (ii) $2:1$

4.6 (i) $1:2$ (ii) $2:1$

4.8 4×10^{-15}

4.10 0.17/day. No difference.

CHAPTER 5

5.1 $90/h$

5.4 $\tau_{\pi_0} \sim 4 \times 10^{-16}$ sec

5.5 $M^* = 2.1$ GeV

5.7 (i) 1.1×10^5 (ii) 24

CHAPTER 6

6.2 1.8×10^{-10} sec on basis of phase-space

6.3 $\mu_{\Omega^-} = -\mu_p$; $\mu_{N^{*++}} = 2\mu_p$

6.5 $-\frac{1}{2}$; $+\sqrt{\frac{3}{2}}$; $-\frac{1}{2}$; $+\frac{1}{2}$

6.6 (i) and (ii) forbidden, (iii) and (iv) allowed

6.7 $1:1/\sqrt{6}:1/\sqrt{6}:1/\sqrt{2}$

6.8 $\sigma_{pp} \sim \frac{3}{2}\sigma_{\pi p}$

CHAPTER 7

7.1 $\Gamma_{+00}/\Gamma_{++-} = 1.26/4 = 0.315.$ Experiment $= 0.31 \pm 0.01.$

7.3 Production reaction $K^- + p \rightarrow Y_0^* + \pi^0$; Kinetic energy at threshold $= 308$ MeV.
  $Y_0^* \rightarrow \Sigma^+ + \pi^-$ and $\Sigma^- + \pi^+$ $(Q = 75$ MeV$)$
  $\rightarrow \Lambda + \pi^+ + \pi^-$ allowed but $Q = 10$ MeV only
  $\rightarrow K^- + p$ forbidden by energy conservation
  $\rightarrow \Lambda + \pi^0$ forbidden by isospin conservation.

7.4 $2\sqrt{E_1^2 - m^2}\sqrt{E_2^2 - m^2} = \pm(M^2 + 2E_1 E_2 - 2M(E_1 + E_2) + m^2)$

7.7 $61°$, 1.06 GeV/c; $0°$, 9.25 GeV/c; 16 GeV^2

7.8 (i) $G = -1$
  (ii) $J^P = 0^-, 1^+, 2^- \cdots$
  (iii) e^{-8t} as in diffraction scattering. A1 meson is example.

7.10 $\left(\dfrac{p}{p_0}\right)^{2l+1} \times \left(\dfrac{E_0}{E}\right)$

Worked Solutions to Selected Problems

2.3

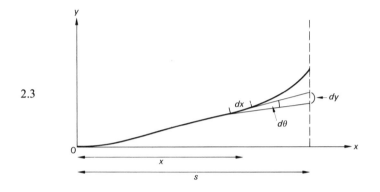

The mean square lateral deflection, projected in the xy-plane, due to scattering in element dx radiation lengths is

$$dy^2 = (s - x)^2 \, d\theta^2 = \frac{1}{2}\left(\frac{21}{p\beta c}\right)^2 (s - x)^2 \, dx; \qquad p\beta c \text{ in MeV}.$$

Integrated over the whole track length s, assuming $p\beta c = $ constant, the rms deflection is

$$y_{rms} = \left(\frac{21}{\sqrt{2}\, p\beta c}\right)\frac{s^{3/2}}{\sqrt{3}}.$$

In the question, $p\beta c$ is not constant. Approximating $\beta = 1$, $pc = E$, we may write, in obvious notation,

$$E(x) = E_0 - \alpha x = \alpha(s - x),$$

where $\alpha = dE/dx = $ constant. Then

$$dy^2 = \frac{1}{2}\frac{(21)^2 \, dx(s - x)^2}{\alpha^2(s - x)^2},$$

or

$$y_{rms} = \left(\frac{21}{\sqrt{2}\, E_0}\right)s^{3/2}.$$

The radial spread is

$$r_{rms} = \sqrt{2}\, y_{rms} = \frac{21}{E_0}s^{3/2}.$$

The mean range of the muons will be

$$R = E_0/(dE/dx) = 0.5 \times 10^6 \, g = \underline{1.6 \, \text{km rock}}.$$

The range measured in radiation lengths is

$$s = R/x_0 = 2 \times 10^4.$$

Then

$$r_{rms} = 59.5 \text{ radiation lengths}$$
$$= 4.95 \, \text{m}.$$

2.6   The threshold energy is obtained by requiring that the available kinetic energy in the CMS be $2m$, in order to create a pair. This means the total CMS energy must be $4m$, corresponding to three electrons and one positron mutually at rest in the CMS frame. If E, p are the energy and momentum of the incident electron in the laboratory frame, the total (four-momentum)^2 of the colliding particles is

$$(E + m)^2 - (p + 0)^2 = 2m^2 + 2mE.$$

This must be equal to the (total energy)^2 in the CMS frame, where the total momentum is zero. Therefore

$$(4m)^2 = 2m^2 + 2mE,$$

or

$$E^{\text{threshold}} = 7m.$$

2.9   Equation (2.2) for the maximum transferable energy is

$$E'_{max} = = 2m\beta^2\gamma^2c^2,$$

where mc^2 is the electron rest energy, and γ is the Lorentz factor of the primary particle. To identify a particle as a pion, it is necessary that $E' > (E'_{max})_K$. For kaons of $5 \, \text{GeV}/c$, $\gamma = 10.2$, $\beta = 1$, so that $(E'_{max})_K = 104 \times 1.02 = \underline{106 \, \text{MeV}}$. For a pion, $(E'_{max})_\pi = 1320 \, \text{MeV}$.
    The probability of observing such a δ-ray in liquid hydrogen is, from (2.1), with $\beta \approx 1$,

$$P(>E') = 2\pi\left(\frac{e^2}{mc^2}\right)^2 \times mc^2 \times N_0 \frac{1}{E'}\left[1 - \frac{E'}{E'_{max}}\left(1 + \beta^2 \log \frac{E'_{max}}{E'}\right)\right]$$

per g/cm^2 traversed. Inserting the numerical values, and with $E' = 106 \, \text{MeV}$, $E'_{max} = 1320 \, \text{MeV}$, one finds

$$P(>106 \, \text{MeV}) = 1.04 \times 10^{-3} \, g^{-1} \, cm^2.$$

Therefore in 1 m path, the probability is 6.2×10^{-3}.

3.13   The $p\bar{p}$ system has C-parity $(-1)^{l+S}$, where l is the relative orbital angular momentum, and S is the total spin (0 or 1).
    The space-parity of the $p\bar{p}$ system is $P = (-1)^{l+1}$, since particle and antiparticle have opposite intrinsic parity. Thus, the initial state has

$$(CP)_{p\bar{p}} = (-1)^{2l+S+1} = (-1)^{S+1} \qquad \text{for all } l.$$

Now let the total angular momentum of the two K^0's be J, where

$$|l + S| \geqslant J \geqslant |l - S|.$$

Measured in the K^0 rest frame, the K_1^0 and K_2^0 mesons have CP eigenvalues $+1$ and -1 respectively. Thus, the final state has

$$CP = (\pm 1)(\pm 1)(-1)^J = (-1)^J \qquad \text{for} \qquad 2K_1^0 \quad \text{or} \quad 2K_2^0,$$
$$CP = (+1)(-1)(-1)^J = (-1)^{J+1} \qquad \text{for} \qquad K_1^0 K_2^0,$$

$\underline{l = 0}$. For annihilation from an S-state, $J = S$, so that the initial state has

$$(CP)_{p\bar{p}} = (-1)^{J+1} \qquad J = 0, 1.$$

Thus, the $K_1^0 K_2^0$ final state is allowed, and $2K_1^0$ or $2K_2^0$ is forbidden.
$\underline{l = 1}$. For the triplet state $(S = 1)$, $J = 0$, 1, or 2. $(CP)_{p\bar{p}}$ is even, so that $J = 0, 2$ allows $2K_1^0$ or $2K_2^0$ in the final state, while $J = 1$ gives $K_1^0 K_2^0$ only. For the singlet state $(S = 0)$, $J = 1$, $(CP)_{p\bar{p}}$ is odd, and only the states $2K_1^0$ or $2K_2^0$ are allowed.

Experimentally, it is found that for annihilations at rest, only $K_1^0 K_2^0$ is observed, proving that annihilation takes place from an $l = 0$ state. On the contrary, annihilation of antiprotons in flight takes place from p-states also, and the modes $K_1^0 K_2^0$, $2K_1^0$, and $2K_2^0$ all appear.

4.10 The reaction rate is given by

$$R = \sigma \phi N,$$

where σ is the cross section per nucleus for neutrino absorption, N is the total number of nuclei in the detector, and ϕ is the neutrino flux $\text{sec}^{-1} \text{cm}^{-2}$.

We take 164 as the molecular weight of C_2Cl_4, and the total mass of liquid as 6×10^8 g; the number of Cl^{37} nuclei is $N = 2.2 \times 10^{30}$. The solar heat flux is $2 \text{ cal cm}^{-2} \text{ min}^{-1}$, or $8.8 \times 10^{11} \text{ MeV cm}^{-2} \text{ sec}^{-1}$. Ten per cent appears as neutrinos, of mean energy 1 MeV. One per cent of the neutrinos are supposed sufficiently energetic to produce a reaction, so that $\phi = 8.8 \times 10^8 \text{ cm}^{-2} \text{ sec}^{-1}$. Thus

$$R = \sigma \phi N = (10^{-45})(8.8 \times 10^8)(2.2 \times 10^{30}) = 1.9 \times 10^{-6} \text{ sec}^{-1}$$
$$= 0.17 \text{ day}^{-1}.$$

4.11 We apply the $\Delta I = 1$ rule by combining the baryon $(I = \tfrac{1}{2})$ with a "spurion" of $I = 1$, to give a final hadronic state of $I = \tfrac{3}{2}$ and $I_3 = \tfrac{3}{2}$ or $\tfrac{1}{2}$.

Referring to Table III of Clebsch-Gordan coefficients, and using the $I = 1$ and $I = \tfrac{1}{2}$ combination, we may write for reaction (i):

$$\phi(1, 1)\phi(\tfrac{1}{2}, \tfrac{1}{2}) = \psi(\tfrac{3}{2}, \tfrac{3}{2}),$$

$$\uparrow \qquad \uparrow$$

$$\text{spurion} \quad \text{nucleon}$$

and for (ii):

$$\phi(1, 1)\phi(\tfrac{1}{2}, -\tfrac{1}{2}) = \frac{1}{\sqrt{3}} \psi(\tfrac{3}{2}, \tfrac{1}{2}) + \sqrt{\tfrac{2}{3}} \psi(\tfrac{1}{2}, \tfrac{1}{2}).$$

If the pion-nucleon system is in a pure $I = \frac{3}{2}$ state, the cross section ratio obtained by squaring the above amplitudes is $\sigma_{(i)}/\sigma_{(ii)} = \frac{3}{1}$. For a $\Delta I = 2$ transition, we use $I = 2$ and $I = \frac{1}{2}$ entry of the table, and find for reaction (i)

$$\phi(2, 1)\phi(\tfrac{1}{2}, \tfrac{1}{2}) = -\sqrt{\tfrac{1}{5}}\,\psi(\tfrac{3}{2}, \tfrac{3}{2}) + \sqrt{\tfrac{4}{5}}\,\psi(\tfrac{5}{2}, \tfrac{3}{2}),$$

and for (ii)

$$\phi(2, 1)\phi(\tfrac{1}{2}, -\tfrac{1}{2}) = \sqrt{\tfrac{3}{5}}\,\psi(\tfrac{3}{2}, \tfrac{1}{2}) + \sqrt{\tfrac{2}{5}}\,\psi(\tfrac{5}{2}, \tfrac{1}{2}).$$

For a final state of $I = \frac{3}{2}$ only, the ratio is then $\sigma_{(i)}/\sigma_{(ii)} = \frac{1}{3}$.

4.12 We assume that the three pions are in a relative S-state. Then, by Bose symmetry, any pair must be in a symmetric isospin state, i.e. $I = 0$ or $I = 2$. Call A and B the amplitudes for an $I = 2$ and $I = 0$ dipion state, respectively. From the $\Delta I = \frac{1}{2}$ rule, we add a "spurion" of $\Delta I = \frac{1}{2}$ to the kaon, of $\Delta I = \frac{1}{2}$, to form states of $I = 0$ or 1. $I = 0$ is forbidden for the three-pion state, which is formed from a dipion of $I = 0$ or 2 and a third pion of $I = 1$. So, we consider a three-pion state of $I = 1$, obtained by adding together $I = 1$ with $I = 0$ or 2. Referring to Table III of coefficients we find, in self-evident notation,

Charged kaon

$$\psi(1, 1) = A[\sqrt{\tfrac{3}{5}}\,\phi(2, 2)\phi(1, -1) - \sqrt{\tfrac{3}{10}}\,\phi(2, 1)\phi(1, 0) + \sqrt{\tfrac{1}{10}}\,\phi(2, 0)\phi(1, 1)]$$
$$+ B[\phi(0, 0)\phi(1, 1)], \tag{a}$$

Neutral kaon

$$\psi(1, 0) = A[\sqrt{\tfrac{3}{10}}\,\phi(2, 1)\phi(1, -1) - \sqrt{\tfrac{2}{5}}\,\phi(2, 0)\phi(1, 0) + \sqrt{\tfrac{3}{10}}\,\phi(2, -1)\phi(1, 1)]$$
$$+ B[\phi(0, 0)\phi(1, 0)]. \tag{b}$$

The next step is to express the various pion combinations in terms of the isospin functions ϕ. The three-pion wave function must be completely symmetric under pion label interchange, as required for identical bosons. So we write the $\pi^+\pi^+\pi^-$ combination as

$$(+ + -) = \frac{1}{\sqrt{6}}(\pi_1^+\pi_2^+\pi_3^- + \pi_2^+\pi_1^+\pi_3^- + \pi_3^-\pi_2^+\pi_1^+$$
$$+ \pi_3^-\pi_1^+\pi_2^+ + \pi_2^+\pi_3^-\pi_1^+ + \pi_1^+\pi_3^-\pi_2^+), \tag{c}$$

the $\sqrt{6}$ factor being to normalize the amplitude to unity. Referring to the 1×1 entry in Table III, treating the first two pions as the "pair" gives

$$(\pi^+\pi^+\pi^-) = \sqrt{\tfrac{3}{5}}\,A; \qquad (\pi^+\pi^-\pi^+) = (\pi^-\pi^+\pi^+) = \sqrt{\tfrac{1}{60}}\,A + \sqrt{\tfrac{1}{3}}\,B.$$

The second result, for example, follows from the fact that the coefficient for combining $I = 1$, $I_3 = +1$ and $I = 1$, $I_3 = -1$ to give $I = 2$, $I_3 = 0$ is $\sqrt{\tfrac{1}{6}}$; and to give $I = 0$, $I_3 = 0$ the coefficient is $\sqrt{\tfrac{1}{3}}$. These factors are then multiplied into the appropriate terms in (a), in order to find $\langle\psi(1, 1) | \pi^+\pi^+\pi^-\rangle$, etc. Adding together all terms in (c) gives us

$$\langle\psi(1, 1) | + + -\rangle = 2\sqrt{\tfrac{2}{3}}\,C \qquad \text{where} \qquad C = \sqrt{\tfrac{4}{15}}\,A + \sqrt{\tfrac{1}{3}}\,B.$$

Similarly, one finds for the other charge combinations

$$\langle \psi(1, 1) \, | \, +00 \rangle = -\sqrt{\tfrac{2}{3}} \, C,$$
$$\langle \psi(1, 0) \, | \, 000 \rangle = -C,$$
$$\langle \psi(1, 0) \, | \, +-0 \rangle = \sqrt{\tfrac{2}{3}} \, C.$$

Squaring these amplitudes, we obtain the ratios of the decay rates

$$\Gamma(K_L \to 3\pi^0) = C^2 = \tfrac{3}{2}\Gamma(K_L \to \pi^+\pi^-\pi^0),$$
$$\Gamma(K^+ \to \pi^+\pi^+\pi^-) = \tfrac{8}{3}C^2 = 4\Gamma(K^+ \to \pi^+\pi^0\pi^0).$$

In order to compare the neutral $(K_L \to \pi^+\pi^-\pi^0)$ with the charged $(K^+ \to \pi^+\pi^0\pi^0)$ decay rate, we need a result from Chapter 4 [Eq. (4.56)]. We have actually calculated the transition for $K^0 \to \pi^+\pi^-\pi^0$. Actually the weak conservation laws only allow half the K^0's to decay in this mode, called K_2^0 or K_L, which has CP eigenvalue -1. The other half is called K_1^0 or K_S, has $CP = +1$, and does not decay to three pions. Thus

$$\langle K^0 | \, T \, | \pi^+\pi^-\pi^0 \rangle = \frac{1}{\sqrt{2}} \langle K_L | \, T \, | \pi^+\pi^-\pi^0 \rangle.$$

Using this result, one obtains

$$\Gamma(K_L \to \pi^+\pi^-\pi^0) = 2\Gamma(K_0 \to \pi^+\pi^-\pi^0) = 2\Gamma(K^+ \to \pi^+\pi^0\pi^0).$$

Experimentally, there is a small deviation from this prediction, indicating that $\Delta I = \tfrac{3}{2}$ as well as $\Delta I = \tfrac{1}{2}$ transitions are involved.

For a more complete discussion of the $K \to 3\pi$ decay modes, the reader is referred to G. Källén, *Elementary Particle Physics*, Addison-Wesley, 1964, Ch. 16; and for a more general treatment of three-pion decays, to the classic paper by C. Zemach, *Phys. Rev.* **133**, B1201 (1964).

5.1   A current of 10 mA of relativistic particles in a ring of radius 10 m corresponds to a circulating charge $q = (2\pi r/c)i$, or, inserting appropriate numbers, $N = 1.3 \times 10^{10}$ circulating electrons or positrons. If the cross-sectional area of the beam is A, the particle density transverse to the beam will be N/A. The reaction rate will therefore be

$$R = \sigma \left(\frac{N}{A} \right)^2 \times Afn,$$

where f is the revolution frequency, and the bunches meet n times per revolution. With $n = 2, f = c/2\pi r, A = 0.1 \, \text{cm}^2$, and $\sigma = 1.5 \times 10^{-30} \, \text{cm}^2$, this formula gives $R \approx 90/\text{h}$.

6.7   The decays (i)–(iv) all have initial states of $U = \tfrac{3}{2}$, so that only $U = 1$ meson states can contribute, if U-spin is to be conserved. The π^0 and η are not pure U-spin states, however; in fact from (6.25)

$$|\eta\rangle = \frac{\sqrt{3}}{2} \, \phi(1, 0) - \tfrac{1}{2}\phi(0, 0),$$

$$|\pi^0\rangle = \frac{\sqrt{3}}{2} \, \phi(0, 0) + \tfrac{1}{2}\phi(1, 0),$$

while

$$|K^0\rangle = \phi(1, 1),$$

where the ϕ's denote U-spin wave functions. The initial states are

$$|N^{*-}\rangle = \phi(\tfrac{3}{2}, \tfrac{3}{2}) = \phi(\tfrac{1}{2}, \tfrac{1}{2})\phi(1, 1),$$

$$|Y_1^{*-}\rangle = \phi(\tfrac{3}{2}, \tfrac{1}{2}) = \sqrt{\tfrac{2}{3}}\phi(\tfrac{1}{2}, \tfrac{1}{2})\phi(1, 0) + \sqrt{\tfrac{1}{3}}\phi(1, 1)\phi(\tfrac{1}{2}, -\tfrac{1}{2}),$$

$$|\Xi^{*-}\rangle = \phi(\tfrac{3}{2}, -\tfrac{1}{2}) = \sqrt{\tfrac{2}{3}}\phi(\tfrac{1}{2}, -\tfrac{1}{2})\phi(1, 0) + \sqrt{\tfrac{1}{3}}\phi(1, -1)\phi(\tfrac{1}{2}, \tfrac{1}{2}),$$

from Table III.

We then obtain, for example,

$$\langle Y_1^{*-} | T | \Sigma^- \pi^0 \rangle$$

$$= A\left\langle \sqrt{\tfrac{2}{3}}\phi(\tfrac{1}{2}, \tfrac{1}{2})\phi(1, 0) + \sqrt{\tfrac{1}{3}}\phi(1, 1)\phi(\tfrac{1}{2}, -\tfrac{1}{2}) \,\middle|\, \phi(\tfrac{1}{2}, \tfrac{1}{2})\left[\tfrac{1}{2}\phi(1, 0) + \frac{\sqrt{3}}{2}\phi(0, 0)\right]\right\rangle$$

$$= A/\sqrt{6},$$

where A is some constant. Similarly, it may be verified that the amplitudes for (i), (iii), and (iv) are A, $A/\sqrt{6}$, and $A/\sqrt{2}$ respectively.

7.2   If we let E, T represent total and kinetic energies of the pions respectively, and Q the total kinetic energy available in the decay, it is readily verified that the conditions are

$$(E_1 E_2 E_3) = \text{maximum} \qquad \text{when } T_1 = T_2 = T_3 = Q/3,$$

$$(E_1 E_2 E_3) = \text{minimum} \qquad \text{when } T_1 = 0, T_2 = T_3 = Q/2.$$

With $x = Q/m$, where m is the pion mass, then

$$(E_1 E_2 E_3)_{\text{max}} = m^3\left[1 + x + \frac{x^2}{3} + \frac{x^3}{27}\right],$$

$$(E_1 E_2 E_3)_{\text{min}} = m^3\left[1 + x + \frac{x^2}{4}\right],$$

so that

$$\varepsilon = \frac{(E_1 E_2 E_3)_{\text{max}} - (E_1 E_2 E_3)_{\text{min}}}{(E_1 E_2 E_3)_{\text{max}}} \approx \frac{x^2}{12(1 + x)}, \qquad \text{where } x \sim \frac{75}{140} \sim \tfrac{1}{2}.$$

Thus $\varepsilon \simeq 0.014$.

7.9   (a) Making a smoothed background subtraction, one can very roughly estimate 18 mb for the $\pi^- p$ cross section from the N^* (1688) state at the peak. This gives $\tfrac{3}{2} \times 18 = 27$ mb for the $I = \tfrac{1}{2}$ cross section. The elastic cross section at the peak is [see Eq. (7.32)]:

$$\sigma_e = 4\pi\lambda^2(J + \tfrac{1}{2})\left(\frac{\Gamma_e}{\Gamma}\right)^2 = 4\pi\lambda^2(J + \tfrac{1}{2})x^2 \qquad \text{where } x = \Gamma_e/\Gamma.$$

The inelastic cross section will be

$$\sigma_r = 4\pi\lambda^2(J + \tfrac{1}{2})\frac{\Gamma_e \Gamma_r}{\Gamma^2} = 4\pi\lambda^2(J + \tfrac{1}{2})x(1 - x),$$

and the total cross section will be

$$\sigma_T = \sigma_e + \sigma_r = 4\pi\lambda^2(J + \tfrac{1}{2}) . x.$$

The next step is to compute the CMS momentum, p, for a pion-nucleon state of invariant mass $M^* = 1688$ MeV; this may be most simply obtained by first deriving the relation for the total energy of the pion in the CMS. One finds

$$E = (M^{*2} - M^2 + m^2)/2M^*,$$

where M, m are the proton and pion masses, and thus $p = 572$ MeV/c, whence $4\pi\lambda^2 = 15$ mb. With $J = \frac{5}{2}$, the above formula gives $x = 0.6$. The agreement with the value in the table in the preface is probably fortuitous.

(b) A better method is to obtain x from examination of the elastic scattering amplitude $f_{el.}$ in the F_{15} wave, as shown in Fig. 7.14(b), computed from a phase shift analysis. For an inelastic resonance, $f_{el.}$ in (7.30) has the form

$$f_{el.}(E) = \frac{\frac{1}{2}\Gamma x}{[(E_R - E) - i\Gamma/2]}.$$

It is seen that, at resonance, $f_{el.} = ix$ and from the figure, $x = 0.6$. Note that x is fairly constant as one goes to either side of resonance.

7.10 The Lorentz-invariant phase-space factor (Appendix D) for a two-body final state is proportional to

$$p^2 \frac{dp}{dE} \times \frac{1}{E_1 E_2},$$

where E_1 and E_2 are the total energies of the particles and p their momentum, in the CMS. The total energy $E = E_1 + E_2$. Writing

$$E = \sqrt{(p^2 + m_1^2)} + \sqrt{(p^2 + m_2^2)},$$

it is found that

$$\frac{dp}{dE} = \frac{E_1 E_2}{p(E_1 + E_2)}.$$

Thus the factor

$$\frac{p^2}{E_1 E_2}\frac{dp}{dE} = \frac{p}{E}$$

enters into the width Γ.

The treatment here refers to an S-wave resonance. For one decaying to two (spinless) particles of orbital angular momentum l, it is necessary also to include a centrifugal barrier factor, familiar from reaction theory in nuclear physics (see, for example, J. M. Blatt and V. F. Weisskopf, *Theoretical Nuclear Physics*, John Wiley, 1952, 320 *et seq.* and Appendix A). This factor is associated with the behavior of partial wave amplitudes near the origin (in the interaction region) which have a radial dependence of approximately $(kr)^l$ when $kr < l$. The width Γ then includes a factor $(pR_0)^{2l}$ where R_0 is some (unknown) range parameter (of order 1 F). Then it is plausible to write

$$\frac{\Gamma(E)}{\Gamma(E_0)} = \left(\frac{p}{p_0}\right)^{2l+1} \times \left(\frac{E_0}{E}\right),$$

where E_0 and p_0 refer to values at the resonance peak. For decay into particles with spin, the formula is more complicated [see, for example, J. D. Jackson, *Il Nuovo Cimento* **34**, 1644 (1964)].

References

Abarbanel, H., S. Drell, and F. Gilman, *Phys. Rev. Lett.* **20,** 280 (1968).

Abashian, A., R. J. Abrams, D. W. Carpenter, G. P. Fisher, B. M. Nefkens, and J. H. Smith, *Phys. Rev. Lett.* **13,** 243 (1964).

Abov, Y. G., P. Krupchitsky, and Y. Oratovsky, *Phys. Lett.* **12,** 25 (1964).

Alff, C., D. Berley, D. Colley, N. Gelfand, U. Nauenberg, D. Miller, J. Schultz, J. Steinberger, T. Tau, H. Brugger, P. Kramer, and R. Plano, *Phys. Rev. Lett.* **9,** 325 (1962).

Alff-Steinberger, C., W. Heuer, K. Kleinknecht, C. Rubbia, A. Scribano, J. Steinberger, M. Tannenbaum, and K. Kittel, *Phys. Lett.* **20,** 207 (1966).

Allaby, J., *et al., Phys. Lett.* **30B,** 500 (1969).

Alston, M., L. Alvarez, P. Eberhard, M. Good, W. Graziano, H. Ticho, and S. Wojcicki, *Phys. Rev. Lett.* **5,** 520 (1960).

Anderson, C. D., *Phys. Rev.* **43,** 491 (1933).

Anderson, C. D., and S. Neddermeyer, *Phys. Rev.* **51,** 884 (1937); **54,** 88 (1938).

Anderson, H., T. Fujii, R. Miller, and L. Tau, *Phys. Rev.* **119,** 2050 (1960).

Augustin, J., J. Bizot, J. Buon, J. Haissinski, D. Labanne, P. Marin, H. Nguyen Ngoc, J. Perez-y-Jorba, F. Rumpf, E. Silva, and S. Tavernier, *Phys. Lett.* **28B,** 508 (1969).

Bailey, J., W. Bartl, G. van Bochmann, R. Brown, F. J. Farley, H. Jostlein, E. Picasso, and R. Williams, *Phys. Lett.* **28B,** 287 (1968).

Barnes, V., *et al., Phys. Rev. Lett.* **12,** 204 (1964).

Bartel, W., B. Dudelzak, H. Krehbiel, J. McElroy, U. Meyer-Berkheut, W. Schmidt, V. Walther, and G. Weber, *Phys. Lett.* **28B,** 148 (1968).

Baton, J., Berthelot, B. Deler, O. Goussu, M. Neveu-Rene, A. Rogozinski, F. Shively, V. Alles-Borelli, E. Benedetti, R. Gasseroli, and P. Waloschek, *Nuov. Cim.* **35,** 713 (1965).

Bell, J. S., *Nucl. Phys.* **8,** 613 (1958).

Bellatini, G., C. Bemporad, P. L. Braccini, and L. Foa, *Nuov. Cim.* **40A,** 1139 (1965).

Bethe, H. A., and J. Ashkin, "Passage of radiations through matter," *Exptl. Nucl. Phys.* **1,** 166 (1953).

Bjorken, J. D., and E. A. Paschos, *Phys. Rev.* **185,**1975 (1969).

Bjorkland, R., W. E. Crandall, B. J. Moyer, and H. F. York, *Phys. Rev.* **77,** 213 (1950).

Blackett, P. M.S., and G. P. Occhialini, *Proc. Roy. Soc.* **A139,** 699 (1933).

Boehm, F., and E. Kankeleit, *Phys. Rev. Lett.* **14,** 312 (1965).

Brabant, J., B. Cork, N. Horowitz, B. Moyer, J. Murray, R. Wallace, and W. Wenzel, *Phys. Rev.* **101,** 498 (1956).

Budagov, I., *et al.*, *Phys. Lett.* **30B,** 364 (1969).

Cabibbo, N., *Phys. Rev. Lett.* **10,** 531 (1963).

Carlson, A. G., J. E. Hooper, and D. T. King, *Phil. Mag.* **41,** 701 (1950).

Cartwright, W. F., C. Richman, M. Whitehead, and H. Wilcox, *Phys. Rev.* **91,** 677 (1953).

Cavenagh, P., J. Turner, C. Coleman, G. Gard, and B. Ridley, *Phil. Mag.* **2,** 1105 (1957).

Chamberlain, O., E. Segre, C. Wiegand, and T. Ypsilantis, *Phys. Rev.* **100,** 947 (1955).

Charpak, G., Bouclier, T. Bressani, J. Favier, and C. Zupancic, *Nucl. Instr. Methods*, **62,** 262 (1968).

Charrière, G., M. Gailloud, Ph. Rosselet, R. Weill, W. M. Gibson, K. Green, P. Tolun, N. Whyte, J. Combe, E. Dahl-Jensen, N. Doble, D. Evans, L. Hoffmann, W. Toner, H. Going, K. Gottstein, W. Puschel, V. Schening, J. Tietge, *Phys. Lett.* **15,** 66 (1965).

Chew, G., S. Frautschi, and S. Mandelstam, *Phys. Rev.* **126,** 1202 (1962).

Christenson, J. H., J. Cronin, V. Fitch, and R. Turlay, *Phys. Rev. Lett.* **13,** 138 (1964).

Clark, D. L., A. Roberts, and R. Wilson, *Phys. Rev.* **83,** 649 (1951); **85,** 523 (1952).

Cnops, A., G. Finocchiaro, J. Lassalle, P. Mittner, P. Zanella, J. Dufey, B. Bobbi, M. Pouchon, and A. Muller, *Phys. Lett.* **22,** 546 (1966).

Coffin, C., N. Dikmen, L. Ettlinger, D. Meyer, A. Saulys, K. Terwilliger, and D. Williams, *Phys. Rev. Lett.* **17,** 458 (1966).

Condon, E. U., and G. H. Shortley, "The Theory of Atomic Spectra" (Cambridge University Press, 1951).

Conversi, M., E. Pancini, and O. Piccioni, *Phys. Rev.* **71,** 209 (1947).

Conversi, M., T. Massam, Th. Muller, and A. Zichichi, *Il Nuovo. Cim.* **40A,** 690 (1965).

Crowe, K. M., and R. H. Phillips, *Phys. Rev.* **96,** 470 (1954).

De Shalit, A., S. Cuperman, H. Lipkin, and T. Rothem, *Phys. Rev.* **107,** 1459 (1957).

Deutsch, M., *Prog. Nucl. Phys.* **3,** 131 (1953).

Dirac, P. A. M., *Proc. Roy. Soc.* **A117,** 610 (1928); also "The Principles of Quantum Mechanics" (Oxford Un. Press, 1947).

Dolen, R., D. Horn, and C. Schmid, *Phys. Rev.* **166,** 1768 (1968).

Dress, W. B., J. K. Baird, P. D. Miller, and N. F. Ramsey, *Phys. Rev.* **170,** 1200 (1968).

Durbin, R., H. Loar, and J. Steinberger, *Phys. Rev.* **84,** 581 (1951).

Erwin, A., R. March, W. Walker, and E. West, *Phys. Rev. Lett.* **6,** 628 (1961).

Fermi, E., *Zeit. Physik* **88,** 161 (1934).

Fermi, E., and C. Yang, *Phys. Rev.* **76,** 1739 (1949).

Feynman, R. P., *Phys. Rev. Lett.* **23,** 1415 (1969).

Feynman, R. P., and M. Gell-Mann, *Phys. Rev.* **109,** 193 (1958).

Fowler, P. H., D. H. Perkins, and K. Pinkau, *Phil. Mag.* **4,** 1030 (1959).

Frauenfelder, H., A. Hanson, N. Levine, A. Rossi, and G. De Pasquali, *Phys. Rev.* **107,** 643 (1957).

Frazer, W., and J. Fulco, *Phys. Rev. Lett.* **2,** 365 (1959); *Phys. Rev.* **117,** 1603 (1960).

Fry, W., J. Schneps, and M. Swami, *Phys. Rev.* **103,** 1904 (1956).

Gell-Mann, M., *Phys. Rev.* **92,** 833 (1953).

Gell-Mann, M., *Phys. Lett.* **8,** 214 (1964).

Gell-Mann, M., and M. Levy, *Nuov. Cim.* **16,** 705 (1960).

Gell-Mann, M., and A. Pais, *Phys. Rev.* **97,** 1387 (1955).

Ghosh, S. K., G. M. Jones, and J. G. Wilson, *Proc. Phys. Soc.* (*Lond.*) **A65,** 68 (1952); **A67,** 331 (1954).

Goldhaber, M., L. Grodzins, and A. Sunyar, *Phys. Rev.* **109,** 1015 (1958).

Gormley, M., E. Hyman, W. Lee, T. Nash, J. Peoples, C. Schultz, and S. Stein, *Phys. Rev. Lett.* **21,** 402 (1968).

Hartill, D. L., B. C. Barish, D. G. Fong, R. Gomez, J. Pine, A. V. Tollestrup, A. W. Marschke, and T. F. Zipf, *Phys. Rev.* **184,** 1415 (1969).

Heisenberg, W., *Zeit. für Phys.* **77,** 1 (1932).

Jackson, J. D., *Rev. Mod. Phys.* **37,** 484 (1965).

Konopinski, E., *Ann. Rev. Nucl. Science* **9,** 99 (1959).

Landau, L., and I. Pomeraneuk, *Dokl. Akad. Nauk. SSSR,* **92,** 535 and 735 (1953).

Langer, L., and R. Moffat, *Phys. Rev.* **88,** 689 (1952).

Lattes, C. M. G., H. Muirhead, C. F. Powell, and G. P. Occhialini, *Nature* **159,** 694 (1947).

Lee, T. C., and C. N. Yang, *Phys. Rev.* **104,** 254 (1956).

Maglic, B., Proc. Lund Int. Conf. El. Part., p. 271. (Berlingska Boktryckeriet, Lund) (1969).

Maglic, B., L. Alvarez, A. Rosenfeld, and M. Stevenson, *Phys, Rev. Lett.* **7,** 178 (1961).

Marshak, R., and H. Bethe, *Phys. Rev.* **72,** 506 (1947).

Marshak, R., and E. Sudarshan, *Phys. Rev.* **109** , 1860 (1958).

Mehlhop, W., S. Marty, P. Bowles, T. Burnett, R. Good, C. Holland, O. Piccioni, and R. Swarson, *Phys. Rev.* **172,** 1613 (1968).

Migdal, A. B., *Phys . Rev .* **103** , 1811 (1956).

Ne'eman, Y., *Nucl. Phys.* **26,** 222 (1961).

Nishijima, K., *Prog. Theor. Phys.* **13,** 285 (1955).

Orear, J., G. Harris, and S. Taylor, *Phys. Rev.* **102,** 1676 (1956).

Pais, A., *Phys. Rev.* **86,** 663 (1952).

Pais, A., and O. Piccioni, *Phys. Rev.* **100,** 1487 (1955).

Pauli, W., *Handbuch der Physik* **24,** 1, 233 (1933).

Panofsky, W. (data of E. Bloom *et al.*) Int. Conf. High Energy Physics, Vienna (1968).

Panofsky, W. K. H., R. L. Aamodt, and J. Hadley, *Phys. Rev.* **81,** 565 (1951).

Plano, R., A. Prodell, N. Samios, M. Schwartz, and J. Steinberger, *Phys. Rev. Lett.* **3,** 525 (1959).

Regge, T., *Nuov. Cim.* **14,** 951 (1959).

Reines, F., and C. Cowan, *Phys. Rev.* **113,** 273 (1959).

Rochester, G. D., and C. C. Butler, *Nature* **160,** 855 (1947).

Rossi, B., and K. Greisen, *Rev. Mod. Phys.* **13,** 240 (1941).

Russell, J. J., R. C. Sah, M. J. Tannenbaum, W. E. Cleland, D. G. Ryan, and D. G. Stairs, *Phys. Rev. Lett.* **26,** 46 (1971).

Sakata, S., *Prog. Theor. Phys.* **16,** 686 (1956).

Shafer, J., J. Murray, and D. Huwe, *Phys. Rev. Lett.* **10,** 176 (1963).

Sonderegger, P., J. Kirz, O. Guisan, P. Falk-Vairant, C. Bruneton, P. Borgeaud, A. Stirling, C. Caverzasio, J. Guillaud, M. Yvert, and B. Amblard, *Phys. Lett.* **20,** 75 (1966).

Sternheimer, R. M., "Interactions of radiation with matter," *Methods of Exptl. Phys.* **5A,** 1 (1961).

Stevenson, M., L. Alvarez, B. Maglic, and A. Rosenfeld, *Phys. Rev.* **125,** 687 (1962).

Street, J. C., and E. C. Stevenson, *Phys. Rev.* **52,** 1003 (1937).

Svensson, B., Proc. 1967 CERN School of Physics, Rattvik, Vol. 2 (CERN 67–24).

Tanner, N., *Phys. Rev.* **107,** 1203 (1957).

Veneziano, G., *Nuov. Cim.* **57A,** 190 (1968).

Von Dardel, B., D. Dekkers, R. Mermod, J. D. Van Putten, M. Vivargent, G. Weber, and K. Winter, *Phys. Lett.* **4,** 51 (1963).

Von Witsch, W., A. Richter, and P. von Brentano, *Phys. Rev.* **169,** 923 (1968).

Weber, G., Proc. 1967 Int. Sym. on Electron and Photon Interactions at High Energies, Stanford, California, p. 59 (1967).

Weber, J., Scientific American **224,** No. 5 (1971).

Wesley, J. C., and A. Rich, *Phys. Rev. Lett.* **24,** 1320 (1970).

Weyl, H., *Zeit. Physik* **56,** 330 (1929).

Wolfenstein, L., *Phys. Rev. Lett.* **13,** 562 (1964).

Wu, C. S., and I. Shaknov, *Phys. Rev.* **77,** 136 (1950).

Wu, C. S., E. Ambler, R. Hayward, D. Hoppes, and R. Hudson, *Phys. Rev.* **105,** 1413 (1957).

Wu, T. T., and C. N. Yang, *Phys. Rev.* **137,** 3708 (1965).

Yukawa, H., *Proc. Phys. Math. Soc. Japan* **17,** 48 (1935).

Zweig, G., CERN Report 8419/Th 412 (1964).

Index